MULTIPLE CRITERIA DECISION ANALYSIS
An Integrated Approach

MULTIPLE CRITERIA DECISION ANALYSIS
An Integrated Approach

VALERIE BELTON
University of Strathclyde
Glasgow, Scotland

THEODOR J STEWART
University of Cape Town
South Africa

Kluwer Academic Publishers
Boston/Dordrecht/London

Distributors for North, Central and South America:
Kluwer Academic Publishers
101 Philip Drive
Assinippi Park
Norwell, Massachusetts 02061 USA
Telephone (781) 871-6600
Fax (781) 871-9045
E-Mail: kluwer@wkap.com

Distributors for all other countries:
Kluwer Academic Publishers Group
Post Office Box 322
3300 AH Dordrecht, THE NETHERLANDS
Telephone 31 786 576 000
Fax 31 786 576 254
E-mail: services@wkap.nl

 Electronic Services <http://www.wkap.nl>

Library of Congress Cataloging-in-Publication Data

Belton, Valerie.
    Multiple criteria decision analysis : an integrated approach / Valerie Belton, Theodor J. Stewart.
        p. cm.
    Includes bibliographical references and index.
    ISBN 0-7923-7505-X
    1. Multiple criteria decision making. I. Stewart, Theodor J. II. Title.

T57.95 .B45 2001
658.4'03--dc21
                                                                                    2001038764

Copyright © 2002 by Kluwer Academic Publishers. Second printing 2003.

All rights reserved. No part of this work may be reproduced, stored in a retrieval system, or transmitted in any form or by any means, electronic, mechanical, photocopying, microfilming, recording, or otherwise, without the written permission from the Publisher, with the exception of any material supplied specifically for the purpose of being entered and executed on a computer system, for exclusive use by the purchaser of the work.

Permission for books published in Europe: permissions@wkap.nl
Permissions for books published in the United States of America: permissions@wkap.com

Printed on acid-free paper.

Printed in United Kingdom by Biddles/IBT Global

This printing is a digital duplication of the original edition.

To our parents

Mona and Harold Phillips
Margaretha and Jack Stewart

who provided us with opportunities
to study not available to themselves

Contents

List of Figures	xi
List of Tables	xiii
List of Example Panels	xv
Preface	xvii
Acknowledgments	xix

1. INTRODUCTION — 1
   1.1   What Is MCDA?   1
   1.2   What Can We Expect from MCDA?   2
   1.3   The Process of MCDA   5
   1.4   Outline of the Book   8

2. THE MULTIPLE CRITERIA PROBLEM — 13
   2.1   Introduction   13
   2.2   Case Examples   16
   2.3   Classifying MCDM Problems   31

3. PROBLEM STRUCTURING — 35
   3.1   Introduction   35
   3.2   Where does MCDA start?   36
   3.3   Problem Structuring Methods   39
   3.4   From Problem Structuring to Model Building   52
   3.5   Case Examples Re-visited   64
   3.6   Concluding Comments   77

4. PREFERENCE MODELLING — 79
   4.1   Introduction   79
   4.2   Value Measurement Theory   85
   4.3   Utility Theory: Coping with Uncertainty   95
   4.4   Satisficing and Aspiration Levels   104
   4.5   Outranking   106

viii   *MULTIPLE CRITERIA DECISION ANALYSIS*

4.6	Fuzzy and Rough Sets	111
4.7	Relative Importance of Criteria	114
4.8	Final Comments	117

5. VALUE FUNCTION METHODS:
PRACTICAL BASICS   119

5.1	Introduction	119
5.2	Eliciting Scores (Intra-Criterion Information)	121
5.3	Direct Rating of Alternatives by Pairwise Comparisons	132
5.4	Eliciting Weights (Inter-Criterion Information)	134
5.5	Synthesising Information	143
5.6	Sensitivity and Robustness Analysis	148
5.7	The Analytic Hierarchy Process	151
5.8	Concluding Comments	159

6. VALUE FUNCTION METHODS:
INDIRECT AND INTERACTIVE   163

6.1	Introduction	163
6.2	Use of Ordinal and Imprecise Preference Information	165
6.3	Holistic Assessments and Inverse Preferences	188
6.4	Interactive Methods based on Value Function Models	193
6.5	Concluding Comments	204
	Summary of main notational conventions used in chapter	207

7. GOAL AND REFERENCE POINT
METHODS   209

7.1	Introduction	209
7.2	Linear Goal Programming	213
7.3	Generalized Goal Programming	220
7.4	Concluding Comments	230

8. OUTRANKING METHODS   233

8.1	Introduction	233
8.2	ELECTRE I	234
8.3	ELECTRE II	241
8.4	ELECTRE III	242
8.5	Other ELECTRE methods	250
8.6	The PROMETHEE Method	252
8.7	Final Comments	258

9. IMPLEMENTATION OF MCDA:
PRACTICAL ISSUES AND INSIGHTS   261

9.1	Introduction	261
9.2	Initial negotiations: establishing a contract	265
9.3	The Nature of Modelling and Interactions	266

9.4	Organizing and Facilitating a Decision Workshop	271
9.5	Working "Off-Line" and in the Backroom	275
9.6	Small Group Interaction and DIY Analysis	279
9.7	Software Support	281
9.8	Insights from MCDA in Practice	283
9.9	Interpretation and Assessment of Importance Weights	288
9.10	Concluding Comments	291

10. MCDA IN A BROADER CONTEXT — 293

10.1	Introduction	293
10.2	Links to Methods with Analytical Parallels	294
10.3	Links to other OR/MS Approaches	310
10.4	Other Methodologies with a Multicriteria Element	320
10.5	OR/MS Application Areas with a Multicriteria Element	327
10.6	Conclusions	329

11. AN INTEGRATED APPROACH TO MCDA — 331

11.1	Introduction	331
11.2	An Integrating Framework	333
11.3	The Way Forward	338
11.4	Concluding Remarks	343

Appendices	
MCDA Software Survey	345
Glossary of Terms and Acronyms	351
References	353
Index	369

List of Figures

1.1	The Process of MCDA	6
3.1	Guidelines for a Post-It exercise	42
3.2	Post-It session in progress	43
3.3	Portion of a cognitive map	49
3.4	Illustrative display from Decision Explorer	50
3.5	Spray diagram for Decision Aid International's location problem	65
3.6	Value tree for Decision Aid International's location problem	66
4.1	Illustration of a "value tree"	81
4.2	Illustration of corresponding trade-offs when $m = 2$	89
4.3	Corresponding tradeoffs for Example Panel 4.2	92
5.1	Value Function for Availability of Staff	125
5.2	Value Function for Accessibility from US	127
5.3	Visual representation of scale	133
5.4	Swing weights - visual analogue scale	137
5.5	Relative Weights (in boldface) and Cumulative Weights (in italics)	140
5.6	Overall evaluation of alternatives – visual displays	145
5.7	Profile graph - top-level criteria	146
5.8	Profile graph – bottom-level criteria	147
5.9	Weighted profile graph - bottom-level criteria	147
5.10	Sensitivity analysis: effect of varying the weight on Availability of Staff	149
5.11	Working with ordinal information on criteria weights	150
5.12	Implication of weight restrictions	151
5.13	Comparison of current weights and potentially optimal weights for Warsaw	152
6.1	Piecewise linear value function for low flow	167

6.2	Categorical scale values for water quality	169
6.3	Outline of the Geoffrion-Dyer-Feinberg algorithm	195
6.4	Outline of the Zionts-Wallenius algorithm	198
6.5	Outline of the Steuer's interactive weighted sums and filtering algorithm	200
6.6	Outline of the generalized interactive procedure	203
8.1	Building an Outranking Relation	238
8.2	Definition of the Concordance Index for Criterion i	244
8.3	PROMETHEE preference functions	253
8.4	GAIA-type plot for Example Panel 8.5	257
9.1	Benefit-cost plot for aerial policing options	285
10.1	Ranges of efficiencies for three DMUs	304
10.2	Graphical display of a simple two-criterion problem	306
10.3	Factor Analysis plotting of the business location data	308
10.4	Scenario-option matrix for illustrative example	314
10.5	A multi-criteria extension of the scenario-option matrix	315
10.6	Alternative evaluations of Chris's preferences	320
10.7	Generic Scorecard Measures	323
10.8	An illustrative interlinked scorecard	324
10.9	The EFQM Excellence Model (adapted from the EFQM website: www.EFQM.org)	326

List of Tables

3.1	Illustrative parameter vales for game reserve example	70
3.2	Payoff table for game reserve example	71
5.1	Intervals on "Accessibility from US" measured by the number of flights per week	126
5.2	Value Function for Accessibility	127
5.3	Beaufort scale	129
5.4	Ratings of alternatives in the business location problem	134
5.5	Swing weights - original and normalised values	138
5.6	Relative and cumulative weights for the example problem	141
5.7	Synthesis of information for the business location case study	144
5.8	Comparison matrix for *quality of life*	154
8.1	Decision matrix for the business location problem	235
10.1	Performance of academic departments taken from Belton and Vickers, 1992	302
A.1	Special purpose MCDA software	347

List of Example Panels

3.1	Use of the CAUSE Framework for the Location Case Study	65
4.1	Illustration of preferential independence violation	88
4.2	Illustration of the corresponding trade-offs condition	91
4.3	Illustration of interval scale property	93
4.4	Illustration of differences between indifference and incomparability	108
5.1	Illustration of local and global scales	122
5.2	Development of a qualitative scale	130
5.3	Illustration of a direct rating process	132
5.4	Conversation between facilitator (F) and decision maker (DM) while assessing swing weights	136
6.1	Illustration of breakpoints	167
6.2	Illustration of categories	168
6.3	Pairwise comparisons in MACBETH	173
6.4	Illustration of inequalities generated from imprecise inputs	178
6.5	The "ϕ_{ijt}" coefficients for the inequality constraints generated in Example Panel 6.4	179
6.6	LP Formulation for consistency checking for the low flow criterion in the land use planning example	182
6.7	LP Formulation to determine bounds on the partial value function for the low flow criterion	185
6.8	Conjoint scaling comparison of two criteria	191
6.9	Application of the Zionts-Wallenius procedure to the game reserve planning problem	199
6.10	Application of the generalized interactive procedure to the game reserve planning problem	205
7.1	Formulation of deviational variables in the game reserve planning problem	216

7.2	Weighted sum and Archimedean goal programming solutions to the game reserve planning problem	217
7.3	Preemptive goal programming solution to the game reserve planning problem	219
7.4	Tchebycheff goal programming solution to the game reserve planning problem	220
7.5	Compromise programming applied to the game reserve planning problem	224
7.6	Application of STEM to the game reserve planning problem	225
7.7	Application of ISGP to the game reserve planning problem	228
8.1	Values of concordance and discordance indices (ELECTRE I) for the business location problem	237
8.2	Building and Exploiting the Outranking Relation for the Business Location Problem	240
8.3	Determining a Ranking of Alternatives Using ELECTRE II	243
8.4	Application of ELECTRE III to the business location case study	248
8.5	Application of PROMETHEE to the business location case study	256

Preface

The field of multiple criteria decision analysis (MCDA), also termed multiple criteria decision aid, or multiple criteria decision making (MCDM), has developed rapidly over the past quarter century and in the process a number of divergent schools of thought have emerged. This can make it difficult for a new entrant into the field to develop a comprehensive appreciation of the range of tools and approaches which are available to assist decision makers in dealing with the ever-present difficulties of seeking compromise or consensus between conflicting interests and goals, i.e. the "multiple criteria". The diversity of philosophies and models makes it equally difficult for potential users of MCDA, i.e. management scientists and/or decision makers facing problems involving conflicting goals, to gain a clear understanding of which methodologies are appropriate to their particular context.

Our intention in writing this book has been to provide a comprehensive yet widely accessible overview of the main streams of thought within MCDA. We aim to provide readers with sufficient awareness of the underlying philosophies and theories, understanding of the practical details of the methods, and insight into practice to enable them to implement any of the approaches in an informed manner. As the title of the book indicates, our emphasis is on developing an *integrated* view of MCDA, which we perceive to incorporate both integration of different schools of thought *within* MCDA, and integration of MCDA *with* broader management theory, science and practice.

It is indeed the integration of the theory and practice of MCDA that has been of central concern to us in writing this book, fuelled by a belief that the continued development of any managerially oriented subject area depends on this synergy. Thus, throughout our discussion of the underlying theory and methods, we have sought to illustrate these by selected case studies, and also by drawing attention to the practical issues of implementation.

We believe that the book can be of value to various reader groups, each with their own objectives:

- *Practising decision analysts or graduate students in MCDA* for whom this book should serve as a state-of-the-art review, especially as regards techniques outside of their own specialization;

- *Operational Researchers or graduate students in OR/MS* who wish to extend their knowledge into the tools of MCDA;

- *Managers or management students* who need to understand what MCDA can offer them.

Certain readers in the second and third groups may wish to omit or to gloss over the more technical algorithmic descriptions in Chapters 6-8 and some parts of Chapter 4. This may be done without losing the main thread of the development, but we would recommend that all readers should at least seek to cover Chapters 1–3; Sections 4.1, 4.2, 4.4 and 4.5 from Chapter 4; the introductory sections of Chaps 5, 7 and 8; and Chapter 9.

We wish you all as much enjoyment and satisfaction in reading this book as we had in preparing it!

VALERIE BELTON AND THEODOR STEWART

Acknowledgments

Many people have contributed directly or indirectly to the emergence of this book as a finished product, through collaborative work, casual conversations or encouraging support.

We would firstly wish to acknowledge the personal and loving support of our partners, Mark and Sheena, without which we would not have been able to see the project through. In the case of Mark Elder this support extended to professional collaboration.

At the risk of overlooking some, we would also like specifically to mention some friends and colleagues whose inputs have been particularly valuable: Fran Ackermann, Euro Beinat, Julie Hodgkin, Ron Janssen, Alison Joubert, Tasso Koulouri, Fabio Losa and Jacques Pictet.

In addition, we acknowledge our debt of gratitude to the many academic colleagues, organizational contacts, students and friends across the world, within and beyond the MCDA community. We have learnt much from discussions, debates and arguments, as well as opportunities to put our ideas into practice.

Finally, we thank Gary Folven of Kluwer Academic Press for his encouragement, and for not giving up on us during this book's long gestation period.

Chapter 1

INTRODUCTION

1.1. WHAT IS MCDA?

Clearly it is important to begin a book devoted to the topic of **multiple criteria decision analysis (MCDA)** with a discussion and definition of what we mean by the term. One dictionary definition of "criterion" (The Chambers Dictionary) is "a means or standard of judging". In the decision-making context, this would imply some sort of standard by which one particular choice or course of action could be judged to be more desirable than another. Consideration of different choices or courses of action becomes a *multiple criteria decision making (MCDM)* problem when there exist a number of such standards which conflict to a substantial extent. For example, even in simple personal choices such as selecting a new house or apartment, relevant criteria may include price, accessibility to public transport, and personal safety. Management decisions at a corporate level in both public and private sectors will typically involve consideration of a much wider range of criteria, especially when consensus needs to be sought across widely disparate interest groups. A number of examples illustrating such situations are presented in the next chapter.

Every decision we ever take requires the balancing of multiple factors (i.e. "criteria" in the above sense) – sometimes explicitly, sometimes without conscious thought – so that in one sense everyone is well practised in multicriteria decision making. For example, when you decide what to wear each day you probably take into account what you will be doing during the day, what kind of impression you want to create, what you feel comfortable in, what the weather is expected to be, whether you want to risk getting that jacket that has to be dry-cleaned dirty,

2 *MULTIPLE CRITERIA DECISION ANALYSIS*

etc. Sometimes you may even try out a number of different alternatives before deciding. Have you ever stopped to build a formal model to analyse this particular decision? The answer is probably "no" – the issue does not seem to merit it, it is not complex enough, it is easy enough to take account of all the factors "in one's head", the consequences are not substantial, are generally short-term and mistakes are easily remedied. In short, the decision "does not matter" that much. However, in both personal and group decision making contexts, we are at times confronted with choices that *do* matter – the consequences are substantial, impacts are longer term and may affect many people, and mistakes might not easily be remedied. It is under these circumstances that the tools and methods presented in this book come into play, since it is well known from psychological research (e.g., the classical work of Miller, 1956) that the human brain can only simultaneously consider a limited amount of information, so that all factors cannot be resolved in one's head. The very nature of multiple criteria problems is that there is much information of a complex and conflicting nature, often reflecting differing viewpoints and often changing with time. One of the principal aims of MCDA approaches is to help decision makers organise and synthesize such information in a way which leads them to feel comfortable and confident about making a decision, minimizing the potential for post-decision regret by being satisfied that all criteria or factors have properly been taken into account.

Thus, we use the expression MCDA as an umbrella term to describe a collection of formal approaches which seek to take explicit account of multiple criteria in helping individuals or groups explore decisions that matter. Decisions matter when the level of conflict between criteria, or of conflict between different stakeholders regarding what criteria are relevant and the importance of the different criteria, assumes such proportions that intuitive "gut-feel" decision-making is no longer satisfactory. This can happen even with personal decisions (such as buying a house), but becomes much more of an issue when groups are involved (such as in corporate decision making, or in other situations in which multiple stakeholders are involved), so that much of our discussion of MCDA in this book will refer to group decision-making contexts.

1.2. WHAT CAN WE EXPECT FROM MCDA?

In answering this question it is important to begin by dispelling the following myths:

Myth 1: MCDA will give the "right" answer;

Myth 2: MCDA will provide an "objective" analysis which will relieve decision makers of the responsibility of making difficult judgements;

Myth 3: MCDA will take the pain out of decision making.

Firstly, there is no such thing as the "right answer" even within the context of the model used. The concept of an optimum does not exist in a multicriteria framework and thus multicriteria analysis cannot be justified within the optimisation paradigm frequently adopted in traditional Operational Research / Management Science. MCDA is an *aid* to decision making, a process which seeks to:

- Integrate objective measurement with value judgement;

- Make explicit and manage subjectivity.

Subjectivity is inherent in all decision making, in particular in the choice of criteria on which to base the decision, and the relative "weight" given to those criteria. MCDA does not dispel that subjectivity; it simply seeks to make the need for subjective judgements explicit and the process by which they are taken into account transparent (which is again of particular importance when multiple stakeholders are involved). This is not always an easy process – the fact that trade-offs are difficult to make does not mean that they can always be avoided. MCDA will highlight such instances and will help decision makers think of ways of overcoming the need for difficult trade-offs, perhaps by prompting the creative generation of new options. It can also retain a degree of equivocality by allowing imprecise judgements but cannot take away completely the need for difficult judgements.

Above all, it is our view that the aim of MCDA should be, and that the principal benefit is, to facilitate decision makers' learning about and understanding of the problem faced, about their own, other parties' and organisational priorities, values and objectives and through exploring these in the context of the problem to guide them in identifying a preferred course of action. This is a view which is shared by many others prominent in the field of MCDA, as illustrated by the following quotes:

> *Simply stated, the major role of formal analysis is to promote good decision making. Formal analysis is meant to serve as an aid to the decision maker, not as a substitute for him. As a process, it is intended to force hard thinking about the problem area: generation of alternatives, anticipation of future contingencies, examination of dynamic secondary effects and so forth. Furthermore, a good analysis should illuminate controversy - to find out where basic differences exist in values and uncertainties, to facilitate compromise, to increase the level of debate and to undercut rhetoric - in short "to promote good decision making"* (Keeney and Raiffa, 1972, pp 65–66)

4   MULTIPLE CRITERIA DECISION ANALYSIS

> ... decision analysis [has been] berated because it supposedly applies simplistic ideas to complex problems, usurping decision makers and prescribing choice!
>
> Yet I believe that it does nothing of the sort. I believe that decision analysis is a very delicate, subtle tool that helps decision makers explore and come to understand their beliefs and preferences in the context of a particular problem that faces them. Moreover, the language and formalism of decision analysis facilitates communication between decision makers. Through their greater understanding of the problem and of each other's view of the problem, the decision makers are able to make a better informed choice. There is no prescription: only the provision of a framework in which to think and communicate. (French, 1989, p1)

> We wish to emphasize that decision making is only remotely related to a "search for the truth." ... the theories, methodologies, and models that the analyst may call upon ... are designed to help think through the possible changes that a decision process may facilitate so as to make it more consistent with the objectives and value system of the one for whom, or in the name of whom, the decision aiding is being practised. These theories, methodologies, and models are meant to guide actions in complex systems, especially when there are conflicting viewpoints. (Roy, 1996, p11)

> The decision unfolds through a process of learning, understanding, information processing, assessing and defining the problem and its circumstances. The emphasis must be on the process, not on the act or the outcome of making a decision ... (Zeleny, 1982)

> ... decision analysis helps to provide a structure to thinking, a language for expressing concerns of the group and a way of combining different perspectives. (Phillips, 1990, p150)

Note that the focus is on supporting or aiding decision making; it is not on prescribing how decisions "should" be made, nor is it about describing how decisions are made in the absence of formal support. There is also a substantial literature, stemming in the main from behavioural psychology, concerned with descriptive decision theory (for example, Mann and Janis, 1985). This work has both informed and fuelled debate amongst those concerned with aiding decision making (e.g. Bell, Raiffa and Tversky, 1988; Hammond, McClelland and Mumpower, 1980). It is an area in which there is substantial practical application, particularly in the field of consumer choice. In Chapter 6 we will see a convergence between descriptive and supportive MCDA in the use of "inverse preference methods" to develop models for choice derived from holistic or historical decision making.

The above brief discussion and comments establish the context in which we believe multiple criteria methods to be practically useful and sets the scene for the book. We would like to emphasise the following points:

- MCDA seeks to take explicit account of multiple, conflicting criteria in aiding decision making;

- The MCDA process helps to structure the problem;

- The models used provide a focus and a language for discussion;

- The principal aim is to help decision makers learn about the problem situation, about their own and others values and judgements, and through organisation, synthesis and appropriate presentation of information to guide them in identifying, often through extensive discussion, a preferred course of action;

- The analysis serves to complement and to challenge intuition, acting as a sounding-board against which ideas can be tested – it does not seek to replace intuitive judgement or experience;

- The process leads to better considered, justifiable and explainable decisions – the analysis provides an audit trail for a decision;

- The most useful approaches are conceptually simple and transparent;

- The previous point notwithstanding, non-trivial skills are necessary in order to make effective use even of such simple tools in a potentially complex environment.

1.3. THE PROCESS OF MCDA

Much of the literature on MCDA can be criticised for adopting a stance of "given the problem" – i.e. taking as a starting point a well defined set of alternatives and criteria, and focusing on evaluation. It is unlikely that in practice any problem will present itself to an analyst in this form. It is more likely that the MCDA process will be embedded in a wider process of problem structuring and resolution, as illustrated in Figure 1.1. This figure, which is deliberately messy in order to convey the nature of MCDA in practice, shows the main stages of the process from the identification of a problem or issue, through problem structuring, model building and using the model to inform and challenge thinking, and ultimately to determine an action plan. This plan may take many forms, for example, to implement a specific choice, to put forward a recommendation, to establish a procedure for monitoring performance,

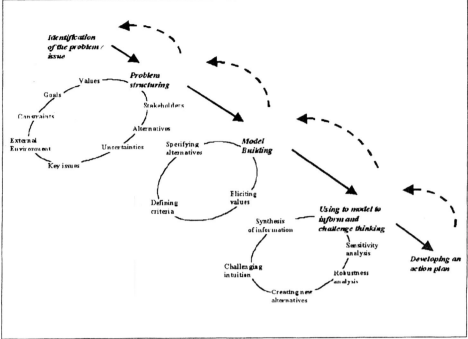

Figure 1.1: The Process of MCDA

or simply to maintain a watching brief on a situation. We group these into three key phases: problem identification and structuring; model building and use; and the development of action plans.

The initial problem structuring phase is one of divergent thinking, opening up the issue, surfacing and capturing the complexity which undoubtedly exists, and beginning to manage this and to understand how the decision makers might move forward. The phase of model building and use represents a more convergent mode of thinking, a process of extracting the essence of the issue from the complex representation in a way which supports more detailed and precise evaluation of potential ways of moving forward. It is possible that the outcome of this phase is a return to divergent thinking, a need to think creatively about other options or aspects of the issue. A phrase which we feel captures the nature of the overall process very effectively, and which represents a recurrent theme throughout the book is: *"Through complexity to simplicity"*. Multiple criteria models often appear very simple, and indeed have been criticized as simplistic (see the quote by French on p.4). However, this neglects the nature of the above process and the fact that the apparently simple model does not deny the complexity, but has emerged from it as a distillation of the key factors in a form which is transparent, easy to work with, and which can generate further insights and understanding.

There are many actors central to the process; these include decision makers, clients, sponsors, other stakeholders, including potential saboteurs, and the facilitators or analysts. As indicated in Figure 1.1 and the above discussion, one can expect iteration within and between the key phases of the process, each of which is subject to a myriad of internal and external influences and pressures. This description of process is generic to the whole of MCDA, although as already mentioned, the emphasis has tended to be on building and using a model. It is here that the different MCDA approaches are distinguished from each other; in the nature of the model, in the information required and in how the model is used. Within that they have in common the need to define, somehow, the alternatives to be considered, the criteria or objectives to guide the evaluation, and usually some measure of the relative significance of the different criteria. It is in the detail of how this information is elicited, specified and synthesized to inform decision making that the methods differ.

Multiattribute value (or utility) theory (MAVT or MAUT) is one of the more widely applied multicriteria methods and it is practitioners and academics in that field who appear to have had most to say about the process of MCDA, in particular about model building and implementation, i.e. use of MCDA in practice. Ever since its origins in the late 1960's, concerns for the practical application of multi-attribute value theory (MAVT) or, more generally, multi-attribute utility theory (MAUT), have influenced developments in the field. For example, concerns about difficulties of using the more complex MAUT models in practice led to the development of SMART (e.g. von Winterfeldt and Edwards, 1986, Chapter 8), a simplified multi-attribute rating approach which now underpins much practical analysis. The field has benefited from the longstanding interests of psychologists, engineers, management scientists and mathematicians which has brought a continuing awareness of behavioural and social issues as well as underlying theory. For example, Decision Conferencing (Phillips, 1989, 1990), described in more detail in Chapter 9, emphasises the importance of the social context in which analysis is conducted. In recent years these issues have become more widely embraced by the MCDA community as a whole, as discussed by Bouyssou et al. (1993) and by Korhonen and Wallenius (1996).

It is important to note that the majority of, if not all, multicriteria analyses reported in the literature are supported by one, or more, analysts or facilitators. In discussing the use of different methods throughout the book we assume that such support is available. The term analyst

8 MULTIPLE CRITERIA DECISION ANALYSIS

tends to be used when there is a strong emphasis on that person working independently to gather information and to capture expertise; a facilitator is more commonly recognised as someone who also brings the skills of managing group processes. Analysts and/or facilitators may be external or in-house consultants, but in either case would be recognised for their expertise in the approach to modelling. In Chapter 9 we explore in more detail the different analytical styles which may adopted.

It is unfortunate that much of the literature on MCDA up to now has tended to be somewhat fragmented, in the sense of concentrating on single methodologies or schools of thought. The view propagated throughout this book is that the process of MCDA needs to be seen and understood in an *integrated* manner. This integration we speak of includes:

- Integration between different approaches to the MCDA problem itself, recognizing that each approach brings particular insights and understanding and may be appropriate at different phases of the analysis;

- Integration between MCDA and other problem structuring and decision evaluation methods from the broader management sciences;

- Integration between tools developed specifically for MCDA and other quantitative tools from OR/MS and statistics.

These issues are discussed at greater length in Chapters 10 and 11.

1.4. OUTLINE OF THE BOOK

In the next chapter we elaborate further on the nature of multiple criteria decision making problems. We distinguish a number of different types or categories of such problems, for example: the choice of a preferred option from a finite and discrete set of possibilities; the task of ranking or categorizing a set of options; or the exploration of an effectively infinite set of possibilities with a view to "designing" a preferred solution. These are illustrated by a number of case studies drawn from or based on our experiences of the use of MCDA in practice. The case studies will be used in later chapters to illustrate the methodologies which we describe, recognizing that different MCDA approaches may be better suited to some problem types than to others. This recognition of the need to match methodologies to problem context is essential to an integrated understanding of MCDA.

As indicated in Section 1.3, MCDM problems will seldom if ever present themselves in terms of a well defined set of alternatives and criteria, ready for evaluation and analysis. Figure 1.1 illustrated the broader process of problem structuring and resolution of which MCDA may be part, and it is the initial stages of this process which form the theme of Chapter 3. We begin by reviewing a number of approaches which have been used widely in practice to facilitate the generation, capture and structuring of ideas, leading to a rich representation and a shared understanding of the perceived decision problems. The emphasis then shifts from problem structuring to the building of a model, establishing and exploring in detail a common framework for MCDA. We illustrate the process by reference to a number of the case studies introduced in Chapter 2.

Chapter 4 provides an overview of a number of different models which have been developed to represent preferences in the context of multicriteria problems. These form the basis of the specific methodologies discussed in more detail in Chapters 5 to 8. It is recognized that the models can be classified into three broad categories, or schools of thought, namely:

1 *Value measurement models* in which numerical scores are constructed, in order to represent the degree to which one decision option may be preferred to another. Such scores are developed initially for each individual criterion, and are then synthesized in order to effect aggregation into higher level preference models.

2 *Goal, aspiration or reference level models* in which desirable or satisfactory levels of achievement are established for each of the criteria. The process then seeks to discover options which are in some sense closest to achieving these desirable goals or aspirations.

3 *Outranking models* in which alternative courses of action are compared pairwise, initially in terms of each criterion, in order to identify the extent to which a preference for one over the other can be asserted. In aggregating such preference information across all relevant criterion, the model seeks to establish the strength of evidence favouring selection of one alternative over another.

The theoretical background to each of these models, including discussion of underlying assumptions concerning preferences and measurement, is presented in Chapter 4. Related issues such as fuzzy and rough set theory are introduced. This chapter also includes a discussion of the concept of the relative importance of different criteria which is central to MCDA. Relative importance is generally represented by means of some form of quantitative importance "weight". We shall argue, together with many other writers, that the meaning of such weights is strongly model-dependent. However, the intuitive sense of the meaning of the notion of relative importance may not necessarily correspond with any of the models. Thus, considerable care needs to be exercised in eliciting weights for use in any multicriteria decision model, a theme which recurs on a number of other occasions in this book.

The next four chapters are devoted to discussion of the practical implementation of the models presented in Chapter 4, supplemented by illustrative examples of the main methodologies. Note that a list of the examples is provided together with lists of tables and figures in the introductory pages of this volume.

Value measurement approaches to MCDA are presented in Chapters 5 and 6. The first of these chapters deals with the explicit construction of value (or preference scoring) functions, typically under the guidance of a facilitator / decision analyst. The larger part of the chapter deals with what is generally termed *multiattribute value theory (MAVT)*, but also includes discussion of the *analytic hierarchy process (AHP)* which is similar in many respects, has been widely applied and has its own substantial following. MAVT and AHP differ primarily in terms of the underlying assumptions about preference measurement, the methods used to elicit preference judgements from decision makers, and the manner of transforming these into quantitative scores.

Chapter 6 is concerned with more indirect methods of building value functions on the basis of partial or incomplete preference statements and/or of inferences drawn from a small number of observed choices. The inferential methods used require some understanding of linear programming principles, so that readers without this background may wish to skip parts of this chapter on first reading. This should not detract from understanding of the ensuing chapters.

Generalized goal programming models are described in Chapter 7. As with the preceding chapter, some of this chapter makes extensive use of linear programming models (which is, in fact, the context in which goal programming was originally developed). Once again, some readers may wish to skip over these sections of the chapter, at least on first reading. The chapter presents first the classical forms of goal programming, which were the first formal multiple criteria models to be found in the management science literature, together with some extensions which have become more accessible with the greater computing power available today. This discussion of classical goal programming principles is then broadened to include the more recent generalizations which have been termed reference point methods.

Both Chapters 6 and 7 contain discussion of so-called "interactive methods" of MCDA. In these methods, a complete preference model is not constructed *a priori* for use in the evaluation of the available

Introduction 11

courses of action. Rather, a sequence of real or hypothetical alternatives is presented to decision makers. As each such alternative is presented, decision makers are invited to indicate preferences between recently seen alternatives and/or to state what improvements they would like to see. Such partial preference statements are used to generate more precise preference models, which in turn guide the search for further (hopefully improved) alternatives to be presented to decision makers for evaluation.

Applications of the outranking approach are described in Chapter 8. The descriptions are provided in a fairly detailed manner for the methods known as ELECTRE I, II and III, in order to trace the evolution in thinking from relatively simple to quite sophisticated implementation. Other more specialized variations of the ELECTRE methods are mentioned more briefly, before turning to discussion of an alternative class of outranking methods termed PROMETHEE.

The final three chapters of the book aim to draw the material together in an integrated fashion. Chapter 9 deals with practical issues in the implementation of MCDA, which are relevant whatever methodology is applied. Much of the discussion relates to working with groups in some form of decision workshop, as it is our belief that most non-trivial MCDA applications occur within this context. We do, nevertheless, give consideration to interactions with single clients, to backroom work done by analysts, and to "do-it-yourself" MCDA performed directly by decision makers themselves. Throughout, the emphasis is on positive guidelines for successful implementation, and on identifying pitfalls or potential misunderstandings that analysts or decision makers may fall into. We emphasize the need for adequate software support for implementation of the models discussed, and for this purpose we have also provided in Appendix A a summary of available MCDA software of which we were aware at the time of writing (early 2001).

Chapter 10 evaluates the role of MCDA within the broader management science context, recognizing four areas in which there exist potential for synergistic link with MCDA:

- Quantitative methods of OR/MS and statistics which have strong parallels with MCDA, such as Game Theory, Data Envelopment Analysis and Multivariate Statistical Analysis;

- Management science approaches which can be used to good effect in conjunction with MCDA, such as cognitive / cause mapping and soft systems methodology;

- Methods of management science which have a strong multicriteria component but which have been developed without any formal links with or reference to MCDA, such as value engineering and the balanced scorecard;

- Methods developed for specific application areas which naturally have a strong multicriteria element.

The final chapter returns to our central concern for an "integrated approach" to MCDA. We suggest means by which the different models and methodologies for MCDA may be drawn together, to provide in effect a meta-approach to MCDA. Such considerations lead naturally to the identification of research needs in the area of MCDA.

Chapter 2

THE MULTIPLE CRITERIA PROBLEM

2.1. INTRODUCTION

What constitutes a "multiple criteria decision making (MCDM)" problem? Clearly there must be some decision to be made! Such a decision may constitute a simple choice between two or more (perhaps even infinitely many) well-defined courses of action. The problem is then simply that of making the "best" choice in some sense. At the other extreme, there may be a vague sense of unease that we (personally or corporately) need to "do something" about a situation which is found unsatisfactory in some way. The decision problem then constitutes much more than simply the evaluation and comparison of alternative courses of action. It involves also an in-depth consideration of what it is that is "unsatisfactory", and the creative generation of possible courses of action to address the situation. Much of the literature on methods of MCDA may tend to suggest that the typical application is in the context of simple choice from amongst a given set of alternatives. In practice, however, most decision making problems of sufficient extent and import to warrant the use of the formal analytical tools discussed in this book will tend to be closer to the unstructured "feeling of unease" category. For this reason, we shall be giving attention to the structuring as well as the analysis and resolution of decision making problems.

By definition, MCDM must also involve consideration of "multiple criteria". In Sections 1.1 and 1.2, we introduced the concept and meaning of multiple criteria in decision making, and indicated some of the reasons why a formally structured analysis may be important, especially in decisions involving multiple stakeholders. In a very real sense, all non-trivial decision making will involve some conflict between different

14   MULTIPLE CRITERIA DECISION ANALYSIS

goals, objectives or criteria (for otherwise no real "decision making" is involved). In not all problems, however, will the multicriteria features, and the need for considering tradeoffs between different goals, be of a sufficient magnitude to warrant the discipline and effort of a full multicriteria analysis as we shall describe it here. Our emphasis will be on those problem contexts in which the multiple criteria nature is a substantial characteristic of the problem. By this we mean that the resolution of the multiple criteria problem is not easily left to intuition, and/or that one or more of the reasons for structured analysis discussed in Section 1.1 and 1.2 may be of particular importance. Later in this chapter we shall illustrate such contexts by a number of case examples.

In much of this book, we will be differentiating between *decision makers* who have responsibility for the decision, and *facilitators* or *analysts* who attempt to guide and assist the decision maker(s) in reaching a satisfactory decision. To a large extent, the tools and methods discussed in this book will be viewed as constituting *multiple criteria decision analysis (MCDA)*, or by some writers *multiple criteria decision aid*. As discussed in Section 1.3, a thread running throughout this book will be the recognition of three key phases of the MCDA process, namely:

1 *Problem identification and structuring:* Before any analysis can begin, the various stakeholders, including facilitators and technical analysts, need to develop a common understanding of the problem, of the decisions that have to be made, and of the criteria by which such decisions are to be judged and evaluated. This phase is discussed in some detail in Chapter 3

2 *Model building and use:* A primary characteristic of MCDA is the development of formal models of decision maker preferences, value tradeoffs, goals, etc., so that the alternative policies or actions under consideration can be compared relative to each other in a systematic and transparent manner. An overview of the different classes of models which have been proposed is given in Chapter 4, while greater details are included in later chapters, particularly Chapters 5 to 8.

3 *Development of action plans:* Analysis does not "solve" the decision problem. All management science, and MCDA in particular, is concerned also with the implementation of results, that is translating the analysis into specific plans of action. We view the process of MCDA not only in terms of the technical modelling and analytical features, but also in terms of the support and insight given to implementation.

Although the methods of MCDA could in principle be utilized by the decision maker directly, it is our view that in the vast majority of non-trivial problems the methods will be implemented by a facilitator or analyst, working with decision makers and/or other interested or responsible parties. This adds a further level of complexity to the range of situations which may be termed MCDM. In some situations, analysts may be tasked with preparing recommendations to be tabled at a later stage before a board or political decision-making body, where the final decision may not be directly facilitated by them. Also, it may occur that the client of the analyst is not (directly) *the* final decision maker, but a group delegated to explore the issue and to make recommendations, or one of many stakeholder groups seeking to influence the final decision maker.

The previous paragraph has identified a number of contexts within which MCDA may be applied, but differing according to the types of output required by the client or user of MCDA. Even when we assume that the alternative courses of action are clearly identified, the outcome may not be the simple choice suggested in the first paragraph of this chapter. Roy (1996) identifies four different *problematiques*, i.e. broad typologies or categories of problem, for which MCDA may be useful:

The choice problematique: To make a simple choice from a set of alternatives

The sorting problematique: To sort actions into classes or categories, such as "definitely acceptable", "possibly acceptable but needing more information", and "definitely unacceptable"

The ranking problematique: To place actions in some form of preference ordering which might not necessarily be complete

The description problematique: To describe actions and their consequences in a formalized and systematic manner, so that decision makers can evaluate these actions. Our understanding of this problematique is that it is essentially a **learning problematique**, in which the decision maker seeks simply to gain greater understanding of what may or may not be achievable.

To these, we might venture to add:

The design problematique: To search for, identify or create new decision alternatives to meet the goals and aspirations revealed through the MCDA process, much as described by Keeney (1992) as "value-focused thinking".

The portfolio problematique: To choose a subset of alternatives from a larger set of possibilities, taking account not only of the char-

16   *MULTIPLE CRITERIA DECISION ANALYSIS*

acteristics of the individual alternatives, but also of the manner in which they interact and of positive and negative synergies.

In the next sections, we shall discuss a number of practical situations in which multiple criteria decision making problems can arise, with the aim of illustrating what we mean by MCDM in a more accessible manner than may be achieved by formal definitions.

The examples illustrate a broad range of problem settings in which the MCDM framework arises quite naturally. They are hypothetical, but are based broadly on case studies from the literature and/or on the experiences of ourselves and colleagues, and are thus realistic. We shall use a number of these case studies throughout the book to illustrate the process and methods of MCDA. Also we hope that all of the examples will serve to promote discussion and reflection in both the practitioner and educational contexts.

2.2.   CASE EXAMPLES

2.2.1.   Discrete Choice Problems

The classical context in which MCDA is often discussed, and the most obvious category of MCDM problem, is that of a simple choice from a set of alternatives. We have emphasized that there is a much wider range of problems to which MCDA may be applied, but it is nevertheless useful to begin by looking at this classical case. Many multicriteria discrete choice problems that nevertheless require a considerable degree of evaluation and analysis have been described in the literature. Some examples include choice of locations for power stations (Keeney and Nair, 1977; Barda et al., 1990); evaluation of tenders and selection of a contractor to develop an information system (Belton, 1985); selection of metro stations in Paris for renovation (Roy et al., 1986); haulier selection (Sharp, 1987); relocation decisions (Butterworth, 1989); selection of an architecture for a defence communication system (Buede and Choisser, 1992); evaluation of public transport options (Bana e Costa and Vansnick, 1997); and selection of a textbook for a course (Weistroffer and Hodgson, 1998).

The impression that may often be gained from a superficial reading of case studies such as those cited above, is that the alternatives and the criteria against which they are judged are more-or-less self-evident, and that the difficult problem is to identify the solution which best satisfies the associated decision-making goals. As we will be emphasizing in the next chapter, such an impression is far from the truth. In most non-trivial situations it is at least as much of a problem to identify suitable alternatives and to establish appropriate criteria, as it is to make the selection from the available alternatives. We now introduce a

The Multiple Criteria Problem   17

typical situation in which such a discrete choice problem may arise in any company or organization. In later chapters, this case study will be used to illustrate the problem structuring process and the analysis of the problem using a number of different techniques. We shall also see that even in such apparently simple choice problems, the purpose of the multicriteria decision analysis is directed much more towards gaining an understanding of the issues involved, than simply towards generating a single "best" choice. For now we look simply at how the problem may first present itself.

An Office Location Problem

A small company called Decision Aid International, based in Washington DC, USA, provides decision support consultancy to the public and private sector. The company employs 5 facilitator / analysts who work directly with clients and are supported by a technical team of 5 software developers and 2 administrative staff. The managing director believes that there is a new and growing market in Europe for services offered by the company and is looking to open an office there. However, she is unsure of the best location and wishes to explore the multitude of possibilities in a structured manner. Many factors relevant to the decision are highlighted in the following conversation between the managing director (MD) and one of the company's analysts:

Analyst: Why do you think it would be advantageous to have a European base? Couldn't the work be handled from Washington?

MD: Well ... firstly I think that developments in the European Union and the growth of the market economy in former Eastern European countries mean that there will be substantial new opportunities for companies such as ours in the near future. Although it would be important to involve the US staff in this work and we could get involved to an extent from our current base, I feel it would be essential to have a permanent presence to be close to clients, to be able to react quickly to opportunities and to keep "in tune" with European cultural issues. Secondly, although occasional trips to Europe would have novelty value for our US employees, too much travel becomes tiring and would undermine morale.

Analyst: Given the need to involve US staff in the European work, then presumably you would be looking for a location which is easily accessible?

MD: Yes, and one which is reasonably attractive to visit for the US staff. It is also important that it is a location which gives good access to

18 *MULTIPLE CRITERIA DECISION ANALYSIS*

other European countries - particularly as the nature of our work means that we tend to spend one day a week with a client over a period of time.

Analyst: Does this restrict you to capital cities - given that they usually have better transport links?

MD: I think we definitely need to be somewhere which has easy access to an international airport - probably with direct flights to the US and a good European network. However, this doesn't necessarily restrict us to capital cities - for example, Glasgow or Manchester in the UK, Milan in Italy or Düsseldorf in Germany all have international airports. These cities may be closer to the county's industrial heart - and may have the advantage of attractive financial packages for new companies locating there.

Analyst: What type of clients are you most interested in?

MD: Currently we have clients in the public sector - particularly government and health - and the private sector. At the moment our private sector clients are primarily in high-tech manufacturing companies - but we have done work in traditional manufacturing industries and in the service sector. However, our expertise is, of course, equally relevant to all organisations. I think we should be located somewhere which gives good access to a range of potential clients, although I think that the public sector is likely to be most important in the immediate future.

Analyst: You mentioned former Eastern European countries earlier - would you consider locating in, say, Warsaw?

MD: That's an interesting question - I really do not feel that I have sufficient knowledge to decide. I would have to be sure that suitable office space was available - when we set up here we initially used an office services company which provided secretarial support, photocopying facilities, etc. - ideally I would like to do the same in Europe with the flexibility of moving to employing our own staff and facilities as we expand. This reminds me that the availability of suitably qualified staff - both consultants and support staff - is, of course, very important. Another issue is the ease of setting up the business - I would want to avoid excessive bureaucracy and legalities - existing links with the local Chamber of Commerce might be useful. It goes without question that there needs to be excellent telecommunications links back to the US. All these comments apply to all potential venues .... it's just that thinking about somewhere unfamiliar brought them to mind.

Analyst: Does thinking about other cities raise concerns in your mind?

MD: Funny you should say that, I was just doing a mental check. I have to say that I have some concerns about language. Although I speak French I'm not sure which of the consultants do - might that be a problem if we located in Paris? None of us speaks any other European language. That might suggest that a UK base would be best. We've hardly touched on financial matters - the high taxes and cost of living in Scandinavian countries is a deterrent, although I have to confess to being attracted to their lifestyle. On the other hand, the food, wine and climate of the southern European countries is rather tempting ... but ... this is meant to be a business decision, isn't it? Where have we got to?

At this point, both the analyst and the MD are beginning to develop some common understanding of the problems which have to be addressed, and of some of the issues (criteria) which need to be taken into consideration. They cannot be sure at this stage that all possible locations have been thought of, nor whether all issues of concern have been identified. Nevertheless, there is enough appreciation of what needs to be done, so that the MD and analyst can start the process of structuring and analyzing the choices which have to be made. They can probably prepare an initial discussion document, to be tabled at a meeting of the critical stakeholders. We shall return to this structuring phase in the next chapter.

2.2.2. Multiobjective Design Problems

The decision problems discussed in the previous section involved a discrete set of explicitly defined alternatives. For example, in the business location problem, the choice would ultimately be made from a list of possible cities in which the company might choose to locate. Frequently, however, alternatives may be more implicitly defined in terms of a variety of components or features which may need to be selected and combined, but where the possible or allowable combinations are subject to a number of feasibility constraints. For example, financial investment options will be constructed from a variety of stocks and shares (and derivatives), property, money market accounts, etc., where choices are restricted by availability of funds and possibly by legal constraints. This process of selecting one or more explicit options from the potentially infinite range of possibilities can be described as a "design problem", in the sense that we are assembling a particular course of action from more fundamental elements. Selection of a satisfactory course of action will in non-trivial instances, just as with any other decision making processes, involve con-

20 MULTIPLE CRITERIA DECISION ANALYSIS

sideration of multiple criteria or objectives, i.e. they are *multiobjective design problems.*

Examples of such multiobjective design problems have included power generation planning (Martins et al., 1996); selection of rootstock in citrus agriculture (Benson et al., 1997); plant layout for manufacturing flexibility (Phong and Tabucanon, 1998); and reservoir operation planning (Agrell et al., 1998).

Multiobjective design problems may emerge at two stages in the broader multicriteria decision making process. As illustrated in the game reserve planning study to be described below, the design problem may follow on from higher level strategic decisions, with the aim of implementing these decisions at a more detailed operational level. In other instances, as will be described in the next section, the design stage may be aimed at generating an explicit set of alternatives for more detailed evaluation before a final decision is reached.

The case study which follows, based broadly on that described by Jordi and Peddie (1988), exemplifies the multiobjective design problem, and will be used later in the book to illustrate a number of MCDA approaches. As the example is developed in later chapters, we will note that it leads naturally to formulation in multiobjective mathematical programming terms. This appears to be a typical feature of many multiobjective design problems, and it is for this reason that, particularly in Chapters 6 and 7, the discussion of MCDA methods will be extended to include mathematical programming (primarily linear programming) approaches. Much of the literature on MCDM problems tends to specialize in either discrete choice or mathematical programming problems, but for an integrated view of MCDA it is important to include both.

Stocking of Game Reserves

In recent years, South Africa has seen the establishment of many private game reserves. The initial strategic management decisions would relate to the size and location of the reserve area, and to which primary game species (typically large mammals) are to be kept in the reserve. The "design" decisions which would follow on from these strategic choices would relate to the operational management of the reserve or game park, and would in particular include decisions regarding:

- Desirable stock levels for each of these species, and sometimes also the allocation of these stocks to different areas or camps (possibly representing different habitat types) within the reserve; and

- The means of disposal of excess stock, as population grows beyond these desirable levels.

The Multiple Criteria Problem 21

Other problems may relate to the balance between predator and prey species, but many smaller reserves do not keep large predators, and for the purposes of this case study we shall assume that the only species under consideration are the larger herbivores (typically a variety of antelope species, perhaps together with zebra, rhinoceros, hippopotamus and elephant). Many different issues need to be taken into consideration, such as the following:

1 The availability of food and water supplies for the animals in various parts of the reserve: This also requires consideration of an adequate balance between various types of herbivores such as browsers and grazers;

2 Direct and indirect economic benefits derived from tourism, which includes game viewing and hunting (generally according to strictly controlled permits): Such benefits come from entry fees, sale of hunting permits, sale of curios, and by provision of accommodation, meals and refreshments;

3 Direct economic benefits from the sale of live animals (to other reserves, or to zoos) or of meat;

4 Maintenance of good relations with surrounding rural communities, often by permitting grazing of cattle and/or free hunting by traditional methods, and/or by providing contributions in cash or kind (e.g. meat) to local villagers: Where these communities do not see a direct benefit to themselves, they may resent the loss of what they often perceive to be their traditional hunting or grazing areas to wealthy financial interests, and this can lead to problems with poaching or even security for visitors.

5 Effects on or contributions to national conservation goals, either out of general conservation interest, or because the potential for spin-offs through public image and general increases in eco-tourism.

Clearly, even within the constraints imposed by the initial strategic decisions, there remains an effectively infinite variety of possible options available, the range of these being defined only implicitly through the constraints placed on the system. This is thus a *design problem* as defined above, the initial evaluation of which may have to be undertaken by a working group or team of experts, rather than directly by the decision makers themselves. This working group must, however, ensure that the management objectives of the decision makers are clearly taken into account. We shall return to more detailed structuring and formulation of their problem in the next chapter.

2.2.3. Mixed Design and Evaluation Problems

As we noted during the discussion of multiobjective design problems in the previous section, some design problems may in fact be preliminary to a more detailed evaluation process. As we shall discuss in later chapters, MCDM problems typically involve many stakeholders who are required to make in-depth value judgments concerning the alternatives before them. In practice, as we shall also be discussing later, the comparative evaluation of alternatives can only applied to a relatively small number of discrete options.

The purpose of the "design" phase is then to generate a suitable shortlist of alternatives from the potentially infinite range of possibilities which may exist, for purposes of detailed evaluation similar to that for the discrete choice problem. In some situations, it would be sufficient to obtain a representative discrete set of alternatives simply by considering all combinations of a few discrete options or levels for each of a number of different actions or interventions. Examples of this approach include river basin planning (Gershon et al., 1982) and nuclear waste management (Briggs et al., 1990). Often, however, the numbers of combinations will be so large that a further selection will need to be made at the design phase, paying careful attention to the underlying decision criteria which will come into consideration at the evaluation stage. The following case study illustrates how such a mixed design and evaluation problem may arise.

Land Use Planning

Particularly in the developing world, policy decisions regarding the allocation of land to new or different uses are inevitably controversial, often generating highly emotional responses, with repercussions way beyond the borders of the country concerned. For example, currently undeveloped land such as virgin forests or grasslands may be used for agriculture or open cast mining, or traditional agricultural or sacred areas may be taken over for industrial development. Decisions are ultimately taken at the political level, but this is typically preceeded by a process of consultation with interested groups, and of analysis and screening of options by consultants and/or state officials. During this process, many preliminary "decisions" will be taken as to what options need to be evaluated, and which of these are to be presented to the political decision makers for final decisions. Ideally, there needs to be an "audit trail" documenting the rationale behind these preliminary decisions, so that the political decision makers are at least informed concerning such choices, both to satisfy themselves that the process has correctly captured the political goals and to have greater understanding of the issues relevant to the final choices they have to make. The "criteria" in this case are closely linked to the values of different interest groups, each of whom may have internally conflicting goals as well. Decision support and aid may be necessary (a) to individual interest groups (in helping them to formulate their own preferences); (b) to group forums involving representatives of each group (in facilitating the reaching of consensus); and (c) analysts seeking to identify the most acceptable courses action to present to the political decision makers.

For purposes of illustration, we focus on the particular types of problem discussed by Stewart and Scott (1995) and Stewart and Joubert (1998). These relate primarily to the controversies in the eastern escarpment areas of South Africa, regarding changes of land use from virgin land and/or traditional farming to commercial forestry plantations and to managed conservation areas, and the construction of reservoirs and/or artificial transfer of water between different river basins. A key concern in these decisions relate to the allocation of increasingly scarce water resources, and the need to ensure proper management of run-off even before it enters defined river courses.

Two levels of choice can be identified as part of the strategic planning processes described above:

1 At early stages of the strategic planning process, the emphasis should be on generating alternatives. The aim should be to have a rich selection of potential courses of action, taking into consideration the values of all interest groups. In this way a very large, or even infinite, number of alternatives is likely to be identified. In some cases the options may be generated implicitly by sets of constraints on key decision variables (e.g. proportions of land in each sub-region allocated to each form of activity). In other cases, the options may consist primarily of proposals from various interests, together with alternatives which combine attractive features of each of the original proposals. Even with a relatively small number of features, each of which may or may not be included in the strategic plan, this can give rise to many hundreds or thousands of alternatives.

In this class of problem it is typically not possible to do a complete evaluation of such large numbers of alternatives, as the evaluations may require public scoping and expert assessments involving explicit consideration of each. The first stage is thus to select a shortlist (of perhaps not more than 10 alternative strategies) for more detailed investigation and evaluation. While this selection is not final (it is possible to re-consider the selection if none of the original selection are adjudged to be satisfactory), it can nevertheless have a profound influence on the final decision, and thus already needs to take all interests into account in some defensible manner. The selection is thus itself a multi-criteria decision, but one that needs to be based on readily available ("objective") quantitative attributes of each alternative.

2 The shortlist of alternatives needs to be evaluated in depth by all interested and affected parties, on the basis of both quantitative and qualitative criteria, some of which may require subjective judgments. The big problem at this stage is to find a course of action which is broadly acceptable to all (or at least to an overwhelming majority of) interest groups. This requires that procedures be developed which ensure adequate communication between groups as to the reasons for their choices. If an acceptable level of consensus is not achieved, then it may be necessary to return to the previous step, using the results of the evaluations by the different groups to refine the selection of the shortlist of alternatives

In summary, we have identified two levels of multiple criteria decision making within the context of land use planning. The first level (the "design" phase) requires an almost automated, but still broadly justifiable and acceptable, procedure for screening out evidently poor alternatives, and for generating a relatively small set of alternative policy scenarios for deeper evaluation. The second level (the "evaluation" phase) involves a search for consensus between possibly highly divergent, and often highly emotive and intangible, interests, taking into consideration many uncertainties and judgmental imprecisions.

2.2.4. Project Evaluation, Prioritization and Selection

Many organisations, in both the public and private sectors, face decisions on how to allocate spending on research and development projects. A number of case studies which illustrate different organisational contexts and different approaches appear in the literature, such as Islei et al. (1991) which describes an approach adopted by the Pharmaceuticals division of ICI (Imperial Chemical Industries) to monitoring R&D expenditure on drug compounds; Stewart (1991) which describes a decision support system developed for the engineering investigations division of a large electricity supply company; Belton (1993) which describes the process adopted to help the Social Services Department of a UK County Council in prioritising new initiatives such as improved care packages or new information systems; and Morgan (1993) which concerns resource allocation in local government.

In order to illustrate this type of application, we describe two specific case studies.

Universities Funding Council:

In 1992 the Universities Funding Council (UFC) in the UK (the government body responsible for the annual allocation of funds for teaching and research at universities in the UK) announced a new initiative which was designed to investigate means of "...using technology to improve the efficiency of teaching and learning in Universities". It was called the Teaching and Learning Technology Programme (TLTP) and was allocated initial funding of £7m per annum for a period of 3 years. The UFC declared its intention of supporting two kinds of project – subject based consortia which should involve a number of Universities working together to investigate the use of technology in the delivery of that subject, and Institutional projects which should be aimed at facilitating or encouraging the use of technology across subject areas within that University. A call for proposals set out in more detail the requirements and constraints of the sponsors and the format to be adopted by submissions. A deadline for submissions was specified.

The situation is one with which anyone who has applied for a grant, from a national research council, a charitable foundation or an international funding body such as the European Union will be familiar. The problem facing the grant awarding body is how best to allocate the available funds.

Many responses to the TLTP call for proposals were received, including both Institutional bids and subject based consortia. The proposed content covered many different subject areas and initiatives. Many of the proposals were to develop computer based teaching / learning materials, some focusing on initial presentation of materials using multimedia and interactive technologies, others on tutorial software, others on alternative means of assessment. Other proposals included the use of video conferencing to give wider access to specific expertise.

We do not propose to discuss here how the decision to allocate the available funds to projects was actually made, but to use the problem to illustrate how multiple criteria analysis could be used as part of the decision making process and to bring attention to some of the contextual issues to which an analyst / facilitator must pay attention.

Let us first consider in more detail the context of the problem and the nature of the desired solution. What would we need to know to be able to help the decision makers? A first question to ask is who are the deci-

26   *MULTIPLE CRITERIA DECISION ANALYSIS*

sion makers in this context? In this case, as in many similar situations, the power to make recommendations lies with a panel of experts recognised for their knowledge of the field. It is interesting to reflect on the nature of the involvement of this group in the decision. The professional involvement of the group in the field should be a guarantee of interest and commitment. However, their personal stake in the problem is limited - perhaps going no further than their personal reputation within the panel of experts - they probably do not have to answer directly to the instigators of the proposals for their decision (in many such situations the feedback received by both successful and unsuccessful candidates is limited) - they are unlikely to be held ultimately responsible for the success or failure of the programme - indeed their identity may never be revealed. How does this compare and contrast with a decision making group within an organisation with a greater degree of responsibility and accountability for their decisions, decisions which may impact directly on the viability and success of the organisation as well as their own progress within that organisation? In what way are these differences likely to impact on decision making?

A second question to ask is the extent of this group's role in the evaluation. Does it include a specification of how the projects are to be evaluated - i.e. what factors are to be taken into account - or are they simply evaluating the projects in the light of a pre-specified agenda?

Other questions include, how do they intend to operate - will they first evaluate the projects individually and then come together to discuss their assessments? Are they aiming to arrive at a consensus view on which projects should be funded or do they intend to work from individually prioritised lists (for example funding each person's 5 preferred projects)? How much time are they able to spend together? How will they communicate at other times?

Let us assume that the panel's role includes determining how the projects are to be evaluated, taking account of the funding council's stated objectives and that their recommendation to the council is to be a list of projects which should be funded. How might they go about this? One approach could be to evaluate each of the projects independently in a way which enables the panel to list them in priority order. Funds could then be allocated beginning with the highest priority and moving down the list until all funds are allocated. What are the potential problems with this approach?

How should projects of different magnitudes be compared? Some of the projects are high cost and likely to yield substantial benefits whilst others are at a smaller scale with correspondingly smaller impact. Is it possible, or reasonable to order the projects in terms of benefits per unit cost? Would this create a bias towards large or small projects?

Are the projects independent? Do some projects have others as a prerequisite? Are there potential synergies or dissynergies? For example, are there projects which are likely to be more successful if others are also selected? Alternatively, are there projects which are broadly similar thus replicating rather than adding overall benefit?

Are there portfolio considerations? For example, is there a wish to achieve a balanced allocation of funds, perhaps across broad subject areas or across Universities? If this is the case then it should be necessary to consider not just the individual projects, but also alternative portfolios of projects. With possibly a 100+ projects, the number of possible combinations which lie within the budget constraint is very large.

However, it would seem that whatever the complexities of the situation it would be sensible to begin by evaluating individual projects – these evaluations could then form the basis for consideration of the portfolio. The following paragraph outlines some of the factors which are considered to be relevant.

The UFC's stated concern was to improve the efficiency of teaching and learning and hence the potential to do so was of significant interest in selecting projects. One aspect of this is the nature of the materials being developed – to what extent might they allow lecturers to spend less time in delivery of materials, or enable delivery to larger groups of students without increasing resources. Another aspect is the potential "reach" of the materials - how many academics would be able to use the materials, how many students? To what extent does the proposal consider ways of achieving maximum reach?

The potential for the project to enhance the effectiveness of teaching is also relevant. This may be viewed as facilitating active learning, facilitating independent learning, giving access to new materials or resources.

The degree of innovation might also be considered - in particular, the UFC would like to encourage exploration of the use of technology in areas which had not previously embraced its use - for example in the Arts, Humanities and Social Sciences. The innovation might also be technical in nature - but it should at the same time be realistic,

The quality of the proposal itself may be an influencing factor: Is it well written? Are the objectives realistic? Does it include a clear schedule? Are there clearly specified deliverables? Is there evidence to suggest that the proposers are able to achieve the objectives (e.g. previous experience and reputation)?

28   MULTIPLE CRITERIA DECISION ANALYSIS

Of course, there are many other factors which the decision making group may want to take into account, not least of all being the cost of the proposed projects.

Problems of this type are in a sense the "inverse" of the one outlined in the previous section, beginning with the "evaluation" of individual projects before moving on to the "design" of a preferred portfolio, once again illustrating the need for an integrated understanding of methods and processes.

Maintenance and Development of Computer Systems:

This case study concerns the Management Information Systems Department of a large banking group. One of the major responsibilities of the department is the maintenance and development of a number of in-house computer systems; these include a number of large scale systems, for example, a cash management system, a customer database, as well as several more specialised DSS. The department also provides support to users, responds to system "emergencies", is involved from time-to-time in the development of new systems, advises other departments of the bank on the purchase of commercial software and advises senior management on issues relating to information strategy. However, it is with the maintenance and development role that this case is primarily concerned. As technology and the business develop so do the needs and expectations of the system users, who include bank staff, corporate and personal customers, creating a constant demand for enhancements to the various systems. In addition to considering requests as they arise, the department tries to be proactive in seeking areas for development, in particular monitoring the customer services offered by competitors.

Each system enhancement is referred to as a project, although these may vary in scope from something which is expected to be feasible in one person day, for example to incorporate a new graphical display in a DSS, through to a major project such as the rewriting of a system in a new language which could absorb several person years.

The estimated resource requirement of the list of potential projects invariably exceeds the resources immediately available, hence projects must be prioritised. A review meeting is held monthly to report on the progress of ongoing projects and to consider new proposals. Occasionally, a new project arising mid-month will be slotted into the schedule if resources permit (for example, if another project has been completed in less time than budgetted or if demands for other services have been lower than anticipated)

The manager of the MIS Department would like to establish a procedure (with appropriate support software) which allows her to:

The Multiple Criteria Problem   29

- Monitor the progress of ongoing projects

- Evaluate newly proposed projects with a view to establishing whether or not they are worth pursuing and what priority they should be allocated

- Allocate resources to projects (principally staff time)

Factors to be taken into account include:

Resources: The main consideration is staff time - measured in person days - although it is not quite so simple, as different staff have different expertise and some projects call for more senior involvement. Larger projects which involve a number of staff also require a project manager. In addition to the resource utilised, the expected duration of the project is also relevant - past experience shows that longer projects are more likely to encounter problems.

Occasional projects may require specialised hardware or software.

Impacts: One consideration in assessing the impact of a project is who will benefit. The number of beneficiaries is an issue, but also whether they are internal to the organisation or customers. If internal, is the benefit to management or operations? If external, what type of customer - corporate or personal?

Of course, the nature of the benefit is also relevant: will the project result in increased efficiency of operations? Will it provide better information for decision making - on a regular basis or in a one-off situation? Will it result in "better" systems - perhaps by moving to a newer, more powerful or more flexible language, or by increasing standardisation across the banks systems? Will it make someone's job easier? Will it generate new customers? Will it improve the service to existing customers?

In common with the previous case, this situation comprises an "evaluation" phase, assessing each project on its own merits, followed by a consideration of how it would fit into the existing portfolio. However, the "design" aspect of the problem is less significant, as the context is one of ongoing monitoring and review, rather than a one-off decision about which projects are selected and which are not.

2.2.5. Classifying Alternatives

In some situations the problem owner is concerned not with choosing a preferred alternative, but with classifying alternatives (i.e. the "sorting problematique" of Roy discussed earlier). It may be a simple two-way

30   MULTIPLE CRITERIA DECISION ANALYSIS

classification - pass / fail, acceptable / unacceptable - or there may be more categories. Some examples have been the evaluation of universities in the UK in terms of research and teaching quality, or the categorization of estuaries or river basins in South Africa into conservation classes.

We now briefly discuss one case study involving the assessment of the quality of suppliers to a high-tech company, categorising them according to whether or not they achieve an acceptable standard.

The company operates in market in which there is a large and in part rapidly changing supplier base of electronic components. For many reasons the company cannot and would not want to restrict its dealings to a single supplier, or even to just a few suppliers - in such a volatile industry it is important not to put all your eggs in one basket. It is not unknown for even reputable and well established suppliers to go out of business overnight. Furthermore, customer loyalty is not strong and suppliers have been known to default on an order (possibly having received a better price from another purchaser). The company has thus to deal with many suppliers - however, they are concerned about supplier quality and would like to develop a rating system which enables them to classify actual and potential suppliers according to whether or not they achieve a specified standard.

Quality is a multi-faceted concept which encompasses issues such as the following:

- Quality of delivered goods - what is the defect rate in goods supplied?

- Reliability of deliveries - do they arrive on time? (too early and too late both present problems)

- Invoicing procedures

- The supplier's own quality procedures (e.g. are they ISO9000 compliant?)

- Reputation of the supplier

The rating system must thus be capable of taking all of the above facets into account when determining a classification for any particular supplier. In short, the rating system addresses a multiple criteria decision making problem. Note that while some of the relevant quality criteria may be assessed on a relatively objective basis, others (such as *reputation*) will remain a primarily subjective judgement.

An important feature of this type of MCDM problem, in common with the one just described, is that the decision analysis is applied not just to a specific instance, a one-off decision, but to the development of the rating system or procedure which will be used on a regular basis for many future decisions. It is important, therefore, that the system be both clearly understandable and broadly acceptable to management and to those who need to implement it. In this case, the rating system must therefore specify both how potential suppliers are to be assessed according to each of the individual criteria, and how these individual assessments are to be aggregated into an overall classification.

2.3. CLASSIFYING MCDM PROBLEMS

The previous case studies illustrate to some extent the range of problems which we can describe as being MCDM problems. In concluding this brief chapter, it is useful to attempt some classification of MCDM, and of MCDA in support of such problems, in terms of the various characteristics which we have identified.

One-off vs. repeated problems: Some decision problems may recur at regular intervals, with essentially the same issues having to be considered at each occasion. For example, many of the project evaluation problems of funding agencies discussed in Section 2.2.4 have to be repeated monthly, quarterly or annually. Other decision problems, however, may be essentially unique in that even apparently similar problems occurring elsewhere will require a fresh look at all the issues. For example, a decision regarding the routing of a new highway will be "one-off" in this sense. While such construction decisions may need to be made on many occasions, each will have substantially unique features, such as criteria to be taken into consideration, stakeholders involved, characteristics of the alternatives, etc.

In the case of repeated decisions, the role of MCDA may be directed more towards the setting up of procedures to be followed each time, than to specific decisions in a particular case. In fact, for project funding decisions in the public sector it may be essential to have the procedure declared up front so that potential applicants are informed of the criteria according to which they will be judged. In this context, it may be particularly important to ensure that the procedure can cope with options well outside the range of those available on the first occasion.

One-off decisions are by definition directed towards the one specific decision, so that the context may be more clearly identified than for repeated decisions, facilitating the analysis. On the other hand, one-off decisions tend often to be of a more strategic nature, so that the criteria may relate to a wider range of fundamental objectives needing deeper levels of exploration.

32 MULTIPLE CRITERIA DECISION ANALYSIS

Number of stakeholders: The "decision maker" may be a single individual, a small group of individuals with more-or-less common goals, or a corporate executive or political decision maker acting on behalf of a large group of interested and affected persons with divergent interests.

For a single individual, or small homogeneous group with shared objectives, the MCDA process can be used to identify the final decision directly, without need to justify or to debate this with other groups.

For decision making involving groups with more divergent objectives, the final decision is likely to involve some form of political negotiation between stakeholders, each of whom may adopt different sets of criteria for evaluating alternatives. The final decision maker will need to take all of these criteria into account in seeking a political consensus or compromise. The analysis may need to be conducted within a group setting involving representatives of all stakeholders, or may be carried out separately for sub-groups as a form of scoping exercise or impact assessment. In this case, the final decision may occur entirely at the political level, not directly facilitated by an MCDA process.

Status and influence of the client: The decision analyst applying MCDA techniques will be providing support to some client. This client may well be the final decision maker(s) themselves, so that the results of the analysis may lead directly to a conclusion reached jointly with the decision maker concerning the course of action to be adopted. In other situations, the client for the MCDA may be a group tasked with preparing recommendations for action by the decision maker, or a board or a council. The MCDA process should then typically generate more than one possible recommendation, with pros and cons for each expressed in terms of the relevant criteria and trade-offs between them.

There may also arise occasions in which the client requesting MCDA assistance has no direct influence on the decision maker. Such a client may, for example, be a particular stakeholder group (see previous point), requiring assistance in understanding their own preferences and in preparing submissions to the final decision making authority.

The "problematique": MCDM problems differ also according to the problematiques identified in the introduction to this chapter. These different problematiques may, however, frequently derive from the number of stakeholders involved, and from the status and influence of the client (i.e. the previous two modes of classification).

Range of available alternatives: In some cases (such as the location problem in Section 2.2.1), the number of alternatives to be considered will naturally be relatively small, and will be identified explicitly. In other cases, the number of alternatives may be very large, or even infinitely many, often defined only implicitly in terms of constraints that the decisions need to satisfy (as, for example, in investment decisions where all possible portfolios of stocks satisfying availability of funds and legal constraints need to be considered). Where the number of alternatives is large or infinite, it may be necessary to focus the more detailed phases of analysis on only a subset, or short-list, of alternatives (as suggested for the land-use planning case study in Section 2.2.3). The selection of the short-list is itself an MCDM problem, so that the analysis is decomposed into two phases, namely selection of the shortlist followed by evaluation of the alternatives in the short-list.

Facilitated vs. DIY analysis: For most of this book, our assumption will be that the MCDA techniques will be used by a facilitator guiding decision makers towards finding the decision which best satisfies their needs. We recognize, however, that in some cases, especially for a single decision maker or a small homogeneous group, decision makers may wish to undertake their own analysis in what we shall term "do-it-yourself" (DIY) mode. This is particularly so in the current day when access to computer facilities and associated decision support systems is widespread. Not all of the MCDA methods discussed in this book are easily applied in DIY mode, but our intention is also to make the methodologies of MCDA accessible, as far as is possible, to decision makers themselves.

For the practice of MCDA, it is important for the analyst to have a clear understanding of the category or categories into which the problem falls. Nevertheless, the need for a clear problem structuring, as will be discussed in detail in the next chapter, in order to identify relevant criteria and to creatively develop value-focused courses of action, is equally important, no matter what the problem classification. Some of the methodologies of MCDA to be introduced in later chapters may be more appropriate to certain types of problem than to others, and we shall from time to time indicate the extent to which a particular method is especially well-suited, or not suited, to specific categories of MCDM problem. It is our belief, however, that most of what we will discuss is applicable to all classes of MCDM problems.

Chapter 3

PROBLEM STRUCTURING

3.1. INTRODUCTION

"A problem well structured is a problem half solved" is an oft-quoted
statement which is highly pertinent to the use of any form of modelling. A mismatch between the problem and model used was the most
frequently cited reason for the failure of OR interventions in a survey
conducted by Tilanus (1983). The validation of a model of human values is a different prospect to that of a physical system, as there is no
external reality against which it can be compared. Indeed, we view the
process as constructive, helping decision makers to understand and to
define their preferences, rather than descriptive, describing what they
do and seeking simply to elicit their preferences. Thus (as we discuss
further in Chapter 4), we do not seek to model a pre-existing "reality"
however abstract. This serves further to highlight the importance of
problem structuring in the context of MCDA.

In Chapters 1 and 2, we distinguished three phases of the process of
MCDA, namely problem structuring, model building and use, and the
development of action plans. In this chapter we focus on the problem
structuring phase and on the link to model building, in particular to
developing the structure of an MCDA model. We begin with a discussion
of the importance of problem structuring in MCDA and then go on to
look at some of the tools, techniques and processes which can help in
this phase of an investigation.

The relevance of problem structuring to MCDA

Problem structuring is the process of making sense of an issue; identifying key concerns, goals, stakeholders, actions, uncertainties, and so

36 *MULTIPLE CRITERIA DECISION ANALYSIS*

on. Rosenhead (1989) defines it as "the identification of those factors and issues which should constitute the agenda for further discussion and analysis."

The process of making sense of the issue may be an informal one; it may be supported by one of a broad range of general managerial tools, such as a "SWOT" (strengths, weaknesses, opportunities, threats) analysis, or by one of the problem structuring methods loosely labelled as "soft OR", for example, SODA (Strategic Options Development and Analysis, Eden and Simpson, 1989) which has been more recently extended to the concept of JOURNEY making (Eden and Ackermann, 1998), SSM (Soft Systems Methodology, Checkland, 1981; 1989), or Strategic Choice (Friend, 1989). From this understanding it may emerge that more detailed modelling of some aspect of the problem is appropriate.

In this book we use the term problem structuring to describe this process of arriving at an understanding of the situation which, if appropriate, provides the foundations for some form of MCDA.

3.2. WHERE DOES MCDA START?

In theory

As mentioned in chapter 2, the majority of the earlier literature on MCDA focuses on the evaluation of alternatives, or the exploration of possible solutions, having identified the boundaries of the problem. The emphasis has been on "solution" rather than "formulation" of the problem, and little has been written on the problem structuring process as defined above. Possibly rightly so, as this stage of an analysis is much broader in scope than MCDA. However, as the field has matured, there is increasing emphasis on model building and methodological issues. This is most developed amongst the advocates and practitioners of multi-attribute value analysis (Von Winterfeldt and Edwards, 1986; Keeney, 1992; Phillips, 1984; Watson and Buede, 1987), with an important contribution from Roy (1985,1996), and increasing attention from researchers/practitioners associated with the outranking school (Bouyssou et al., 1993) and multi-objective programming (Korhonen and Wallenius, 1996). Others (for example, Belton et al., 1997; Bana e Costa et al., 1999) have looked to the now well established field of problem structuring methods (e.g., Rosenhead, 1989) to provide this support for MCDA.

In practice

As discussed in chapter 2, all problems and decisions are multicriteria in nature; a multicriteria analysis begins when someone feels that the issue matters enough to explore the potential of formal modelling. The nature of a problem, and the extent to which it has been defined, will of course have a significant effect on the point of departure of an analysis, as indicated by the following illustrative scenarios.

Starting point: a complex and unstructured one-off problem. The realisation of the potential relevance of MCDA and the engagement of an analyst or facilitator may occur very early in the process, before any attention has been given to problem structuring, and long before it is clear what the potential courses of action are. For example, an organisation may have expressed a wish to explore an issue defined in fairly general terms, such as one of the following:

How can we improve the quality of teaching in our University?

Why are sales of product X decreasing and what can we do about it?

How can we increase the use of MCDA methods in practice?

How can we reduce traffic congestion in the city?

How do we encourage greater use of public transport rather than private cars?

How can we encourage recycling of domestic waste?

Should we be seeking to expand the company?

Such issues clearly involve multiple, potentially conflicting goals, and it is likely that there will be alternative courses of action, or strategies, to be considered. However, we are a long way from the well defined input required for any multiple criteria assessment tool; first of all it is important to establish a good understanding of the situation.

Starting point: a complex one-off decision. Consideration of an issue has progressed to the point of having identified a need or desire to take action, but there are a number of possibilities, which must be evaluated in detail. It may be necessary to take account of the views of different stakeholders and to consider alternative futures. The options may be well defined or they may need to be designed. Examples of such scenarios are:

The evaluation of tenders to develop a complex computer system

The evaluation of alternative road schemes

The evaluation of potential landfill sites

An expanding company evaluating potential acquisitions or mergers

The appointment of a new Managing Director

38   *MULTIPLE CRITERIA DECISION ANALYSIS*

The location of a new call centre / distribution hub / manufacturing facility

Although the alternative courses of action are more clearly defined in this situation, the potential for formal modelling arises because of the complexity of the decision, the significant consequences and possibly the need for accountability.

Starting point: a recurring decision or evaluation.   Rather than being faced with a one-off problem, an organisation wishes to develop a more structured way of handling decisions or evaluations which occur on a regular basis. It may be the case that the individual decisions are not in themselves of significant consequence, but the totality is. A few examples of such situations are:

Evaluation of research grant proposals

Monitoring of research and development programmes (e.g. in a pharmaceutical company)

Personnel selection in large companies

Performance measurement

Supplier evaluation

In such circumstances there may be considerable experience and expertise in making the particular decision. The emphasis here is on making explicit and formalising that process for purposes of coherency and consistency, whilst still leaving room for and encouraging ongoing reflection and review.

The three "categories" of problem described above apparently range from the very messy and unstructured to the reasonably well defined. The need and potential role for problem structuring is most easily seen in the first category, with the emphasis shifting more to model building in the third category. However, although the decisions outlined in the third category are apparently much more clearly defined, these decisions must be taken in the context of an organisational strategy and the extent to which such has been defined may influence the ease with which the operational strategy necessary to formalise this kind of decision making can be specified. Seeking to better understand the organisational context could thus necessitate taking a step back to consider problem structuring from a more general perspective.

Both the theory and practice of MCDA seem to suggest that the MCDA does not start until some degree of problem structuring has been carried out. However, if MCDA is to have a real impact on decision making in practice then analysts must develop the skills which are necessary to handle this broader process. To neglect to do so by shunning involvement in this stage of an analysis is to forfeit significant opportunity for involvement in more significant issues. To neglect to do so by treating an issue as well-defined from the start risks over-simplification and trivialisation of both the problem and the methods. There are, of course, other ways of achieving impact on decision making in practice – for example, by working cooperatively with those with expertise in problem structuring, by convincing these and other OR/MS practitioners that MCDA is a useful addition to their toolbag, and for the future by influencing the education of potential practitioners.

Whatever the point of departure of an analysis, we advise that the process begins with a stage of free-thinking around the issue. This should be a process which surfaces values, beliefs, priorities, facts, points of view, constraints, consequences, causes, ... a process which starts to reveal all which is relevant to an issue. The question is, how can we as analysts or facilitators aid our clients at this stage of the process?

3.3. PROBLEM STRUCTURING METHODS

As already indicated, there is much to be learned about problem structuring from the body of work stemming from the fields of Operational Research and Systems in the UK, collectively referred to as "Soft OR" or Problem Structuring Methods (see Rosenhead, 1989). Included under this umbrella term are the following approaches, which pay attention to multiple objectives and multiple perspectives in a more or less formal way: SODA (Strategic Options Development and Analysis) developed by Eden (e.g., Eden and Simpson, 1989), and more recently extended to the concept of JOURNEY making (Eden and Ackerman, 1998), using the Decision Explorer (formerly COPE) software for cognitive mapping; the Strategic Choice approach, developed by Friend and Hickling (1987); and Soft Systems Analysis developed by Checkland (1981, 1989). Each of these approaches has something to offer problem structuring for MCDA.

From the MCDA literature, the most directly relevant contributions are Keeney (1992) on "value focused thinking" and Po Lung Yu (1990) in his discussion of the manner in which our "habitual domains" affect the manner in which problems are perceived. In addition, the general literature on creativity, for example, the many books by De Bono (e.g. 1985, 1990), also offer useful ideas on how to stimulate thinking. Van Gundy's (1988) book on "Techniques of Structured Problem Solving" is a compendium of approaches designed to aid problem definition, idea

40   *MULTIPLE CRITERIA DECISION ANALYSIS*

generation and evaluation; we have found variants of many of these useful in the process of decision support.

Before going on to consider some ideas for stimulating thinking, let us pause for a moment and consider the context in which this is likely to be happening. It may be a group of people working together to explore an issue of common concern, or it may be an individual working alone. They may be helped by one, or more, facilitators; for the purpose of this chapter, we assume that at least one facilitator is involved, although the issues are of equal relevance in the case of unsupported ("D.I.Y") work. Facilitators must pay attention not only to ways of stimulating thinking, but also to how to capture the ideas which are generated in a way which will be supportive of the continuing process. Furthermore, when working with a group of people, they need also to be sensitive to group process issues, a matter which will be discussed in chapter 9.

3.3.1.   Idea generation and capture

By this we mean the process of making explicit and recording notions of relevance to the problem, however defined. In addition to providing a mechanism to allow participants to contribute everything they know and feel about a problem, the idea generation process should encourage the surfacing of tacit knowledge and beliefs, and should foster creative thinking.

The first step is to identify the issue under consideration and to ensure that all participants in the process share a broad understanding of the concern. It is helpful to write this down somewhere (e.g. on a flip chart) as a reminder of the focus. An example might be any of the issues listed under the scenarios described in Section 3.2, for example, *"How do we encourage greater use of public transport rather than private cars?"*.

Facilitating the ensuing process of idea generation and capture can be done in many different ways. Different facilitators will prefer different ways of working and different approaches may be required when working with an individual, as opposed to a group. Furthermore, external constraints such as the physical environment, the time available, the availability of computer support, etc., may influence the way of working.

At one extreme the facilitator may simply engage in a conversation with the decision making group, eliciting ideas, information and values, but not seeking initially to record these in detail. A structured representation of the problem in appropriate form is written down either as the conversation progresses, or to summarise it. This may be a list of key concepts, stakeholders, uncertainties, goals, possible actions or it may be an initial model structure. This way of working captures the essence of the issue, but the detail, although surfaced, is not recorded. It places a lot of emphasis on the skill and memory of the facilitator and is probably not workable for large, complex issues. There is a danger that the representation of the issue is perceived as belonging to the facilitator rather than the group.

At the other extreme, the process may be designed to capture as much detail as possible and to fully engage the decision making group in doing so. One of the most successful and widely used "tools" to support such a way of working is the humble "Post-It" (adhesive notelets), or more sophisticated variants on this theme such as Oval Mapping (Eden and Ackermann, 1998) and Hexagons (Hodgson, 1992). In the group context this approach enables the simultaneous generation and recording of ideas from a number of participants, initially in an unstructured format, although the ideas are captured in a form which facilitates display and subsequent structuring. The ease with which ideas can be moved around to generate structure means that it can be helpful to use Post-Its also when working with individuals. The requirements for, and conduct of a Post-It session are as described in Figure 3.1 and illustrated in Figure 3.2 which shows a Post-It session in progress.

Of course, these two extremes are not the only ways of working. The facilitator might capture a conversation as it unfolds using Post-Its, or a second facilitator might concentrate on capturing the ideas whilst the first manages the discussion.

A process such as this has the following key benefits:

- As individuals are working initially in parallel, many ideas are generated and captured in a relatively short time

- The information is recorded for longer term reference in a form which is instantly accessible to everyone

- The use of Post-Its, rather than writing directly on a whiteboard or flipchart, allows for the ideas to be moved around and clustered.

- The use of a large wall space means that a lot of information can be captured and viewed simultaneously

- All participants contribute their ideas on an equal footing and have an equal chance to contribute - the process is not dominated by more eloquent or more powerful individuals and there is a degree of anonymity

In some circumstances it may be useful to capture the Post-It exercise in electronic form. This can be done simply in a word processor; however, it may be important to capture the structure of ideas and to preserve the visual impact. The Decision Explorer software (formerly known as COPE, and available from Banxia Software in the UK) provides a

42 MULTIPLE CRITERIA DECISION ANALYSIS

REQUIREMENTS
You will need: a block of Post-Its and pen (one which encourages large writing) for each participant: a large, flat wall space (e.g. a whiteboard or a wall covered with flip-chart paper), to which the Post-Its will stick: space in front to allow participants access to the wall to stick up their own Post-Its and to read the ideas posted by others: a flexible seating arrangement which allows participants to easily move around the room.

INDIVIDUAL IDEA GENERATION
Each participant is asked to write down their ideas / comments / questions – one idea per Post-It, using up to 7-10 words – relevant to the issue under consideration. This stage of individual reflection is something a Post-It session shares with many group idea generation processes (e.g. the nominal group process described in Delbeq et al., 1975, and in Moore, 1994). Subsequent to an initial period of individual reflection participants are free to chat and to discuss ideas.

SHARING IDEAS
After a short time interval participants are invited, and should be encouraged, to stick up their Post-Its on the wall - initially randomly, but as similar ideas start to emerge clustering these together. Everyone should be encouraged to read others' ideas and to continue contributing new ideas as long as they come to mind.

STRUCTURING IDEAS
The nature of appropriate structuring will depend on the specific issue under consideration, ideas may be clustered according to areas of concern, key issues, action areas, areas of responsibility, ...

Figure 3.1: Guidelines for a Post-It exercise

vehicle for capturing, displaying and also (as discussed later) analysing the structure of ideas. Electronic capture may be done by the second facilitator as ideas emerge, or if a facilitator is working alone it may be done after the session for use in subsequent meetings, or an individual may work directly with the software.

"Hi-tech" systems for idea generation

Technological developments over the past decade or so (such as the growth of networks, wireless LANs, enhanced screen resolution and display facilities) have facilitated the development of electronic parallels to the Post-It exercise. To date these have mostly utilised local area networks, which have allowed participants working together in the same room to input their ideas via their own computer keyboard and screen. All ideas are then displayed on a central screen and can be accessed by

Figure 3.2: Post-It session in progress

each individual. The Group Systems software (Nunamaker and Dennis, 1993) was one of the first such systems. The more recent growth in use of the World-Wide-Web and accompanying development tools has spawned similar systems allowing dispersed and asynchronous working.

Electronic versions of the process offer a number of potential advantages over a manual approach, for example, immediate capture of ideas in electronic form and complete anonymity. However, in comparison with "writing on the wall" they are relatively restricted in their display capabilities, being constrained by the amount of information which can be reasonably displayed on a screen. Early versions of the software simply displayed ideas in lists, which made it difficult to see which ideas were linked and to conceptualise an emergent structure. More recently facilities for grouping ideas have been added to the Group Systems software and a multi-user version of Decision Explorer (known as Group Explorer), providing full structuring and analysis capabilities, has been developed.

The technology itself can, of course, operate as a motivator or a deterrent, depending on the attitudes of the participants. Also, its use

44 MULTIPLE CRITERIA DECISION ANALYSIS

has implications for room layout and flexibility, an issue related to the practicalities of implementing MCDA discussed in Chapter 9.

Stimulating idea generation

It is usually the case that a large number of ideas emerge from the initial process, as described above, particularly if working with a group. The subsequent stage of clustering and linking concepts, as discussed below, will stimulate further discussion. However, it may also be worthwhile for the facilitator to introduce specific activities to focus thinking, in order to ensure that all aspects of the situation have been considered. A number of such activities are now described.

Checklists. A checklist serves to ensure that key aspects of an issue are not overlooked. Perhaps the most widely known is the one denoted by Kipling's six thinking men - "Who and How and Where and What and Why and When". This provides a useful framework for gathering information about a situation and as a basis for idea generation. Checkland's "CATWOE" analysis (Customers, Actors, Transformation, Worldview, Owners and Environment), widely acclaimed amongst systems thinkers, focuses on the nature of the system and the stakeholders involved. We suggest that problem structuring for MCDA may usefully be guided by consideration of CAUSE – Criteria, Alternatives, Uncertainties, Stakeholders and Environmental factors or constraints.

Let us illustrate how the above checklists may be used in creating structure for addressing a question which might arise in transportation planning, such as *"How do we encourage greater use of public transport rather than private cars?"*

Kipling's six thinking men:

> *Who do we want to encourage?*
> Commuters? Shoppers? Leisure visitors? Tourists?

> *How can we encourage use?*
> Ban cars from the city centre? Increase cost of car parking? Improve public transport?

> *Where is public transport used?*
> In areas of low car ownership. In areas of high traffic congestion.

> *When do people use public transport?*
> When their cars are at the garage. When they want to drink. When they are going to the city centre. When they don't want to drive.

What is public transport?
Scheduled buses. Underground. Taxis. (what about cyclists?)

Why is public transport not used more?
Inconvenient. Crowded. Dirty. Unsafe. Difficult to carry shopping. Expensive?

This is only the start of the idea generation process, but it is already starting to throw up reasons for non-use of public transport and some thoughts on potential actions to encourage greater use. The thinking might be further informed by the following CATWOE analysis.

Checkland's CATWOE: In carrying out a CATWOE analysis it is perhaps best to start by thinking about the nature of the system – the Transformation – but we follow the order of items in the acronym.

C – Customers i.e. the people who benefit (or are victims of) the system:
Users – actual and potential

A – Actors i.e. the people who deliver the system:
The operators (drivers, conductors, ..), the mechanics, the management

T – Transformation i.e. the purpose of the system:
A shared or mass transportation system which people can pay to use to travel from one destination to another

W – Worldview i.e. how is transportation system viewed?:
A means of moving people? A means of encouraging social interaction? Contribution to national economic efficiency? Increasing access to touristic and commercial centres?

O – Owners i.e. the people who have the power (who could sabotage the system):
Drivers (by striking, not driving safely, etc.); Users (by withdrawing their use); Vandals

E – Environment i.e. external constraints and demands:
Provision of bus-stops, bus lanes, taxi-ranks, etc.; Subsidies

CAUSE: We can view the same problems in terms of the CAUSE checklist:

C – Criteria:
Economic benefits and costs. Convenience of access. Time delays. Safety and security. Adequate service to disadvantaged or underprivileged. Air pollution.

46 *MULTIPLE CRITERIA DECISION ANALYSIS*

A – Alternatives:
Transportation options (large and small buses; minibuses; light rail). Legislative options (restrictions on access to and parking in city centres). Incentives (free decentralized parking, subsidized fares). Publicity and advertising.

U – Uncertainties:
Future transport demand. Reactions of people to restrictions and/or incentives. Changes in population profiles and/or economic conditions.

S – Stakeholders:
Local users of the system (current and potential). Tourists. Operators of current systems. The motor industry. Residents on proposed new transportation routes

E – Environment factors and constraints:
Available investment funding. Levels of disposable income of local users. Availability of space and/or widths of roads

The CAUSE checklist is suggested here as it is closely linked to the basic concepts of the MCDA models which are developed in this book.

Thinking about specific actions or alternatives.   Selecting a specific course of action and focusing attention on its strengths, weaknesses and interesting or unusual characteristics (cf. De Bono's (1995) "Plus - Minus - Interesting Analysis") can be a useful way of surfacing values.

Comparing alternatives can also be helpful. One possibility is to take two randomly selected alternatives, to note how they differ and whether or not those differences would be relevant if you had to chose between them. Along similar lines, the Repertory Grid Approach, as described by Eden and Jones (1984), randomly selects three alternatives, divides them into a pair and a single and asks how the single differs from the pair. This approach was formally incorporated in one of the earliest software systems available to support MCDA, namely MAUD (multi-attribute utility decomposition) developed in the early 1980's by the Decision Analysis Unit at London School of Economics (Wooler, 1982).

Keeney (1992) cautions, however, against over-emphasis on alternative focused thinking, distinguishing this from value focused thinking. In the latter, the process focuses on eliciting the decision makers' values prior to identifying alternatives. In the former, alternatives are identified at an early stage in the decision process and the focus is on distinguishing and choosing between the alternatives. Keeney suggests that decision analysis should be driven by value focused thinking as alternative focused thinking tends to " ... anchor the thought process, stifling creativity and innovation" (p48). However, in practice it is often the case that decision makers are faced with a situation in which well defined alternatives already exist and in such circumstances the alternatives can provide a useful stimulus for thinking (as suggested here and as Keeney himself describes - p57), provided that a focus on values is preserved.

Adopting alternative perspectives. Participants are asked to imagine themselves in the role of a specific stakeholder in the problem. For example, suppliers might be asked to imagine themselves in the role of purchaser, nurses as patients, teachers as students or parents. They might be asked to think generally about the stakeholders values, or more specifically about how they would expect the stakeholder to react if a particular action were pursued, and why they would expect them to behave in that way.

Positive and negative reviews. Participants are asked to consider, on the one hand, what could go wrong if a particular course of action were followed – What are the potential pitfalls? What are the possible negative consequences? What barriers might be encountered? On the other hand, what are the anticipated benefits? Are there other possible spin-offs?

Positive and negative reviews correspond to De Bono's (1990) "yellow hat" and "black hat" thinking. He comments that "black hat" thinking is relatively easy because we are used to behaving cautiously; avoiding danger and mistakes is a natural part of the survival instinct. In contrast "yellow hat", or optimistic thinking calls for significant conscious effort. The notion of positive and negative review is also encapsulated in the Idea Advocate approach, countered by a Devil's Advocate, described by Van Gundy (1988, p232). Here specific individuals in the group take on the roles of positive and negative advocate. This approach was adopted in the case study described by Belton (1985) to prompt further thinking when it proved difficult to choose between two options with complementary strengths and weaknesses.

Barriers and constraints. Are there any minimum standards that must be achieved? Are there any limits that must be adhered to? Is there anything which might prevent a particular course of action being implemented, or would any opposition be anticipated?

As mentioned earlier, there are many other "aids to thinking"; those discussed above are ones we have found particularly useful in the context

48   *MULTIPLE CRITERIA DECISION ANALYSIS*

of MCDA, but each intervention is unique and a facilitator should be able to draw on many different ways of prompting thinking, selecting the most appropriate in a given context.

3.3.2.   Structuring of ideas

The aim of the structuring process is to identify key areas of concern, to organise ideas in a way which clarifies goals and actions, and to highlight any gaps in the picture. As suggested in the description of a Post-It session, some structuring of ideas may take place as they emerge. As clusters of related ideas begin to form, Post-Its should be moved to a position close to others to which they relate. The facilitator may wish to maintain control of this, or they may wish to encourage members to the group to actively participate. The process of building clusters will stimulate discussion amongst participants, furthering shared understanding. For example, do similarly worded ideas reflect shared concepts or quite different perspectives? does a particular idea fit better in one cluster or another?

In building clusters it is useful to position the most general concepts at the top of the cluster, cascading down to more specific detail at the bottom. A name which captures the unifying concept of a cluster should be sought; it may already exist as one of the contributed ideas, but it may be necessary to add a new concept.

If the focus of the idea generation process was to identify criteria for the evaluation of specified alternatives, then a clustering exercise such as this may be an adequate basis from which to move directly to building a multicriteria model. However it may be worthwhile structuring the ideas in a more formal way, particularly if the initial problem statement was a very general one, or if the issue is a particularly messy one. One way of doing so, which looks potentially useful for MCDA (Belton et al., 1997, Bana e Costa et al., 1999) is to make use of cognitive or cause mapping, as described by Eden (1988), Eden and Simpson (1989), Eden and Ackermann (1998), and outlined very briefly below.

Cognitive mapping. Cognitive Mapping (Eden, 1988) was developed as a way of representing individual 'construct systems' based upon Kelly's Personal Construct theory (Kelly, 1955), which asserts that "people try to make sense of their environment through comparing and contrasting new experiences with previous ones in order to try to predict and therefore manage their future". A cognitive map aims to represent the problem/issue as a decision maker (participant) perceives it, in the form of a means-ends network-like structure. For example, as part of structuring the location decision problem described in Section 2.2.1, some ideas regarding constraints on choice of location may be linked as illustrated in Figure 3.3.

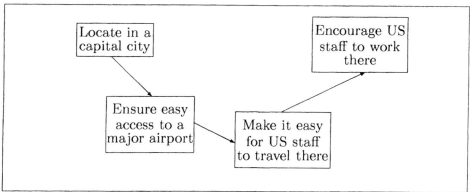

Figure 3.3: Portion of a cognitive map

The arrows in Figure 3.3 should be read as "leads to" or "implies". Often concepts are elaborated by making explicit an opposing pole. For example, the alternative to "locate in a capital city" may not simply be "do not locate in a capital city", but more specifically, "locate in a major regional city". In such situations, the opposing pole will usually be indicated by expressing the concept in the form *Locate in a capital city ... Locate in a major regional city*, where the dots should be read as "rather than". Note the action oriented phraseology in the above examples – each of the concepts contains a verb – this is a particular characteristic of Eden's approach to the use of cognitive mapping for strategy development.

Possible initial components of a more comprehensive cognitive map for a land-use planning problem as described in Section 2.2.3, are illustrated in Figure 3.4. The underlying problem was that of allocating land to exotic forest plantations, discussed in Stewart and Joubert (1998). This map was developed by using the *Decision Explorer* software referred to earlier, which is an extremely useful tool for constructing cognitive maps in an interactive manner on a computer screen. Note that a number of the arrows in Figure 3.4 are associated with negative signs to indicate a counter-effect, i.e. a move to the opposite pole. For example, the negative link from *controlled forestry ... uncontrolled forestry* to *deterioration in water quality ... preservation of water quality* represents the fact that increasing control of exotic forests will counter deterioration in water quality, leading to the opposing (in this case preferred) pole of preservation of water quality.

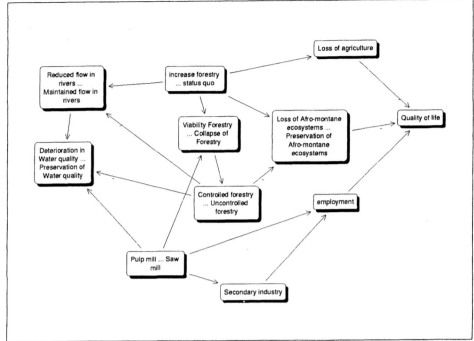

Figure 3.4: Illustrative display from Decision Explorer

A cognitive map is a representation of an individual's cognition; individual maps can be merged to produce a group map, or a process such as the post-it exercise described above can be used to directly generate the group map. To build the map from ideas generated, the facilitator works with the group to cluster and link concepts. The most general concepts, representing the more fundamental values of contributors, should be placed at the top of the clusters. Other ideas, elaborating on those values and describing ways and means of achieving them should be grouped below. Effort should be made to integrate all ideas into the overall structure. If it is not clear how an idea might be linked, posing questions such as "what might this action lead to?" or "how can we achieve this aim?" may help. This process of linking ideas encourages further discussion, leading to greater level of shared understanding and ownership of the model. Within each cluster, a hierarchical structure of ideas should begin to emerge. Of course, we would expect there to be links across clusters, particularly at lower levels as actions can be expected to impact on multiple goals. However, strong links at higher levels in the hierarchy of ideas might suggest that the current structure, or formulation of the problem, is not the best for applying multicriteria

Problem Structuring 51

analysis. Perhaps this might suggest a re-structuring of the problem in other ways, although we need to concede that MCDA is not the panacea for all complex decision problems; there may exist complex interactions or dynamics which it might not be possible to capture in a multicriteria analysis. Nevertheless, in many problem settings, the use of cognitive mapping together with the hierarchical clustering of concepts described above may lead to the emergence of key elements which map well on to the components of a multi-criteria analysis (criteria, alternatives, uncertainties, stakeholders and environmental factors, i.e. CAUSE).

Maps can include several hundred ideas, thus the software tool Decision Explorer allows the contents and associated structure to be captured in a flexible and graphical manner. The software allows the user to focus on a particular part of the map, perhaps a particular key issue, and to explore the ideas relating to that issue. It also supports analysis of the structure, for example: heads and tails (ideas with nothing "above" or "below" them), or orphans (ideas not linked to any other idea) can be identified; cluster analysis enables the detection of emergent themes, by indicating well elaborated and central ideas; feedback loops (vicious or virtuous circles) can be revealed: and goal systems can be developed. Different styles (colour and font) can be used to distinguish goals, actions, descriptive detail, etc. The forms of analysis which are available are described in detail by Eden and Ackermann (1998).

A potential benefit of having the detailed model available in electronic form is its use as an organisational memory - a way of capturing detailed thinking which can be referred to, reviewed and added to over the longer term. This is particularly useful in planning and decision making processes which evolve over an extended period of time, especially since the MCDA model itself (representation in terms of alternatives and criteria) tends to be more concise, sparser in information, than what may be captured in the full cognitive map.

Other ways of representing and structuring ideas. A cognitive, or causal map is a useful way of capturing ideas in a dynamic form which permits further analysis. However, it is not the only way of organising ideas and representing them in a structured form. Simple spray diagrams capture linkages between ideas; Buzan's (1993) mind maps are similar, but more graphical representations; Checkland's rich pictures capture the essence of an issue in pictorial form.

Keeney (1992) discusses in detail the use of fundamental objectives hierarchies and means-end objective networks. The former capture the decision makers' values with respect to a decision, whilst the later capture factual information and causality. Both are important in under-

52 MULTIPLE CRITERIA DECISION ANALYSIS

standing an issue. A cognitive / causal map, as described above would capture both aspects, together with other relevant material, in the single representation.

3.4. FROM PROBLEM STRUCTURING TO MODEL BUILDING

At some stage the emphasis must move from problem structuring to model building, the development of a framework for the evaluation of alternatives. Model building should be regarded as a very a dynamic process, informed by and informing the problem structuring process, and interacting with the process of evaluation. It may involve much iteration, search for new alternatives and criteria, discarding, reinstating and redefining of old ones, and further extensive discussion amongst participants. Moving from a broad description of the problem, whether it be simply a clustering of ideas, a fully elaborated group map, or some other representation of the issue, to a preliminary definition of a model for multicriteria analysis, is one which requires a good understanding of the chosen approach to multi-criteria modelling. The nature of the model which is sought will differ according to the nature of the investigation, whether alternatives are explicitly or implicitly defined, and the particular approach selected for analysis.

The key elements of the model framework, as highlighted by the CAUSE framework introduced earlier (p.45), are recalled here and will each be considered in detail:

- The alternatives (options, strategies, action plans, ...) to be evaluated

- The model of values (criteria, objectives, goals, ...) against which they will be evaluated

- Key stakeholders, their perspective on the decision and how this will be taken into account

- Key uncertainties – both internal and external – and how these will be modelled

These elements are, of course, inter-dependent and cannot be considered in isolation as will become clear in the following discussion.

3.4.1. Identifying alternatives

The evaluation, or exploration, of alternatives – choices, action plans, strategies, etc. – is the essential focus of MCDA. Those alternatives may be relatively few and explicitly defined, as in the evaluation of tenders for a contract, or submissions to a grant awarding body, or they may be infinitely many and implicitly defined, for example, how to best allocate available resources across competing needs. On occasions the alternatives to be evaluated may appear to be clearly defined. In other circumstances the definition, or discovery of alternatives may be an integral part of a study. Sometimes the challenge may be to find *any* suitable alternatives; sometimes it may seem to be impossible to manage the overwhelming complexity of options. Although the rationale for MCDA may appear to be the evaluation of *given* alternatives, equal emphasis should be given to the potential for *creating* good alternatives. Pruzan and Bogetoft (1991) describe the process of planning as "... a search and learning process successively increasing our awareness of our objectives, the alternatives which can be considered, and the relationships between alternatives and objectives". The implication is that alternatives are not givens, but that they evolve. Van der Heijden (1996) talks about option evaluation in the context of scenario planning, saying that "Most work associated with strategic decisions is concerned with redesigning proposals and options such that the upsides are maximised and the downsides are minimised. .... The purpose is not primarily to decide between acceptance or rejection, but to work towards improving the proposal, such that outcomes are as robust as possible over a range of possible futures." If MCDA is to be useful in these contexts design must go hand in hand with evaluation. What guidance do we have for working with alternatives?

Once again, until the work of Keeney (1992), who devotes two chapters of "Value Focused Thinking" to the process of generating alternatives, there has been little in the multiple criteria literature addressing this aspect of the decision process. A number of authors have commented on the importance of generating alternatives, Zeleny (1982) has a chapter on the invention of alternatives and conflict dissolution and Von Winterfeldt and Edwards (1986) mention some experimental work in the area. Both recognise the importance of creative thinking and refer to literature in that area, with important contributions from De Bono (1985), Ackoff (1978) and Yu (1990). The work on Strategic Choice (Friend and Hickling, 1987) also has a useful contribution to make on the subject of designing alternatives. Walker (1988) gives a brief overview of work on generating alternatives.

If an intervention begins with a general approach to problem solving, then it is likely that potential alternatives, or the basis for such will emerge from this process. It may be necessary to ask the group to elaborate these in order to create well defined options for evaluation. It may also be appropriate to conduct an initial crude prioritisation

54 MULTIPLE CRITERIA DECISION ANALYSIS

exercise (such as that described by Belton et al, 1997) to identify those which are of greatest interest.

Combinations of choices

Often the "alternatives" for evaluation are a complex combination of independent choices or actions. For example, an alternative may define a sequence of actions over time, or it may represent a portfolio of investments or activities. Most multicriteria evaluation methods are designed for the evaluation of independently defined alternatives; however, to explicitly enumerate and evaluate all possible combinations of alternatives in such circumstances would be prohibitively time consuming. The Analysis of Interconnected Decision Areas (AIDA) which is embedded in the Strategic Choice approach focuses on defining feasible combinations of actions prior to evaluation. Some similar concepts are also discussed by Stewart and Scott (1995).

EQUITY, a decision support tool for resource allocation, is used to assess different levels of investment across a number of decision areas (e.g. departments of an organisation, countries, project teams) where each individual element is assessed on a cost and benefit scale (possibly derived from a detailed multi-criteria evaluation). Following work which evaluated individual projects which make up a portfolio (Belton, 1993) we proposed the use of meta-models to evaluate project portfolios, where projects are initially evaluated individually in terms of contribution and costs, but in assessing portfolios of projects new criteria come into play, for example, a balance between large and small projects and across activity areas. (See also Stewart, 1991.)

Screening alternatives

Sometimes the problem is not one of generating alternatives, but one of identifying an appropriate and manageable set for detailed evaluation from a much larger set of possibilities – a screening process. The appropriate set may be made up of "good" alternatives, or it may be a set which is representative of the range of possibilities. Walker (1988) discusses two strategies for screening alternatives, firstly bounding the space of promising alternatives and secondly using a simplified assessment model. DEA (Data Envelopment Analysis) could be used as a way of identifying promising alternatives from a long list of possibilities (cf. Belton and Stewart, 1999). It has also been suggested that an outranking method, such as ELECTRE, might be used to draw up a shortlist of alternatives for more detailed evaluation.

It certain situations it may be necessary to first screen alternatives in order to exclude ones which are "non-compliant" in that they do not meet certain minimum specifications. This should always be done with care, making sure that a degree of non-compliance on one criterion could not be compensated for by exceptional performance elsewhere.

Framing issues

An issue of some concern in defining a set of alternatives for evaluation is the potential influence of "phantom" alternatives (Farquar and Pratkanis, 1993). A phantom is an alternative which is apparently available, but subsequently turns out not to be. On the one hand, phantom alternatives can help to promote creativity, but research has shown that the inclusion of alternatives with particular characteristics can also significantly affect choice behaviour. For example, the consideration of a phantom which is particularly attractive in some respect can have two effects: firstly, the contrast effect, which tends to lead to the lowering of attractiveness of other options on the "focal" attribute: and secondly, the importance shift, whereby greater weight tends to be given to the focal attribute in deciding between available options. The presence of phantoms can also have disturbing effects if a policy of screening for promising alternatives is adopted. It is unlikely that such effects can be eliminated, but an awareness of their possibility is the first line of defence.

3.4.2. Identification of criteria

Without exception, all multicriteria methods call for the identification of the key factors which will form the basis of an evaluation. These are referred to variously as: values, (fundamental) objectives, criteria, (fundamental) points of view. The extent to which, and the way in which, these key factors are elaborated in the model structure differs between the methodologies (or "schools" of MCDA) which we shall be discussing in the next few chapters. Multiobjective (goal) programming methods focus on identifying a small number of objectives (goals) which must be measurable on a quantitative scale. Methods based on the use of value functions, including the Analytic Hierarchy Process, generally seek to identify a hierarchy of criteria, or a "value tree". Outranking methods also tend to focus on a few, key criteria although each may be a complex construct derived from a number of sub-criteria.

An initial candidate set of key factors, which we shall term *criteria* should emerge from the problem structuring process. In identifying these the following considerations are relevant to all MCDA approaches.

Value relevance: Are the decision makers able to link the concept to their goals, thereby enabling them to specify preferences which relate

56   *MULTIPLE CRITERIA DECISION ANALYSIS*

directly to the concept? To give a simple example, in comparing models of car, size may have emerged as a criterion. However, it is not clear how size relates to values – it could be that size is important because it relates to the amount of luggage space (which we would like to maximise) or to the number of passengers who can be carried (which should be greater than 3) or to the perceived status of the car (the bigger the better). Structuring the problem as a cognitive map would have helped here by ensuring the linking of the concept to higher level goals.

Understandability: It is important that decision makers have a shared understanding of concepts to be used in an analysis. The absence of such can lead to confusion and conflict rather than constructive discussion and mutual learning. It is, however, always possible that a misunderstanding does not emerge until later in the process - for example, in evaluating potential house purchases, a couple agree that distance from the station is an important factor, but only when evaluating specific options does it become clear that one person prefers to be close (to minimise the distance walked each day) whilst the other prefers to be more distant (away from the disturbance from noise and commuters' cars). This is another illustration of a concept which has not been linked to values.

Measurability: All MCDA implies some degree of measurement of the performance of alternatives against specified criteria, thus it must be possible to specify this in a consistent manner. It is usual to decompose criteria to a level of detail which allows this. However, different approaches call for different levels of precision and different degrees of explicitness, as will be seen in subsequent chapters.

Non-redundancy: Is there more than one criterion measuring the same factor? When eliciting ideas often the same concept may arise under different headings. If both are included in the analysis then it is likely that as a consequence the concept will be attributed greater importance. One can easily check for criteria which appear to be measuring the same thing by calculating a correlation coefficient if appropriate data are available, or carrying out a process of matching as associated with analysis of repertory grids (see Wooler, 1982; Eden and Jones, 1984). Remember that criteria which reflect similar performances over the set of options currently under consideration may not do so in general. As a general rule it is better to combine similar criteria in a single concept. However, there may be good reasons to refer to similar factors, or even the same criterion, in different parts of an analysis as it reflects different values in the different contexts. For instance, Butterworth (1989) describes a multi-attribute value analysis of a bank's decision to relocate one of its head office functions. The top level of the value tree was split into staff acceptability and bank acceptability, both of which had as sub-criteria factors relating to unemployment. Not only would the importance of this be likely to be different for the staff and for the bank, but for staff low unemployment is preferable reflecting concerns about the ease with which other family members could obtain jobs, whilst the opposite is true for the bank, reflecting a desire to be able to recruit new staff without difficulty. Roy (1996) also discusses the splitting of a single dimension into multiple criteria (for example, in evaluating the quality of a service operation waiting times may be important - however, the problem owners may be concerned about average waiting time and about maximum waiting time). Keeney (1992) suggests that double counting can be avoided by differentiating between means objectives and fundamental (ends) objectives and ensuring that only the latter are incorporated in the value tree to be used for evaluating alternatives (although a separate tree of means objectives may be constructed to facilitate understanding of the situation).

In a similar way to double counting, the greater the level of detail pertaining to an objective, possibly reflected in the number of sub-criteria in a value tree, the more likely it is that it will be attributed a high level of importance. Weber et al. (1988) and von Nitzsch and Weber (1993) have done research work which supports these statements. It is advisable to be aware of these issues, however, there are no hard and fast guidelines as to how they should be handled.

Judgmental independence: Criteria are not judgmentally independent if preferences with respect to a single criterion, or trade-offs between two criteria, depend on the level of another. For example, in evaluating alternative job offers someone might feel that "... if I take the job in Greece I would prefer more annual leave to an increased salary, but if I take the job in Alaska I would prefer the salary to the holiday". That is, the trade-off between salary and leave is dependent on location. It may be possible to overcome judgmental dependency by redefining criteria (see, for example, Keeney, 1981). The theoretical validity of value-function approaches (see Chapters 5 and 6) requires judgmental independence, and some studies (Stewart, 1996b) have indicated that violation of this condition can be quite critical in practice. Whilst outranking methods (see Chapter 8) do not in principle demand judgmental independence, proponents of this school of

58 *MULTIPLE CRITERIA DECISION ANALYSIS*

thought (see Roy, 1996, p229) point to the complexity and problems which arise in its absence. Similar comments no doubt apply to the application of goal programming methods (Chapter 7), although some simulation studies (Stewart, 1999) have indicated that these methods may be robust to violations of judgmental independence.

Balancing completeness and conciseness: Keeney and Raiffa (1976) note that desirable characteristics of a value tree are that it is complete, i.e. that all important aspects of the problem are captured, and also that it is concise, keeping the level of detail to the minimum required. Roy (1996) also cites exhaustiveness as a requirement of a coherent family of criteria. Clearly completeness or exhaustiveness and conciseness are potentially conflicting requirements; how can an appropriate balance be achieved? Phillips' (1984) concept of a requisite model is a potentially useful one; he defines a model as requisite when no new insights are generated. However, judging when this stage has been reached may not be easy.

Operationality: Associated with the need to achieve a balance between completeness and conciseness, it is also important that the model is usable with reasonable effort - that the information required does not place excessive demands on the decision makers. The context in which the model is being used is clearly important in judging the usability of a model; for example, if the time available is restricted to a one day workshop then the facilitator must take account of this in guiding the initial specification of a model.

Simplicity versus complexity: As discussed in Chapter 1, the value tree, or criteria set is itself a simple representation, capturing the essence of a problem, which has been extracted from a complex problem description. Notwithstanding that, some representations will be more or less simple than others as a consequence of the degree of detail incorporated and in the nature of the specific structure. In our experience the modeler should strive for the simplest tree which adequately captures the problem for the decision maker. However, in practice, the initial representation tends often to be more detailed than either necessary or operationally desirable, and it is only through the process of attempting to use the model that this becomes apparent, leading to further refinement.

A note on building value trees: top-down vs bottom-up

Von Winterfeldt and Edwards (1986) and Buede (1986) both describe two distinct approaches to the structuring of value trees, the *top-down* and *bottom-up* approaches. The top-down approach tends to be objective led, beginning with a general statement of the overall objectives and expanding these initial values into more detailed concepts which help to explain or clarify the former. Expansion should continue until it is felt that the emergent criteria are measurable. In contrast, the bottom-up approach is alternative led, beginning with elicitation of detail stimulated by thinking about the strengths and weaknesses of available alternatives. The top-down and bottom-up approaches reflect Keeney's value focused and alternative focused thinking and are conceptually relevant in the context of any MCDA analysis, whether or not it makes use of value trees or hierarchies. It should be noted that there is no one "right" way of doing things and in practice one perspective should inform the other. A more general approach to problem structuring and model building such as described above can encompass value-focused and alternative-focused thinking as well as top-down and bottom-up approaches to model building. The outcome of this stage of the analysis should be an initial family of criteria, or a value tree which captures the decision makers' values.

3.4.3. Stakeholders

There is a substantial literature on stakeholder identification, analysis and management (see, for example Eden and Ackermann, 1998), which it is impossible to cover here in any depth. However, in a multicriteria analysis, as in any form of problem analysis, it is important to recognise the existence and potential impact of both internal and external stakeholders.

As a first step the stakeholders must be identified. A dictionary definition (The Chambers Dictionary, 1998 edition) of a stakeholder is as "someone with an interest or concern in a business or enterprise", noting, however, the extension of this definition to the "economy or society" as a whole. Checkland's CATWOE analysis, discussed in Section 3.3.1, can act as a useful prompt in identifying stakeholders. However, the extent to which the decision making group would wish to take account of a particular stakeholder's views is likely to differ according to the influence which that stakeholder could exert and in particular, their power to sabotage a decision. Eden and Ackermann use a two-dimensional grid to classify stakeholders according to their level of interest and power with respect to an issue. They further highlight the importance of not only considering stakeholders in isolation, but also taking into account their propensity to influence other stakeholders as well as the potential for coalitions. Such an analysis is useful in identifying which stakeholders'

60   MULTIPLE CRITERIA DECISION ANALYSIS

views it is important to consider; there is then the question of how to do so.

A question to be asked is whether it is desirable and/or possible to involve representatives of the relevant stakeholder groups in the decision process. This, of course, will depend on the aim of the process – is it to arrive at decision which everyone can buy into or is it explicitly to explore different perspectives? Involvement should ensure that all parties views are fully considered and, if the group has a shared sense of purpose, can be the way to develop ownership of and commitment to a way forward.

If particular stakeholders are not represented in the decision making group then role play may be a helpful way of encouraging consideration of an issue from different perspectives. The group, or sub-groups, can be asked to think generally about the issue from a perspective other than their own, or they might be asked to anticipate how a particular stakeholder would react to specific decision. Role play can also be useful in a supportive environment in building understanding of the perspectives of other group members.

The enforced consideration of different perspectives serves both to encourage more open thinking about the problem and as a means of anticipating the reactions of stakeholder groups.

These processes serve to surface issues relating to different stakeholders and to encourage the broader consideration of an issue. Following on from this is the question of how this might then be incorporated in the multicriteria modelling. A number of approaches have been adopted. In the context of multi-attribute value analysis two approaches can be distinguished by whether or not the model structure, the value tree, is shared – i.e. the same for all stakeholders – or different for each stakeholder. In the case of a shared model the same criteria and structure are adopted by everyone, but different stakeholders have the opportunity to express their individual preferences through the values entered into the model (Brownlow and Watson, 1987). A different model for each stakeholder may be appropriate if they have very different concerns, manifested in different criteria sets. Butterworth (1989) is an example of such an analysis. The decision under consideration was the relocation of a department of the Bank of England; the analysis models separately the interests of the management of the bank and the bank staff, combining these at the top level of the value tree.

3.4.4.   Handling Uncertainty

Uncertainty is inherent to some degree in most decision situations. It can, however, take on many different forms which may be most appropriately handled in correspondingly different ways. Increased effort in problem structuring, as well as in data gathering and analysis, can lead to a better understanding of the nature of that uncertainty and may in some instances reduce it, but cannot eliminate it. Thus, any multicriteria analysis should pay attention to the issue of uncertainty, and especially if, how, and to what extent it should formally be incorporated in the analysis.

The nature of uncertainty

Uncertainties in problem structuring and analysis take a number of different forms, and arise for a number of different reasons. A number of authors have proposed categorisations of uncertainty which might prove useful in thinking about a problem. For example, Friend (1989), as part of the Strategic Choice methodology, classifies uncertainty in terms of uncertainties about values, about related decision areas, and about the environment. French (1995) identifies 10 different sources of uncertainty which may arise in model building for decision aid, which he classifies into three groups, referring broadly to uncertainties in the modelling (or problem structuring) process, in the use of models for exploring trends and options, and in interpreting results.

For purposes of multicriteria decision aid, it may be useful to differentiate between *internal uncertainty*, relating to the process of problem structuring and analysis, and *external uncertainty*, regarding the nature of the environment and thereby the consequences of a particular course of action, as described below.

Internal uncertainty. This refers to both the structure of the model adopted and the judgmental inputs required by those models. It is the first of these two types of internal uncertainty which is most relevant in the context of problem structuring and model building. It can take many forms, some of which are resolvable others which are not. Resolvable uncertainties relate to imprecision or ambiguity of meaning – for example, what exactly is meant by the concept "quality of life"? However, it may be the case that a number of alternative value trees, or criteria sets, emerge from the problem structuring process, with no clear indication of which is most appropriate. Or, it may not be clear what level of detail should be incorporated in a model. Similarly, if the alternative courses of action are not clearly defined, then it may not be obvious which ones should be chosen for more detailed analysis. Uncertainties of this nature are unresolvable, hence it is important that the MCDA process allows the opportunity to go back, restructure, and consider the

62   MULTIPLE CRITERIA DECISION ANALYSIS

problem from a different angle. The structuring process should ensure that such iteration is both catered for and encouraged.

The second kind of internal uncertainty impinges most significantly on the elicitation of values and use of a model than on problem structuring and model building. Such uncertainty, often referred to as imprecision, relates largely to the more quantitative assessments required for preference modelling, such as the assessment of performances with respect to specified criteria, acceptable trade-offs between performance on different criteria, indifference and preference thresholds. Uncertainty of this nature, and ways of handling it, will be discussed in Sections 5.6 and 6.2 (in the context of value function methods), and in Chapter 8 (when dealing with outranking methods).

External uncertainty. This refers to lack of knowledge about the consequences of a particular choice. Friend and French both highlight distinctions between uncertainty about related decision areas, and uncertainty about the environment, as described below.

- *Uncertainty about related decision areas* reflects concern about how the decision under consideration relates to other, interconnected decisions. For example, suppose a company which supplies PCBs to computer manufacturers is looking to invest in a management information system. They would like their system to be able to communicate directly with that of their principal customers, however, at least one of these customers is planning to install a new system in the near future. The customer's decision could preclude certain of the options open to the supplier and would certainly have an impact on the attractiveness of options. The appropriate response to uncertainty of this kind may be to expand the decision area to incorporate interconnected decisions, or possibly to collaborate or negotiate with other decision makers.

- *Uncertainty about the environment* represents concern about issues outside the control of the decision maker. Such uncertainty may be a consequence of a lack of understanding or knowledge (in this sense it is similar to uncertainty about related decision areas) or it may derive from the randomness inherent in processes (for example the chance of equipment failure, or the level of the stock market). For example, the success of an investment in new production facilities may rest on the size of the potential market, which may depend in part on the price at which the good will be sold, which itself depends on factors such as the cost of raw materials and labour costs. A decision about whether or not to invest in the new facilities must take all of these factors into

Problem Structuring 63

account. This kind of uncertainty may be best handled by responses of a technical nature such as market research, or forecasting.

There are a number of ways in which consideration of uncertainty about the environment has been integrated with multiple criteria analysis, as outlined below:

Scenario Planning is one approach which is used to focus attention on uncertainty about the environment. Scenario Planning was developed and has been widely used in strategic planning in Shell International Petroleum. The approach, which is described in detail in the very readable book by van der Heijden (1996) on "Scenarios: The Art of Strategic Conversation", requires decision makers to identify a number of, usually three, scenarios relevant to the decision context. There is no attempt to explicitly model the likelihood of the different scenarios happening, or to describe all possible scenarios. Alternative actions, or strategies, under consideration are summarily evaluated for each of these scenarios. As noted above, the emphasis in scenario planning is on defining good strategies which are robust over a range of possible futures. It is an approach which could be effectively integrated with MCDA. Bear in mind, however, that the development of scenarios is itself a substantial task which would significantly expand the process of problem structuring.

Decision Theory provides another approach to modelling uncertainty in the environment. Decision Theory is based on the use of probability to describe the likelihood of uncertain events, using *utility* to model the decision maker's attitude to risk. A decision maker may be risk averse, risk seeking or risk neutral, depending on the context of the decision and on personal factors. Consider, for example, an investment decision in which the only objective is to maximise financial return – there are a number of possible outcomes and although it is not known for certain which will happen, the probability of occurrence of each is known. If the stakes are small but the potential gains are large then many people would be willing to invest even though the Expected Monetary Value (EMV) of investing is negative – i.e. they would be risk seeking. National Lotteries and other similar gambles rely on such behaviour. However, if the potential losses are substantial even though the EMV of an investment may be positive, many people are reluctant to invest – i.e. they are risk averse. This concept is extended to multi-attribute utility to model decision maker preferences in situations which require attention to be given to both multiple objectives and environmental uncertainty. Multi-attribute utility theory (MAUT), which is described in more detail

64 *MULTIPLE CRITERIA DECISION ANALYSIS*

in Section 4.3 of the next chapter, is the only approach to multiple criteria analysis which seeks to model attitude to risk in this way.

Risk as a criterion: Another way to take account of uncertainty is to include level of risk (represented perhaps by a probability of success or failure in achieving certain desired ends) as one of the criteria in a multiple criteria analysis (e.g., Stewart, 1997). Such an approach implies that decision makers may be willing to accept a certain level of risk in return for increased benefits or reduced costs in terms of the other criteria.

3.5. CASE EXAMPLES RE-VISITED

In this section, we extend the discussion of three of the case examples of Section 2.2 to provide some illustration of results which may be expected to emerge from the structuring process described earlier in this chapter. These three case studies, in particular, will be used to illustrate a number of the MCDA methods discussed in the later chapters.

3.5.1. Location Decision

The problem introduced in Section 2.2.1, regarding the selection of a European office by Decision Aid International, would involve a relatively small number of stakeholders. The hypothetical discussion between the analyst and managing director, as recorded in Section 2.2.1, already represents a substantial move towards structuring of the problem. It is clear from that discussion that the alternatives are not clearly defined at the outset. In order not to close out options too early, it would be helpful to follow the value-focused thinking approach of Keeney (1992). The analyst might thus facilitate a discussion, or decision workshop (described more fully in Chapter 9), involving both staff and management, on the issue of opening a new office. Some of the factors which need to be taken into account have already emerged in the original discussion, but use of any of the various techniques discussed in the previous sections could further enrich this list. Example Panel 3.1 illustrates what might possibly emerge from a discussion using the CAUSE framework.

Relevant factors such as those illustrated in Example Panel 3.1 could be captured manually using Post-Its or electronically using software such as Decision Explorer, as illustrated by the spray diagram shown in Figure 3.5. This would facilitate initial clustering by the analyst, in discussion with the decision makers, leading to the emergence of a more fully structured value tree such as that illustrated in Figure 3.6.

Once the criteria illustrated in Figure 3.6 have been identified, participants engaged in the decision workshop would be encouraged to think

Criteria	costs, attractiveness of location, ease of operating, communication links, size of local market, do US staff want to work there, ...
Alternatives	whether or not to open a European office? Why not the Far East? Australia? Which city?
Uncertainties	there is some uncertainty about most of the criteria noted above
Stakeholders	company staff (in particular those expected to work at or visit the new office), managers (who need to deal with issues relating to set up and operation), shareholders (concerned about the financial viability of the company, existing customers concerned about quality of support), ...
External Factors	closely linked to the uncertainties noted above — nature of the market for services, political environment in different countries, nature of competition.

Example Panel 3.1: Use of the CAUSE Framework for the Location Case Study

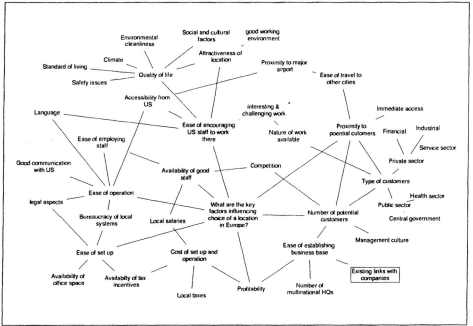

Figure 3.5: Spray diagram for Decision Aid International's location problem

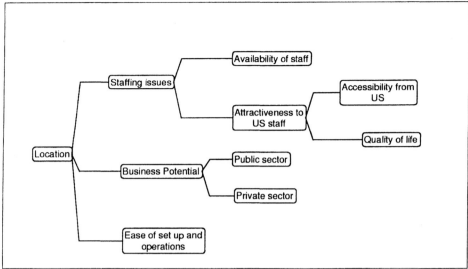

Figure 3.6: Value tree for Decision Aid International's location problem

creatively about which cities are particularly good as regards each of the criteria in turn (i.e. using the value-focused thinking approach). Some of these may eventually need to be rejected, but gradually a short-list of alternative cities would emerge, agreed by the group to be deserving of further investigation in greater detail. For later use of this case study as an example to illustrate some of the MCDA methodologies, let us suppose that the following seven cities have been identified as being good candidates in terms of some if not all of the criteria: Paris, Brussels, Amsterdam, Berlin, Warsaw, Milan, and London.

Decision Aid International will need, sooner or later, so associate some form of assessment of the degree to which each city contributes to satisfaction of each of the criteria in the value tree. For some of the criteria, it may be possible to base the assessments on available quantitative measures (or "attributes"), for example:

- Information may be obtained from recruitment agencies on numbers of suitably qualified applicants typically responding to advertisements for posts in each of the candidate cities, at salary ranges likely to be offered by Decision Aid International. This could serve as a measure of performance on the criterion *availability of staff*.

- The numbers of flights per week out of the local airports to major US destinations, might serve as a useful measure of *accessibility from US*.

Many of the other criteria are more qualitative in nature, and it may be necessary to construct relevant scales of measurement. We delay discussion of this process to Chapter 5.

We shall make extensive use of this case study, to illustrate the methodologies discussed particularly in Chapters 5 and 8.

3.5.2. Stocking of Game Reserves

For the game reserve planning problem described in Section 2.2.2, consideration of the initial strategic issues (such as location of the proposed game reserve, and selection of which species to stock) would follow a similar process to that described for the office location decision. We shall thus focus here on the ensuing "design problem", namely the decisions regarding stocking levels (numbers of each species to be maintained), the distribution of the animals across different areas of the game park (taking into consideration available water and grazing, as well as demands on these resources by local communities), and the means of disposal of excess population growths (e.g. by traditional or commercial hunting, or by live animal sales).

Those involved at this stage might be representatives of investors or funding agencies; game reserve management and game wardens; tour operators; and representatives of local communities close to the boundaries of the game reserve. The process would start by inviting all stakeholders to public discussion and information meetings, out of which representatives may be identified to participate in more detailed planning workshops. Much as in the previous case study, the first task would be to develop some form of value tree, representing the criteria which need to be taken into consideration. At this stage, it is probable that a small expert working group would need to go away and develop tentative solutions which are technically feasible but also in some sense best satisfy the criteria which have been established. The process would be iterative. The initial solutions would be discussed in further workshops, and any dissatisfaction with the solutions tabled would reveal either criteria not previously identified, or a misinterpretation of the fundamental values underlying the criteria. For example, initial solutions might allow too much commercial hunting, in the sense that it would become difficult to market the game park as a conservation area, revealing a further criterion of image in the ecotourism market. The expert working group would then need to repeat their analyses with the new information.

For purposes of using this case study for later examples (especially in Chapters 6 and 7), we shall focus on the problems as faced by the expert working group, in creating policies to present to the full workshop. At this stage of the process, such design problems are often easily formu-

68 MULTIPLE CRITERIA DECISION ANALYSIS

lated in algebraic terms, i.e. in terms of a *mathematical programming* structure, which we now describe.

Let S be the number of different species to be stocked, which we can index by $s = 1, 2, \ldots, S$. The first problem may be that the proposed game reserve is not one homogeneous whole, but may be bisected by large rivers or man-made structures such as roads. The total reserve area will then have to be viewed as consisting of a number of more-or-less independent areas which we shall term *camps*. Let C be the number of camps identified, indexed by $c = 1, 2, \ldots, C$. The first decision in which we are interested is thus to determine values for X_{sc} for each s and c, being the number of animals of species s which are to be stocked in camp c.

The other component of the design problem relates to the means adopted for removing excess stocks which arise through population growth. The problem structuring process may have identified three possible means, *viz.* sales of live animals, culling (with the resulting meat being sold or distributed to workers or local communities), and commercial hunting (i.e. by the sale of hunting licenses for the shooting of a specified number of animals in each species). Thus, let Y_{sj} be the numbers of species s which are to be disposed of in manner j (for $j = 1, 2, 3$) during each planning period, with the implication that $\sum_{j=1}^{3} Y_{sj}$ should equal the natural population growth over that same period. In other words, if the proportional rate of uncontrolled population growth per planning period of species s is α_s, then for sustainable management we require:

$$\sum_{j=1}^{3} Y_{sj} = \alpha_s \sum_{c=1}^{C} X_{sc} \tag{3.1}$$

for each species s.

The structuring process undertaken during the workshops may also have identified direct constraints which need to be placed on choice of the primary decision variables X_{sc} and Y_{sj}. At least two other types of constraint may emerge. Firstly, various critical resources (such as food and water) may be identified, imposing constraints on the numbers of animals in each camp. Suppose that there are R such resources, indexed by $r = 1, \ldots, R$. In order to keep this case study simple, we shall ignore complications of seasonality, and suppose that resource r is available at a constant rate of A_{rc} per time period in camp c, and that each animal of species s continuously consumes β_{rs} of this resource per time period.

This implies the following restriction for each camp (c) and resource (r):

$$\sum_{s=1}^{S} \beta_{rs} X_{sc} \leq A_{rc} \qquad (3.2)$$

Secondly, there may be requirements that a minimum number (say M_s) of certain species need to be stocked (for example to maintain a viable breeding stock for an endangered species, not stocked elsewhere). These impose constraints of the form:

$$\sum_{c=1}^{C} X_{sc} \geq M_s \qquad (3.3)$$

The problem is thus to choose nonnegative values for the X_{sc} and the Y_{sj}, subject to a set of constraints such as those above. The set of feasible decisions is thus the solution set of a linear programming problem, superficially similar to many encountered in standard management science texts. The number of alternatives is infinite, and the set of alternatives is defined only implicitly by the constraints, which contrasts with the situation in the other case examples in which a discrete (generally finite) set of alternatives is explicitly defined.

In addressing such problems with a mathematical programming structure, it is quickly realized that there is not a single well-defined objective function which needs to be maximized or minimized. Objectives such as maximizing direct economic benefits, maximizing contributions to national conservation goals, maximizing tourist benefits and maximizing benefits to local rural communities will all have been identified by the problem structuring workshops, but will to a greater or lesser extent be in conflict with each other. If each objective can be expressed as a linear function of the X_{sc} and the Y_{sj}, then we have in principle a number of linear programming problems, but the solution of any one of these in isolation from the others will in general produce highly unbalanced answers. We shall, in fact, demonstrate in a later chapter (see Example Panel 7.2 on page 217) that simple composites of the different objectives (e.g. maximizing a weighted average of them) also exhibit erratic and unbalanced solutions. More care is needed, in other words, to address what is the *multiple objective linear programming problem*.

In order to crystallize this case example for use later in the book, let us introduce a simplified version of the problem with specific numerical details. We shall suppose that the numbers of animals involved are sufficiently large to be able to ignore the fact these should be integers, and for numerical convenience, we shall re-scale the X_{sc} and Y_{sj} to be

70 MULTIPLE CRITERIA DECISION ANALYSIS

Table 3.1. Illustrative parameter valeus for game reserve example

	Resource:	
	1	2
Consumption per 100 animals:		
Species 1 (β_{r1})	1	2
Species 2 (β_{r2})	3	1
Species 3 (β_{r3})	8	4
Availability in:		
Camp 1 (A_{r1})	14	12
Camp 2 (A_{r2})	25	30

in terms of hundreds of animals. Suppose that we have $S = 3$ species to be stocked in $C = 2$ camps. The growth rates per time period (the α_s) are 0.4, 0.3 and 0.2 respectively for the three species. Suppose that there are also $R = 2$ critical resources, and that the consumptions by each species (β_{rs}) and the availability (A_{rc}), are as in Table 3.1.

Finally, a lower limit of 3 (i.e. 300 animals) is placed as a hard constraint on the stock of species 3. Also for illustration purposes, we define four objectives below, which can be viewed respectively as maximizing revenues from sales and hunting (z_1); maximizing contribution to national conservation effort (z_2); maximizing food value to local communities (z_3); and maximizing viewing pleasure for visitors to the reserve (z_4). The full multiple objective linear programming formulation is then specified as follows:

$$
\begin{aligned}
\text{Max } z_1 &= 2Y_{12} + 2Y_{13} + 3Y_{22} + 6Y_{23} + Y_{32} + 8Y_{33} \\
\text{Max } z_2 &= 5Y_{12} + 3Y_{22} + 2Y_{32} \\
\text{Max } z_3 &= 20Y_{11} + 15Y_{21} \\
\text{Max } z_4 &= 3X_{11} + 15X_{21} + 6X_{31} + X_{12} + 5X_{22} + 2X_{32}
\end{aligned}
\tag{3.4}
$$

Subject to:

$$\begin{aligned}
-0.4X_{11} - 0.4X_{12} + Y_{11} + Y_{12} + Y_{13} &= 0 \\
-0.3X_{21} - 0.3X_{22} + Y_{21} + Y_{22} + Y_{23} &= 0 \\
-0.2X_{31} - 0.2X_{32} + Y_{31} + Y_{32} + Y_{33} &= 0 \\
X_{11} + 3X_{21} + 8X_{31} &\leq 14 \\
2X_{11} + X_{21} + 4X_{31} &\leq 12 \\
X_{12} + 3X_{22} + 8X_{32} &\leq 25 \\
2X_{12} + X_{22} + 4X_{32} &\leq 30 \\
X_{31} + X_{32} &\geq 3
\end{aligned} \tag{3.5}$$

A useful first step is simply to maximize each objective in turn, recording the impacts on the other objectives in a "payoff" table as illustrated by Table 3.2. The tendency to extreme values, which is a typical feature of linear programming formulations, is quite evident, so that compromise solutions are not immediately obvious.

Table 3.2.   Payoff table for game reserve example

Objective being maximized	Values obtained for:			
	z_1	z_2	z_3	z_4
z_1	16.80	30.00	0	33.00
z_2	12.60	31.20	0	33.00
z_3	4.80	0	120.00	33.00
z_4	4.80	0	22.50	77.67

The initial analysis of a multiple objective linear programming problem will typically need to be carried out by analysts having access only to relatively limited information concerning decision maker preferences. The aim would then be to generate a shortlist of potentially optimal solutions for more detailed evaluation. The present case study will serve to illustrate the use of so-called interactive methods based on value functions (Chapter 6), and linear goal programming (Chapter 7) for this purpose. It will be seen that naive methods of solution can lead to highly misleading results, but that use of appropriate MCDA methodologies can lead to robust and defensible recommendations.

3.5.3.   Land Use Planning

As indicated in Section 2.2, problems of this nature need to be viewed as occurring in two phases, namely:

1 An initial screening of a potentially large number of possible land-use plans, typically carried out by a technical team (i.e. not the decision

72   *MULTIPLE CRITERIA DECISION ANALYSIS*

makers or representatives of interested parties), in order to produce a shortlist of development plans (or "policy scenarios" as we shall call them below) for more detailed evaluation;

2 Evaluation of the plans included in the shortlist by the various interest groups, taking into consideration the political and value judgmental aspects, with the aim of finding broad consensus.

The main reason for the two-phase approach is that the evaluation and comparison of policy alternatives require many political and value judgements to be made, many of which are not easily captured in quantitative measures. This is particularly true when some of the interested parties are unsophisticated technically or scientifically (as is often the case in less developed communities). For example, impacts on living conditions in rural communities might have to be evaluated by community representatives by examining visual displays such as videos. In practice, only a small number of alternatives can be evaluated in this way, so that it is becomes necessary to restrict the shortlist to perhaps no more than 7-9 alternative policy scenarios (the "magic number 7 plus or minus 2"). This is a multicriteria choice in its own right, as effort is necessary to ensure that the alternatives included in the shortlist do include potentially good compromise solutions.

It is worth noting that the alternatives selected in this way for more detailed evaluation will in general not be defined to the finest level of detail. For example, a policy alternative which specifies that a certain percentage of the land in a particular sub-region needs to be retained as a conservation area might not specify in detail how this area will be managed. The alternatives are then perhaps better termed *policy scenarios*, in the sense that details of implementation will only be worked out at a later stage, possibly as part of the evaluation in the second stage. All that is required is that there is sufficient detail so that these policy scenarios can be distinguished in terms of degree of satisfaction of the various goals and interests. In fact, an important reason for adopting the two-phase approach is precisely because the evaluation of policy scenarios needs to give consideration to the finer levels of implementation detail which cannot be assessed for all possible policy alternatives.

Identification of Policy Scenarios

As indicated, the first phase in the decision structuring process is to identify a short-list of possible policy actions and ranges of action. This phase would start with some form of workshop involving primarily planners from relevant state departments and technical experts. One of the key tasks of this initial workshop would be to identify the stakeholders and other national or regional interests which may be relevant, so that representatives of such groups may be contacted. This initial workshop would be followed up by a series of public meetings, advertised broadly in the local press, to allow scope for all interested parties to contribute views and concerns at an early stage. Much as in the case of the game reserve planning study, the intention is that a representative working group may emerge from the public participation meetings, who would take the structuring process further in more detailed decision workshops.

The issues first to be considered by the working group would be the development of the value tree, and also of the possible actions or interventions that can be considered. For example, in the case studies reported by Stewart and Scott (1995) and by Stewart and Joubert (1998), the main such actions identified were associated with:

- *Direct land use options* such as agriculture, commercial forestry or conservation for each subregion under consideration;

- *Binary choices* such as constructing or not constructing reservoirs at each of one or more designated sites;

- *Imposition of restrictions* such as various levels of limitation on abstraction of water from rivers or dams for irrigation purposes.

Policy alternatives, or policy scenarios, would then be constructed as combinations of feasible options for each action. For example, one policy scenario might be (a) to restrict commercial forestry to specified regions only; (b) not to build any reservoirs; and (c) to allow dry-land farming only (i.e. not to allow any irrigation). In principle, there may well be an infinite variety of feasible combinations, for example when land can be subdivided more or less continuously between possible land uses, in which case the situation then becomes similar to the multiobjective mathematical programming problem illustrated by the previous case example (game reserve planning). In many land use planning problems, however, the quantitative consequences of any specific combination of actions will require the use of quite complex models (e.g. extensive simulations), so that the mathematical programming approaches described for the game reserve planning problem may become impracticable. The problem has then to be discretized into a finite number of options for each action, leading to a finite (but still potentially quite large) number of discrete feasible combinations, as illustrated in the next paragraph. (Interpolation between these discrete alternatives may nevertheless be justified in some cases, to give an approximation expressed in non-linear mathematical programming terms.)

74   *MULTIPLE CRITERIA DECISION ANALYSIS*

As an illustration, suppose that the structuring process leads to a recognition of two critical land uses, viz. commercial forestry and conservation, subject to restrictions that:

- the area allocated to forestry should be between 15% (the status quo) and 40% of the land, which may be considered in steps of 5%, giving 6 possible forestry options;

- the area allocated to conservation should be between 0 and 30% of the land, again viewed in steps of 5% (giving 7 options); while

- the total of forestry and conservation should not exceed 50% of the land (to leave space for other traditional activities).

In view of the last restriction, not all 42 combinations of forestry and conservation options are feasible; in fact, it turns out that only 32 feasible combinations remain. These we view as defining the feasible land-use scenarios. In addition, however, two additional actions need to be considered for each of these 32 land use scenarios:

- *Construction of reservoirs:* Major reservoirs can be constructed at one or both of two identified sites in the region (giving 4 possibilities);

- *Forestry products processing:* Here there are three options, viz. (a) felling only, with transport of logs out of the region for processing elsewhere; (b) local small saw mills, to reduce the logs to sawn timber for export to other regions; or (c) a chemical pulp mill to produce paper.

Combining the land use scenarios with the other options generates a total of 384 possibilities, for each of which the relevant simulation or other models will need to be run in order to evaluate predicted consequences.

The key interest and stakeholder groups which emerged from the quoted case studies included community representatives (interested in socio-economic upliftment of the area), the forestry industry, other agricultural groups, and conservation groups (both governmental and nongovernmental). Ultimately, the value judgements of these groups will need to be incorporated into the planning process, but as we have indicated above, this will require a substantial level of direct subjective comparisons of policy scenarios, which cannot be done on 384 alternatives. It is for this reason that the short-list needs to be created. In order to assist planners in creating such a short-list, we need to start with a set of relatively well-defined system attributes, i.e. consequences of the alternative policy scenarios, which can be agreed by the different interest and stakeholder groups (at the workshop) as representing their aspirations at least for purposes of initial screening of the options. Typical attributes which have emerged at this stage include:

- New employment generated in the region

- Return on investment from forestry industry

- Public sector investment costs

- Gross regional product

- Profitability of the forestry industry

- Area of sensitive ecosystems permanently lost due to flooding by the reservoirs

- Quality of ecosystem preservation in conservation areas (possibly qualitatively assessed on a categorical scale by ecologists)

- Low flows in rivers during dry seasons

- Water quality in rivers (possibly also qualitatively assessed on a categorical scale by ecologists)

In principle, these attributes can be calculated for each alternative scenario. In most cases, the direction of preference is self-evident (where, "all other things being equal", we would ideally wish to maximize employment, gross regional product, profitability of the forestry industry, number of ecosystem types preserved and river flow in dry seasons, while minimizing investment costs and area of sensitive ecosystems lost). Conflicts may, however, arise when communities on the banks of the rivers might seek to minimize river flood levels while ecologists might argue that regular floods are essential to the ecosystems. In some cases, interest or stakeholder groups may be able to specify levels of achievement for some of these attributes (either minimal requirements or ideal aspirations).

The MCDM problem at this stage is to identify a small number of widely dispersed alternatives (within the set of 384 potential policy scenarios) which perform well according to all attributes. As with the design phase of the game reserve planning case study, the process of generating the short-list of representative alternatives (i.e. finding "solutions" to this MCDM problem) will largely be delegated to a specialist team, who would produce recommendations to the full working group.

76 *MULTIPLE CRITERIA DECISION ANALYSIS*

Evaluation of Policy Scenarios

Suppose now that the first phase MCDM problem, as described above, has been "solved" in the sense that a shortlist of perhaps not more than 7-9 alternative policy scenarios has been identified. As indicated above, these need to be evaluated in depth by the interest and stakeholder groups (jointly or separately) on the basis of both quantitative (objective) and qualitative (subjective) criteria. Depending on the sizes of the groups involved, and the degree of conflict between them, the decision analysts/ facilitators may either run a series of separate decision workshops with each of the stakeholder groups (reporting results eventually back to the full working group), or work directly with the full group. In either case, the MCDM problem evaluated by the group (separately and jointly) will in essence be similar to that of the location decision problem presented as the first case example, and would be approached in much the same way.

Where the different groups do meet separately, each such stakeholder group would develop its own set of criteria, and would thus arrive at a preference ordering for the alternatives which would in general be different for that of other groups. In this case, the search for consensus between the interest or stakeholder groups may well be seen as yet a further MCDM problem, in which the "criteria" are essentially the interests of each group (which could be viewed as the highest level criteria in a "value tree", as will be discussed in Section 4.1). One important difference between the within-interest and between-interest MCDM problems, however, is that whereas a single interest group might well be able to agree on trade-offs between criteria (i.e. the extent to which good performance on one criterion can compensate for poorer performance on other criteria), this might well not be true between interests.

It is important to re-emphasize that the two phases of the problem will in general be iterative. Complete consensus might not be achievable in the first iteration, but it may be possible to exclude a certain number of the policy scenarios included initially in the shortlist. This preference information can be fed back to the first phase evaluation, leading perhaps to an improved shortlist generation. See Stewart and Scott (1995) for further discussion.

In completing the discussion of this case example, it is worth noting that an additional complication may be the existence of substantial uncertainties in the outcomes associated with each alternative. In some cases, such uncertainties may be represented by scenarios defined in terms of external uncertainties (for example, economic growth rates, or environmental trends such as changes in rainfall patterns). At some stage, the analysis will need to come to grips with such problems, as discussed above in Section 3.4.4.

Aspects of this case study will be used to illustrate value function methods based on imprecise preference information (Section 6.2), in which, for example, it is possible to identify all alternatives which are at least potentially optimal in the light of the available information.

3.6. CONCLUDING COMMENTS

In this chapter, we have (together with many other authors) stressed that decision "problems" do not present themselves in a structured form, complete with lists of alternative courses of action and decision making objectives, ready for systematic analysis. The structure is a human construct which emerges as decision or policy makers struggle to gain a shared understanding of the situation which is perceived to be sufficiently disquieting that some form of intervention or action is necessary. There is no one right structure to describe the situation facing the decision makers, but we believe that some processes are better suited than others to developing clarity of understanding and confidence in the final decisions, and it is these which we have described.

We have reviewed a number of different procedures to assist planners or decision makers in developing a useful and coherent structure within which to analyze future courses of action. The discussion here has necessarily been brief, but we have referred extensively to a wide and rich literature concerned with problem structuring. As the present book is concerned with *multiple criteria decision aid*, i.e. with providing aid in the types of decision situations described in Chapter 2, we have focused particularly on structuring in terms of *alternatives* and *criteria*. These key components of a problem structure are central to the methodologies of MCDA which are described in the next five chapters. Chapter 4 surveys different models which have been developed to represent preferences of decision makers within the structure of multicriteria decision problems, and from these models emerge the specific tools and methodologies. In the present chapter, we have also presented a number of case examples in which the problem structuring leads naturally to the multicriteria formulation, and these same examples will be used to illustrate the various decision aiding tools which are presented.

Chapter 4

PREFERENCE MODELLING

4.1. INTRODUCTION

At this stage, let us assume that the problem structuring phases have generated a set of alternatives (which may be a discrete list of alternatives, or may be defined implicitly by a set of constraints on a vector of decision variables), and a set of criteria against which these alternatives are to be evaluated and compared. In order to provide aid or support to decision makers in their search for satisfactory solutions to the multicriteria decision problem, it becomes necessary to construct some form of model to represent decision maker preferences and value judgements. Such a preference model contains two primary components, viz.:

1 *Preferences in terms of each individual criterion*, i.e. models describing the relative importance or desirability of achieving different levels of performance for each identified criterion.

2 *An aggregation model*, i.e. a model allowing inter-criteria comparisons (such as trade-offs), in order to combine preferences across criteria.

The concept of "modelling" is central to all of management science or operational research. In multicriteria decision analysis, however, we also model the value judgements and preferences of the decision maker. This differs substantially from the modelling of essentially external realities, such as the operation of a production facility or the behaviour of an economic or ecological system. In such cases, although many different perceptions of the reality may exist, and no model will capture all aspects of the reality, the real world remains in principle observable as a standard against which the model can be tested and validated. In other words, predictions of the model can be tested against actual system be-

80 MULTIPLE CRITERIA DECISION ANALYSIS

haviour. When we turn to the modelling of decision maker preferences, however, the situation is dramatically different. Decision makers seek assistance from a decision aiding procedure precisely because they have difficulties in understanding what they really want to do, and what options will best satisfy their aspirations in the long run. In a very real sense, the preferences and value judgements do not exist (or at least are very incompletely formed) at the start of the decision analysis, and are formed at least partially as a result of the decision aiding process. The model is thus a mechanism whereby decision makers are enabled to learn about their own preferences.

The purpose of modelling for multicriteria decision aid is thus to *construct* a view or perception of decision maker preferences consistent with a certain set of assumptions, so as to give coherent guidance to the decision maker in the search for a most preferred solution. The modelling approaches described in the sections which follow differ in terms of both the type and the strength of assumptions made. In comparing these approaches, however, it is important to bear the constructive (as opposed to a descriptive) intention in mind. One set of assumptions cannot be compared with another on the basis of how well they *describe* actual decision making (although behavioural research on decision making should not be ignored), but rather on the basis of the insights and guidance provided to the decision maker. The view of the present authors is that such insight is best achieved by constructing the simplest possible models. These may appear on the surface to imply strong sets of assumptions, but it is always possible to carry out extensive sensitivity studies as a means both to weaken the effects of the assumptions and to facilitate learning.

Before turning to discussion of the different forms of preference modelling, it is useful to review some terminology and basic principles. We suppose that m criteria have been identified. In the process of structuring the problem (cf. Chapter 3), it is possible (even likely) that the criteria may have been constructed hierarchically in terms of a "value tree", an example of which (based broadly on the land use case example introduced in Sections 2.2.3 and 3.5.3) is illustrated by the diagram of Figure 4.1. Relatively broad general interests (social, economic and environmental concerns in Figure 4.1) are represented at the "top" of the tree (shown, however, on the left of the figure for ease of presentation). These are increasingly broken down into more specific *criteria* such as employment, agricultural output, water quality, etc. at the lowest level (shown to the right of the figure). The assumption is that these lowest level criteria are defined in such a way that a more-or-less unambiguous ordering of the alternatives can be stated in terms of each criterion. If

Preference Modelling 81

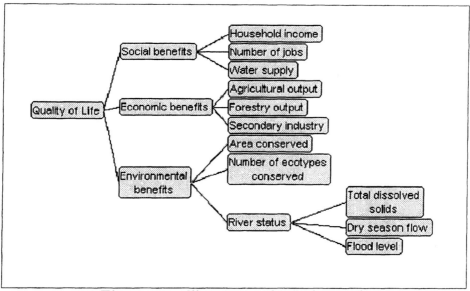

Figure 4.1: Illustration of a "value tree"

this is not true, then further structuring is necessary in order to identify the underlying conflicts, which may lead either to a further splitting of the criterion into more elementary components, or perhaps to a complete re-definition of the criteria.

After preferences have been expressed in terms of these lowest level criteria, the aggregation step can be applied either across all the m criteria in a single operation (even if the original structuring was hierarchical), or may be applied hierarchically by aggregating at each level of the value tree across those criteria which share the same parent criterion at the next highest level in the tree. In the latter case, the aggregation will need to be repeated as often as is necessary, moving from one hierarchical level to the next, until the overall aggregation is achieved. For example, in the value tree of Figure 4.1, preferences between alternatives would first be assessed separately for each of the $m = 11$ lowest level criteria shown to the right of the figure. In some MCDA approaches, an attempt might immediately be made to aggregate these 11 separate sets of (usually conflicting) preferences into a single set of preferences in terms of the overall "quality of life" criterion. Alternatively, one might start by aggregating the preferences in terms of the three criteria "total dissolved solids", "dry season flow" and "flood levels" into a statement of preferences in terms of "river status". Thereafter, aggregation across "area conserved", "number of ecotypes conserved" and "river status" would generate corresponding preferences in terms of "environmental

82   MULTIPLE CRITERIA DECISION ANALYSIS

benefits". This process would continue until "quality of life" preferences are finally obtained by aggregation of social, economic and environmental benefits.

In the remainder of this chapter we shall examine a number of different means by which such aggregation of preferences across criteria can be achieved. For purposes of exposition, we shall often describe these methods of aggregation at a single level only, i.e. as if the aggregation is performed across all m criteria simultaneously (in what may be termed the *extensive* form of analysis). The same principles will, however, apply just as well to aggregation in hierarchical form.

In some (but by no means all) instances, the very definition of a criterion may imply a (more-or-less) objective ordering of the set of alternatives in terms of this criterion, the direction of preference on which may be self-evident. (Consider, for example, a criterion such as investment cost.) This typically occurs when performance levels in terms of the criterion can be associated with an objectively quantifiable attribute. For example, if sulphur pollution from a power station is a criterion relevant to choice of technology, then SO_2 concentration at a certain defined point in the smokestack may be the associated attribute; and an alternative with lower levels of this concentration can immediately be stated as being more preferred in terms of this criterion to another with a higher concentration. Where such a well-defined measure of performance exists, we shall represent the performance level or attribute value of the alternative a according to criterion i by $z_i(a)$. For ease of discussion (but without loss of generality), we shall suppose that all attributes are defined in such a way that *increasing* values are preferred. For example, the sulphur pollution attribute above may be defined as the amount by which SO_2 concentrations are reduced below some fixed level.

Where a criterion is associated with a measurable performance level as described in the previous paragraph, there may be a tendency to conclude that in this case the first component of the preference model is trivial. Such a conclusion is naive and potentially dangerous (i.e. capable of generating extremely misleading or biassed results). In all the approaches discussed below, we shall stress the importance of explicitly modelling *value* judgments, rather than to rely on seemingly "objective" measures, since preferences are rarely linearly related to attribute values. As a simple example, consider the use of an attribute such as fuel consumption in choosing a car. In the USA, fuel consumption will typically be expressed as miles per gallon, and if three alternative cars have consumptions under standard conditions of (say) 23, 27 and 31 mpg respectively, a naive view may be that the gain in moving from the first to the second is essentially the same as the gain in going from the second to the third. In other parts of the world, however, consumption may be measured in terms of litres per 100km, giving for the same three cars 10.3, 8.8 and 7.7 litres per 100km respectively. Now the gain in going from the first to the second appears markedly larger than the gain in going from the second to the third. A more substantial form of non-linearity arises if there exist clear preference thresholds, either minimal levels of achievement below which the decision maker will not be prepared to go, or desirable levels of achievement beyond which further improvements have little additional value to the decision maker. Non-linearities as described above can in some cases result in substantial impacts on the solutions obtained (as discussed, for example, in Section 5.2.2, and in Stewart, 1993, 1996).

Even if no natural attribute is directly linked to a criterion, it may still be possible to construct a scale to summarize performance, possibly an ordinal or an ordered categorical scale, for each criterion. We shall use the nomenclature $z_i(a)$ in this case as well. Whether the $z_i(a)$ derives from a natural attribute, or whether it is simply a constructed summary, it remains a measure of performance for criterion i, and can be viewed as a *partial preference function* in the sense that alternative a is strictly preferred to b in terms of criterion i if and only if $z_i(a) > z_i(b)$. No other properties of such performance measures or partial preference functions are implied at this stage. It will sometimes be convenient to adopt the notation $\mathbf{z}(a)$ to represent the *vector* of performance measures for alternative a, i.e. $\{z_1(a), z_2(a), \ldots, z_m(a)\}$.

Once partial preference functions have been associated with each criterion, it is often useful to perform an initial check for the existence of any pairs of alternatives a and b for which a is at least as good as b on *all* criteria (i.e. $z_i(a) \geq z_i(b)$ for all i), and is strictly preferred to b on *at least one* criterion (i.e. $z_i(a) > z_i(b)$ for at least one i). If this occurs, then the vector of performance measures $\mathbf{z}(a)$ is said to *dominate* $\mathbf{z}(b)$. A vector $\mathbf{z}(a)$ is defined to be "non-dominated" (or "Pareto optimal") if there is no other alternative b such that $\mathbf{z}(b)$ dominates $\mathbf{z}(a)$; in this case the alternative a is said to be "efficient". It is assumed in many MCDA approaches that further analysis can be restricted to efficient alternatives. Note that this is only a valid assumption provided that the intention is to select one unique decision alternative (cf. the discussion of problematiques in Chapter 2), and that the set of criteria being used is complete. It is ill-advised to discard dominated alternatives if there are any reservations concerning the validity of the above two conditions.

The purpose of this chapter is primarily to review three different classes of preference model which have been adopted in considering multiple criteria decision problems, viz. value measurement (Section 4.2),

84   *MULTIPLE CRITERIA DECISION ANALYSIS*

satisficing and aspiration-based methods (Section 4.4), and outranking (Section 4.5). The descriptions provided here are quite brief, aimed at providing a working understanding of the key ideas needed for understanding the strengths and weaknesses of the associated MCDA methodologies to be described in the ensuing chapters. More detailed discussions may be found in the following:

- *Value measurement theory:* Keeney and Raiffa (1976); Roberts (1979); von Winterfeldt and Edwards (1986, Chapters 7-10); French (1988).

- *Satisficing:* Simon (1976); Lee and Olson (1999) and Wierzbicki (1999) (the latter two references being Chapters 8 in 9 in Gal et al., 1999).

- *Outranking:* Roy and Bouyssou (1993); Roy (1996).

The discussion of value measurement models is extended in Section 4.3 to include the notion of "utility theory" for dealing with uncertainties in MCDA. This is included as it represents the only formalized theory for explicit modelling of uncertainty in MCDA, but is aimed perhaps more at the MCDA researcher than at the practitioner. The discussion is also rather more technical in nature, so that it may be skipped by less technically inclined readers.

In Section 4.6 we also briefly review some concepts from fuzzy and rough set theories which have been applied to MCDA. We deal with these concepts separately, as they do not represent a separate methodology for MCDA *per se*, but are tools which can be (and have been) applied within each of the broad approaches identified above. As with the section on utility theory, this more technical discussion may be skipped by readers less concerned with technical details.

Finally, in Section 4.7 we address the critical issue of the relative importance of the different criteria, generally expressed in terms of some form of weight parameter. We shall emphasize that the interpretation of such weight parameters differs from one preference model to another, but that none of these interpretations need necessarily coincide with the intuitive sense of "importance weight" understood by decision makers or other stakeholders. Furthermore, this intuitive sense can well be influenced by the manner in which the problem is formulated or structured. We argue, therefore, that considerable care needs to be taken when implementing MCDA models, to ensure that the interpretation of weights is consistent with the model being used and understood by all participants in the process.

4.2. VALUE MEASUREMENT THEORY

The intention in this approach is to construct a means of associating a real number with each alternative, in order to produce a preference order on the alternatives consistent with decision maker value judgments. In other words, we seek to associate a number (or *"value"*) $V(a)$ with each alternative a, in such a way that a is judged to be preferred to b $(a \succ b)$, taking all criteria into account, if and only if $V(a) > V(b)$, which also implies indifference between a and b $(a \sim b)$ if and only if $V(a) = V(b)$. It is worth noting immediately that the preference order implied by any such *value function* must of necessity constitute a complete weak order (or preorder), i.e.:

- *Preferences are complete:* For any pair of alternatives, either one is strictly preferred to the other or there is indifference between them (i.e. either $a \succ b$, or $b \succ a$, or $a \sim b$).

- *Preferences and indifferences are transitive:* For any three alternatives, say a, b and c, if $a \succ b$ and $b \succ c$, then $a \succ c$, and similarly for indifference.

The value measurement approach thus constructs preferences which, in the first instance, are required to be consistent with a relatively strong set of axioms. It is important to realize, however, that in practice value measurement will not (and should not) be applied with such a literal and rigid view of these assumptions. The construction of a particular value function does impose the discipline of coherence with these "rationality assumptions", but the results of and conclusions from the value function will be subjected to extensive sensitivity analyses, as will be described in Chapter 5. The end result will generally be much less rigidly precise than may be suggested by the axioms.

Within the value measurement approach, the first component of preference modelling (measuring the relative importance of achieving different performance levels for each identified criterion) is achieved by constructing "marginal" (or "partial") value functions, say $v_i(a)$, for each criterion. A fundamental property of the partial value function must be that alternative a is preferred to b in terms of criterion i if and only if $v_i(a) > v_i(b)$; similarly, indifference between a and b in terms of this criterion exists if and only if $v_i(a) = v_i(b)$. Thus the partial value function satisfies our previous definition of a preference function. However, as we shall shortly be describing, the partial value functions will in addition need to model strength of preference in some sense, so that stronger properties than simple preservation of preference ordering will in general be needed. When criterion i is associated with an attribute $z_i(a)$ as

86   MULTIPLE CRITERIA DECISION ANALYSIS

described in the introduction to this chapter, then $v_i(a)$ is of necessity a non-decreasing (but not necessarily linear) function of this attribute. We may then describe the marginal value function as a function of this attribute directly, say $v_i(z_i)$, without reference to any specific alternative a. For ease of notation, we shall use the same notation $v_i(\cdot)$ whether the argument is an alternative (a) or a performance level in terms of an attribute (z_i); in other words $v_i(a) = v_i(z_i(a))$. This should not cause any confusion. Of course, the means by which the marginal value function is constructed will depend upon whether values are assessed directly on the alternatives, or indirectly via an attribute value. We shall return to this point in Chapters 5 and 6.

A central feature of value measurement is that the properties required of the partial value functions and the form of aggregation used are critically interrelated. For the most part in this section, we shall examine the required properties underlying additive aggregation, i.e. in which $V(a)$ is constructed in the form:

$$V(a) = \sum_{i=1}^{m} w_i v_i(a).$$

Our reason for emphasizing additive aggregation is that it is the form most easily explained to and understood by decision makers from a wide variety of backgrounds, while not placing any substantially greater restrictions on the preference structures than more complicated aggregation formulae.

It is generally advisable to standardize the partial value functions $v_i(a)$ in a well-defined manner. Typically, "best" and "worst" outcomes need to be described for each criterion (either locally, being simply the best and worst of the available alternatives, or more globally as the best and worst conceivable outcomes in similar contexts). This is most easily done for criteria associated with measurable attributes, but can be done qualitatively in other cases. The partial value functions can then be standardized to 0 at the worst outcome and to some convenient value (e.g. 100) at the best. With these conventions, the "weight" w_i represents the relative importance of criterion i, in the sense of being a measure of the gain associated with replacing the worst outcome by the best outcome for this criterion.

Two observations may be useful at this stage:

- If all criteria are associated with measurable attributes z_i (i.e. defined on cardinal scales), then it is sufficient to work in terms of a value function $V(\mathbf{z})$, where $\mathbf{z}$ represents the vector of attribute values, without reference to any specific alternative. The additive assumption can then be expressed in the form:

$$V(\mathbf{z}) = \sum_{i=1}^{m} w_i v_i(z_i).$$

For the purposes of this section, we shall work in terms of the more general form $V(a)$, which allows that the assessments on one or more criteria may represent subjective values not directly linked to quantitative system attributes. The principles naturally apply to the special case in which $V(\mathbf{z})$ is defined.

- As indicated in the introduction to this chapter, the additive aggregation has been expressed above in what we have the termed extensive form, i.e. over all "lowest level" criteria simultaneously. This approach is adopted for ease of discussion only, and does not represent any fundamental restriction. For example, suppose that the value tree consisted of two hierarchical levels, with M upper level criteria (labelled by $I = 1, \ldots, M$). Let L_I be the set of lower level criteria which have been derived as subcriteria of upper level criterion I. In Figure 4.1, for example, if $I = 1$ represents the social benefits criterion, the $L_1 = \{1, 2, 3\}$, i.e. {household income; number of jobs; water supply}. The additive expression for $V(a)$ can then be written in the form:

$$V(a) = \sum_{I=1}^{M} \sum_{i \in L_I} w_i v_i(a) = \sum_{I=1}^{M} W_I \sum_{i \in L_I} w_{i|I} v_i(a)$$

where $W_I = \sum_{i \in L_I} w_i$ is a measure of the importance of the I-th upper level criterion, and $w_{i|I} = w_i/W_I$ is the importance of lower level criterion i, *relative to* the other subcriteria of the same upper level criterion. In some practical situations, it may be convenient to assess the w_i directly, while in other situations it may preferable to assess the W_I and $w_{i|I}$ separately (as will be discussed in Chapter 5). It is evident, however, that extensive and hierarchical aggregation formulae are algebraically equivalent, and must therefore imply identical assumptions and restrictions which we shall now discuss.

As indicated above, the simplicity of additive aggregation is particularly appealing. We need to ask, however, what additional assumptions, above those required for the existence of a value function, are required in order to justify this additive aggregation. Surprisingly, almost, it turns out that only relatively minor additional assumptions are needed and even these relate primarily to the need for care in defining criteria, and

88 MULTIPLE CRITERIA DECISION ANALYSIS

in the interpretation of the marginal value functions and the weights (w_i). These additional requirements are easily understood from the algebraic properties of the additive form, as described in the following.

Preferential independence: Suppose that two alternatives a and b differ only on $r < m$ criteria; let D be the set of criteria on which the two alternatives differ. Since, by definition, $v_i(a) = v_i(b)$ for $i \notin D$, it follows that a is preferred to b if and only if:

$$\sum_{i \in D} w_i v_i(a) > \sum_{i \in D} w_i v_i(b)$$

irrespective of the performances of these two alternatives according to criteria not in D (i.e. on which their performances are identical). The somewhat trivial case in which $r = 1$ implies merely that preference orderings in terms of one criterion should not depend on the levels of performance on other criteria, provided only that they remain fixed.

When $r > 1$ (implying that $m > 2$), it follows from the above that for an additive aggregation to apply, the decision maker must be able to express meaningful preferences and trade-offs between levels of achievement on a subset of criteria, assuming that levels of achievement on the other criteria are fixed, *without* needing to be concerned with what these fixed levels of achievement are. This is the property which we term *preferential independence* of the criteria, illustrated in Example Panel 4.1.

Suppose that three criteria relevant to a water development scheme in an arid region were identified as investment cost (from taxpayers' money), person-days of recreational facilities provided, and number of invertebrate species conserved. From a political viewpoint it may be much more difficult to justify restrictions on recreational access to achieve higher numbers of invertebrate species conserved if investment costs have been high (i.e. high demands on taxpayers) than if these costs are low.

If this is true, then the pair of criteria relating to person-days of recreational facilities provided and to number of invertebrate species are not preferentially independent of investment cost. Some re-formulation will be necessary to justify use of an additive value function model; perhaps using person-days of recreation and invertebrate species conserved *per unit of investment cost* may be better.

Example Panel 4.1: Illustration of preferential independence violation

For $m = 2$, the required independence property takes on a slightly different form. This is most easily understood for the situation in which both criteria are associated with measurable attributes, say z_1

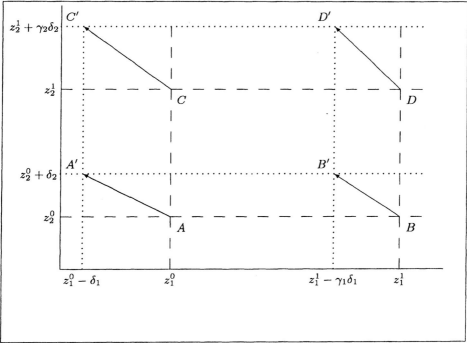

Figure 4.2: Illustration of corresponding trade-offs when $m = 2$

and z_2. Any alternative can thus be represented by a point in a two-dimensional plane, as illustrated in Figure 4.2. Consider a situation represented by the pair of values z_1^0, z_2^0, i.e. point A in the figure. Suppose that the decision maker expresses a tradeoff between the attributes in the sense that a loss of δ_1 units in z_1 is just compensated by a gain of δ_2 units in z_2 (implying indifference between the outcomes represented by points A and A'). The value function must thus satisfy the following equality:

$$w_1 v_1(z_1^0) + w_2 v_2(z_2^0) = w_1 v_1(z_1^0 - \delta_1) + w_2 v_2(z_2^0 + \delta_2)$$

Now consider higher values for each attribute, say $z_1^1 > z_1^0$ and $z_2^1 > z_2^0$, and suppose that trade-offs are also assessed at the points B (z_1^1, z_2^0) and C (z_1^0, z_2^1). There is no reason to suppose that the tradeoffs should be the same at these points (although they may be). At B, suppose that a gain of δ_2 in z_2 is judged just to compensate for a loss of $\gamma_1 \delta_1$ units in z_1 (where $\gamma_1 \neq 1$ in general). Similarly, suppose that at the outcome represented by C, a loss of δ_1 in z_1 is just compensated by a gain of $\gamma_2 \delta_2$ units in z_2. These tradeoffs imply indifferences between B and B', and between C and C', so that the

90   *MULTIPLE CRITERIA DECISION ANALYSIS*

value function must satisfy the following equalities:

$$w_1 v_1(z_1^1) + w_2 v_2(z_2^0) = w_1 v_1(z_1^1 - \gamma_1 \delta_1) + w_2 v_2(z_2^0 + \delta_2)$$

$$w_1 v_1(z_1^0) + w_2 v_2(z_2^1) = w_1 v_1(z_1^0 - \delta_1) + w_2 v_2(z_2^1 + \gamma_2 \delta_2)$$

Simple algebraic elimination of terms between the above three equalities (adding the last two, and substituting from the first) leads to the further implication:

$$w_1 v_1(z_1^1) + w_2 v_2(z_2^1) = w_1 v_1(z_1^1 - \gamma_1 \delta_1) + w_2 v_2(z_2^1 + \gamma_2 \delta_2)$$

This last equality has an important implication for the preferences being modelled, as it implies indifference between the outcomes represented by the points D and D', *no matter what the form of the functions* $v_i(z_i)$ *are*, i.e. that at the point D, a loss of $\gamma_1 \delta_1$ units in z_1 is just compensated by a gain of $\gamma_2 \delta_2$ units in z_2. This result can be seen from two perspectives, viz.:

- For any fixed value of z_2, the size of the loss in z_1 which is compensated by a specified gain in z_2 changes by the same factor (γ_1) when moving from z_1^0 to z_1^1, irrespective of this fixed value for z_2.

- For any fixed value of z_1, the size of the gain in z_2 which compensates for a specified loss in z_1 changes by the same factor (γ_2) when moving from z_2^0 to z_2^1, irrespective of this fixed value for z_1.

This is termed the *corresponding tradeoffs*, or *Thomsen*, condition, as illustrated in Example Panel 4.2.

The preference and tradeoff properties described above are necessary for the use of an additive aggregation of partial value functions. In fact these preferential independence conditions are also *sufficient* to prove the existence of an additive value function, i.e. there will always exist an additive model representing preferences which are consistent with these conditions (see Keeney and Raiffa, 1976, Chapter 3). Before proceeding with the use of an additive value function modelling approach in MCDA, it is thus necessary to confirm that the criteria are so defined that decision maker preferences are consistent with these properties. In the event of serious doubt about the validity of these assumptions, one should return to the structuring process.

It is perhaps important at this stage to differentiate between the concept of preferential dependence, and that of structural or statistical dependence between performance levels for different attributes. It may well happen that the structure of the decision space is such that

Suppose that in a choice of routes for a highway, two criteria are total investment cost and mean transit time. Suppose further that in considering one alternative for which the cost is 18 million US Dollars and mean transit time is 20 minutes (the point A in Figure 4.3), the city manager states that the council would accept an increase of 2 million US Dollars in investment cost if this could reduce the mean transit time by 2 minutes (i.e. a move from A to A' in Figure 4.3). The following conversation between the city manager and the analyst might then ensue:

Analyst: Suppose that the investment cost was actually somewhat lower, say 12 million US Dollars, would you still be prepared to pay 2 million Dollars to reduce the mean transit time from 20 to 18 minutes?

City Manager: No, from a lower base, I'd need to be more cautious; probably only about one million Dollars (i.e. a move from B to B').

Analyst: OK, fair enough. But let's try another hypothetical scenario. Suppose the cost is still 18 million Dollars, but that the transit time is now 15 minutes (point C in Figure 4.3). Would an additional 2 million Dollars in investment be worth while if this could be reduced to 13 minutes.

City Manager: No again. I don't think that reducing transit times much further would be worth that expense. For 2 million Dollars more, I would want to see a much larger reduction, say 4 minutes, i.e. a reduction to 11 minutes (point C'). Does that make sense?

Analyst: Well, it's your values I'm trying to measure, so it's not for me to argue. But to be consistent with the models we are using, if you had an option that cost 12 million Dollars and gave a mean transit time of 15 minutes, you should be prepared to accept an additional one million Dollars in cost to reduce the transit time by a further 4 minutes (a move from D to D'). Does *that* sound reasonable.

City Manager: Umm ... yes, that does sound about right.

Example Panel 4.2: Illustration of the corresponding trade-offs condition

improved performance in terms of one criterion is strongly associated with poor performance in terms of another. This does not in any way preclude the possibility that decision maker preferences may still satisfy the preferential independence properties.

Interval scale property: It is clear that the addition or subtraction of a constant term to each of the $v_i(a)$ (i.e. independent of a) will not affect the preference ordering implied by the resultant $V(a)$. The addition of such a term simply re-defines the (generally arbitrary) level of performance which is allocated a "zero value" so as to provide a reference point against which other performances are measured. Ratios of partial values will accordingly not be meaningful in general (since addition of constants to each $v_i(a)$ will change their ratios), except in

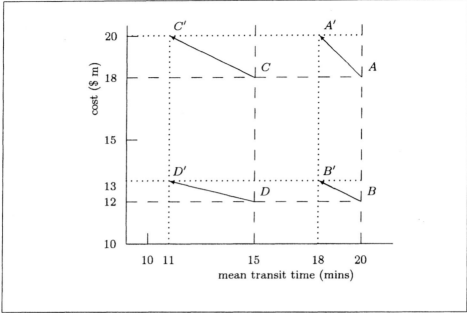

Figure 4.3: Corresponding tradeoffs for Example Panel 4.2

special circumstances under which a natural and unambiguous zero value point exists, precluding such addition or subtraction of constant terms. Consequently, only ratios of differences between the $v_i(a)$ (i.e. the relative magnitudes of differences) have any absolute meaning independent of the generally arbitrary choices of zero point and scaling. This is the defining property of an *interval scale* of preferences.

The implication of the interval scale property is best seen by looking again at tradeoffs between criteria which are associated with measurable attributes. For criterion i, consider two pairs of values for the associated attribute z_i, viz. $z_i^1 > z_i^0$ and $z_i^3 > z_i^2$. Suppose that the corresponding value differences are equal, i.e. that $v_i(z_i^1) - v_i(z_i^0) = v_i(z_i^3) - v_i(z_i^2)$. Clearly, if an increase from z_i^0 to z_i^1 just compensates for specified losses in other criteria, then an increase from z_i^2 to z_i^3 will equally well just compensate. In constructing partial value function models, therefore, it is important to check that equal increments in $v_i(z_i)$ do indeed represent equal tradeoffs with other criteria, as illustrated in Example Panel 4.3.

Weights as scaling constants (trade-offs): The weight parameters w_i have a very specific algebraic meaning. It is important to emphasize this, as people are often willing to specify the relative impor-

Consider again the problem of choice of routes for a highway, with the two criteria: total investment cost and mean transit time. In Chapter 5 we shall look at means by which value functions may be assessed, but for now suppose that we have established a partial value function for mean transit time as $v_2(z_2) = (z_2 - 30)^2$ for $10 < z_2 < 30$. Note that in this case, the value function *decreases* with increasing values of the attribute. It is easily confirmed that $v_2(21) - v_2(25) = 56$ and $v_2(15) - v_2(17) = 56$. These value differences are illustrated in the example below by the gaps between the pair of horizontal dotted lines, and between the pair of horizontal broken lines.

Thus, whatever the value function for investment cost is, if decision makers are prepared to accept an increase in investment cost from 20 to 21 million dollars in order to reduce mean transit time from 25 to 21 minutes, then for any additive model to hold, they should equally well be prepared to accept an increase in investment cost from 20 to 21 million dollars in order to reduce mean transit time from 17 to 15 minutes.

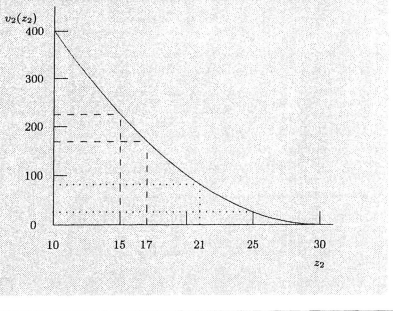

Example Panel 4.3: Illustration of interval scale property

tances of criteria on numerical scales, but these may not necessarily coincide with the algebraic meaning in the additive value function. The weights here are simply scaling constants which render the different value scales commensurate. Suppose that partial value functions for two criteria have been constructed, satisfying the preferential independence and interval scale properties. Suppose further that two alternatives a and b, differing only in terms of the two criteria r and

94   *MULTIPLE CRITERIA DECISION ANALYSIS*

s, are judged by the decision maker to be equally preferred. Since the alternatives are equally preferred, it follows that $V(a) = V(b)$, implying (since the alternatives differ on two criteria only) that $w_r v_r(a) + w_s v_s(a) = w_r v_r(b) + w_s v_s(b)$, i.e. that $w_r \Delta v_r = w_s \Delta v_s$, where $\Delta v_r = v_r(b) - v_r(a)$ and $\Delta v_s = v_s(a) - v_s(b)$. The weights are thus required to be in the ratio: $w_r/w_s = \Delta v_s/\Delta v_r$. Any method of assessing weights must be consistent with this requirement.

The above conditions for the validity of an additive value function model place relatively modest constraints on the preference structures being modelled, apart from those implied by the existence of the value function itself. They do, however, impose some discipline on the manner in which the model is constructed, primarily as regards definitions of the criteria (to ensure the necessary preferential independence), the construction of the partial value functions (to ensure the interval scale property), and the assessment of the weights (to ensure the trade-off property). We shall return to these issues in Chapters 5 and 6.

Other forms of aggregation in value function preference modelling are certainly possible. A multiplicatively aggregated preference function of the form:

$$V(a) = \prod_{i=1}^{m} [v_i(a)]^{w_i}$$

has been popular in some MCDA approaches. Taking logarithms converts the above into an additive form, so that the properties discussed above still apply, except that the interval scale properties apply to $\log v_i(a)$ rather than to $v_i(a)$. The implication is that the partial value functions in this case must have a ratio scale interpretation, and some workers (e.g. Barzillai, 1987, and Lootsma, 1997) have argued in effect that preferences are often naturally perceived in ratio scale terms, so that the product form is more natural. A true ratio scale demands a natural zero point, which certainly exists in assessments of properties such as mass or length, for example. Many attributes which arise in decision analysis do not appear, however, to exhibit an unambiguous and clear-cut zero-value point, so that apparently natural ratio-judgement expressions (e.g. "A is twice as good as B") seem more likely to be expressions of ratios of gains relative to an unstated and essentially arbitrary reference level. In this case, these natural ratio statements may be more characteristic of an interval rather than a ratio scale. With this in mind, we shall describe the implementation of value function methods in the context of additive value functions, noting that due care needs to be taken in assessing the partial value functions to ensure that these are on an interval scale.

The conditions described above are necessary to ensure that an additive aggregation faithfully models preference orders. It must be re-emphasized that the only properties that have been assumed up to now for the global value function, is that it preserves preference ordering ($a \succ b$ if and only if $V(a) > V(b)$). If stronger properties are desired (for example, that the interval scale property is preserved in the aggregate function), then additivity may not hold even with the above assumptions. We shall not pursue this point here (see, however, von Winterfeldt and Edwards, 1986, Chapter 8, for further discussion), but the issues are closely related to the problem of defining "utility functions", which are preference models permitting the application of an expectation operator under conditions of uncertainty. These are discussed in the next section.

4.3.   UTILITY THEORY: COPING WITH UNCERTAINTY

Utility theory can be viewed as an extension of value measurement, relating to the use of probabilities and expectations to deal with uncertainty. Although, for the most part, we shall not in this book be making use of probability theory or statistical expectations to model uncertainties in the decision process, it seems useful to include an outline of multiattribute utility theory (MAUT) as part of our survey of preference models. This is included primarily for completion, in order to provide analysts and researchers in MCDA with a readily available overview of the principles behind utility theory. A secondary result of the discussion, however, will be a demonstration that in many applications it may not be necessary to invoke the substantially more complicated multiattribute utility theory models. The simple additive value functions of the previous section, linked to sensitivity analysis, provide in most instances essentially the same results and insights, and it is for this reason that we shall largely concentrate on the simpler models in later chapters.

The discussion of MAUT is of necessity at a relatively higher technical level than was needed for the simpler value function models of the previous section. However, as this section is primarily aimed at developers of decision support systems and processes, it can without serious loss be omitted by less technically-minded readers.

This presentation of MAUT is also relatively brief, attempting to capture the key concepts only. For a more detailed and rigorous discussion, the reader is referred to Keeney and Raiffa (1976).

4.3.1. Basic utility principles

For purposes of developing multiattribute utility theory, we shall need to assume that each criterion is directly associated with a quantitative attribute measured on a cardinal scale. The attribute values (z_i) are, however, not fully determined by choice of the alternative, but may also be influenced by unknown exogenous ("random") factors. The consequences of each alternative is thus described in terms of a probability distribution on the attribute vector $\mathbf{z}$.

We now seek a function, say $U(\mathbf{z})$ (which we shall term a *"utility function"*), such that alternative a is preferred to alternative b if and only if the *expectation* of $U(\mathbf{z}^a)$ is greater than that of $U(\mathbf{z}^b)$, i.e. $\mathrm{E}[U(\mathbf{z}^a)] > \mathrm{E}[U(\mathbf{z}^b)]$, where $\mathbf{z}^a$ and $\mathbf{z}^b$ are the (random) vectors of attribute values associated with alternatives a and b respectively. This condition is called the *expected utility hypothesis*. Since this property must also hold for deterministic outcomes (which are simply degenerate random variables), it must also follow that the (deterministic) outcome described by $\mathbf{z}^1$ will be preferred to that described by $\mathbf{z}^2$ if and only if $U(\mathbf{z}^1) > U(\mathbf{z}^2)$. Thus the utility function $U(\mathbf{z})$ also satisfies the properties of a value function as defined in the previous section. It is not necessarily, however, identical to the additive or multiplicative forms of value function $V(\mathbf{z})$ defined there, although the functions must be related to each other via some strictly increasing transformation (since $U(\mathbf{z}^1) > U(\mathbf{z}^2)$ if and only if $V(\mathbf{z}^1) > V(\mathbf{z}^2)$). The possibility of a distinction between $U(\mathbf{z})$ and $V(\mathbf{z})$ arises because the requirements of the expected utility hypothesis and those of seeking simple forms of aggregation (such as additive) may be incompatible.

The fundamental question at this stage is: Does there exist *any* function $U(\mathbf{z})$ satisfying the expected utility hypothesis? Establishing the existence of a utility function in this sense turns out to be a non-trivial problem even for the single attribute case, which is the context in which von Neumann and Morgenstern (1947) first developed expected utility theory. Suppose that alternatives are evaluated on the basis of a single measure of reward, z, but that for each alternative, the reward is a random variable having a known probability distribution over some set of possible rewards Z. We can then associate probability distributions on Z with each action, and can speak in terms of preferences between these probability distributions, often termed *lotteries*. For any pair of lotteries P and Q, we shall use the notation $P \succ Q$ to denote that P is preferred to Q. The question then becomes whether there exists a function $u(z)$ such that such that $P \succ Q$ if and only if $E[u|P] > E[u|Q]$, where $E[u|P]$ denotes the statistical expectation of $u(z)$ when z is generated from the probability distribution P. For complete and transitive preferences on

Z, the function $u(z)$ would need to be continuous and strictly increasing. The existence of such a function then also implies the existence of the *certainty equivalent (CE)* of the lottery, namely the unique reward z_{CE} such that $u(z_{CE}) = E[u|P]$.

It is convenient to introduce here the notion of a *mixture lottery* denoted by $\{P, \alpha, Q\}$. This will denote a two-stage lottery, in which a random selection is first made between two lotteries P and Q, with the probability of selecting P being given by α, after which the reward (z) is generated according to whichever of P and Q is chosen in the first stage.

The expected utility theory developed by von Neumann and Morgenstern is based firstly on the assumption that preferences are to be defined over a set of lotteries which includes degenerate lotteries (i.e. deterministic outcomes) and which is closed under mixing as defined above. Apart from the completeness and transitivity assumptions of value measurement theory discussed in the previous section, a critical further assumption is the *independence axiom*:

> Preferences between two mixture lotteries $\{P, \alpha, R\}$ and $\{Q, \alpha, R\}$ depend only on preferences between P and Q, and are independent of the common outcome represented by R. In formal terms: $\{P, \alpha, R\} \succ \{Q, \alpha, R\}$ if and only if $P \succ Q$.

It can be shown that the independence axiom, together with the assumptions of completeness, transitivity and a technical assumption regarding continuity of preference, are necessary and sufficient for preferences to satisfy the expected utility hypothesis, viz. the existence of a function $u(z)$ such that $P \succ Q$ if and only if $E[u|P] > E[u|Q]$. Note that the function $u(z)$ is not uniquely defined, since the preference ordering implied by the expectations will be unchanged if $u(z)$ is replaced by $a + bu(z)$, for any a and any $b > 0$. In other words: It is always possible to arbitrarily select two values, say z^0 and z^1 with $z^0 < z^1$, such that $u(z^0) = 0$ and $u(z^1) = 1$. Apart from this arbitrary choice for the origin and scale, however, the utility function is unique, as more general monotonic transformations will lead to changes in orderings for some lotteries at least.

Although the independence axiom has a normative appeal, it has been shown in many contexts to be systematically violated by many people (i.e. it is not descriptively valid). An example of violation of the independence axiom is provided by *Allais' Paradox* (e.g. Howard, 1992), which can be described in terms of the following lotteries:

P: $1m for sure

98 MULTIPLE CRITERIA DECISION ANALYSIS

Q: \$5m with probability 10/11 (and nothing otherwise)

R: nothing (for sure)

Suppose now that for $\alpha = 0.11$, the decision maker is faced with the following two choices:

- Choice 1: Select between

  ▷ $\{P, \alpha, P\}$ i.e. the degenerate lottery in which P, giving \$1m for sure, is the certain outcome; and
  ▷ $\{Q, \alpha, P\}$ which gives \$5m with 10% probability, \$1m with 89% probability, and 0 with 1% probability.

- Choice 2: Select between

  ▷ $\{P, \alpha, R\}$ which gives \$1m with 11% probability (nothing otherwise).
  ▷ $\{Q, \alpha, R\}$ which gives \$5m with 10% probability (nothing otherwise).

It has been found that many people prefer the \$1m for sure option in the first case, but $\{Q, \alpha, R\}$ (giving \$5m with 10% probability) in the second case. This is in violation of the independence axiom, which would require $P \succ Q$ to be consistent with the first choice, but $Q \succ P$ to be consistent with the second choice. It can, in fact, be demonstrated directly that there is no utility function which is consistent with both preferences.

A number of explanations for the above behaviour have been suggested, for which we do not have space here to do justice. One explanation which we find appealing, however, is that of a "reference level", i.e. an outcome which the decision maker perceives as neutral in the sense that any lesser values are perceived as losses, while larger values are perceived as gains. The framing of the two choices is such that in the first instance the reference level is likely to be the winning of \$1m, while in the second instance it is likely to be the *status quo*. As a consequence the first choice includes both an outcome which is perceived to be a loss (of \$1m) and one which is perceived to be a gain (of \$4m); while the second choice only involves perceived gains. There should be no expectation that risk preferences should be the same in both cases. The practical problem is that it may difficult to know what the appropriate reference level may be in any one situation.

The descriptive failure of expected utility theory, however, does not lessen the value of constructing preferences according to this model.

Substantial insight can be provided to the decision maker through the discipline of maintaining consistency with the stated axioms, especially if a plausible reference level can be agreed upon. Such an approach is particularly relevant when handling uncertainty in multicriteria decision support, as few other useful models exist for this case.

4.3.2. Problems of Aggregation

Our primary interest here is on the extension of these concepts to the multiple criteria problem, i.e. the manner in which utility functions for each of a number of different criteria (say $u_i(z_i)$ for $i = 1, \ldots, m$) can be aggregated into an overall multiattribute utility function (say $U(\mathbf{z})$, where $\mathbf{z} = z_1, \ldots, z_m$) in such a way that the expected utility hypothesis applies for multivariate gambles over $\mathbf{z}$. Key questions include whether the partial utility functions $u_i(z_i)$ which model preferences between lotteries are equivalent to the partial value functions $v_i(z_i)$ defined in Section 4.2; and whether additive aggregation of the $u_i(z_i)$ is justified.

The added complications of dealing with uncertainties are evident even for the case of $m = 2$, as may be illustrated by a simple example. Suppose that lotteries involve two outcomes only $(z_i^0 < z_i^1)$ for each of the two attributes $(i = 1, 2)$. Without loss of generality, the partial utility functions can be standardized such that $u_1(z_1^0) = u_2(z_2^0) = 0$ and $u_1(z_1^1) = u_2(z_2^1) = 1$. Suppose that we propose to model preferences for bivariate lotteries by means of an additive utility function of the form $w_1 u_1(z_1) + w_2 u_2(z_2)$. Consider then the expected utilities for the following two lotteries:

- The lottery giving equal chances on $(z_1^0 \; ; \; z_2^0)$ and $(z_1^1 \; ; \; z_2^1)$; and

- The lottery giving equal chances on $(z_1^0 \; ; \; z_2^1)$ and $(z_1^1 \; ; \; z_2^0)$.

It is easily verified that both of these lotteries yield an expected utility of $(w_1 + w_2)/2$. The model thus suggests that the decision maker should always be indifferent between these two lotteries. There seems, however, to be no compelling axiomatic reason for forcing indifference between the above two options. Where there is some measure of compensation between the criteria (in the sense that good performance on one can compensate for poorer outcomes on the other), the second option may be preferred as it ensures that one always gets some benefit (a form of multivariate risk aversion). On the other hand, if there is need to ensure equity between the criteria (if they represent benefits to conflicting social groups, for example), then the first lottery (in which loss or gain is always shared equally) may be preferred.

100 *MULTIPLE CRITERIA DECISION ANALYSIS*

4.3.3. Aggregation across two attributes

For multicriteria decision making with certain outcomes, the assumption of *preferential independence* turned out to be necessary and sufficient for additive aggregation across criteria to apply. Recall, however, that the intention in that case was only to establish an aggregate function preserving preference ordering between outcomes. Preferential independence is, however, not sufficient to justify additive aggregation of utilities if it is required that the aggregate utility function $U(\mathbf{z})$ should also satisfy the expected utility hypothesis for uncertain outcomes. Two possibly stronger properties than preferential independence have been suggested as follows for the case of two attributes:

Additive Independence: Two attributes are additive independent if preferences between lotteries depend on the marginal probability distributions only, i.e. the CE for z_1 depends only on the marginal distribution of z_1, and not on its joint distribution with z_2. This assumption leads to additive aggregation of the partial utilities, but as we have seen in the above example, this may turn out to be too strong an assumption in some contexts at least.

Utility Independence: The attribute z_1 is utility independent (UI) of z_2 if preferences between lotteries on z_1 for *fixed* z_2 do not depend on this fixed value, i.e. the CE for z_1 *when z_2 is fixed* does not depend on this fixed value of z_2.

If z_1 is UI of z_2 and z_2 is UI of z_1, then z_1 and z_2 are said to be *mutually utility independent.*

For convenience, suppose that we select values $z_i^0 < z_i^1$ for each attribute, and standardize both the partial and the joint utility functions so that:

$$U(z_1^0, z_2^0) = u_1(z_1^0) = u_2(z_2^0) = 0$$

and

$$U(z_1^1, z_2^1) = u_1(z_1^1) = u_2(z_2^1) = 1.$$

Some messy but quite simple algebra shows that if z_1 and z_2 are mutually utility independent, then $U(z_1, z_2)$ is related to $u_1(z_1)$ and $u_2(z_2)$ in the following manner:

$$U(z_1, z_2) = k_1 u_1(z_1) + k_2 u_2(z_2) + k k_1 k_2 u_1(z_1) u_2(z_2) \qquad (4.1)$$

where

$$k = \frac{1 - k_1 - k_2}{k_1 k_2}$$

i.e. a *bilinear* aggregation of $u_1(z_1)$ and $u_2(z_2)$.

If $k = 0$, then the bilinear form is simply additive. Otherwise, by multiplying both sides of (4.1) by k, and adding 1 to both sides, we obtain an alternative multiplicative representation of the aggregation of the partial utilities:

$$1 + kU(z_1, z_2) = [1 + kk_1 u_1(z_1)] [1 + kk_2 u_2(z_2)] \qquad (4.2)$$

where (by considering the case of $z_1 = z_1^1$ and $z_2 = z_2^1$):

$$1 + k = [1 + kk_1] [1 + kk_2]$$

It is instructive to examine the results of applying the above aggregation formulae to the simple two point lottery example used above:

- The lottery giving equal chances on $(z_1^0 \; ; \; z_2^0)$ and $(z_1^1 \; ; \; z_2^1)$ has expected utility $\frac{1}{2}(k_1 + k_2 + kk_1 k_2)$; while

- The lottery giving equal chances $(z_1^0 \; ; \; z_2^1)$ and $(z_1^1 \; ; \; z_2^0)$ has expected utility $\frac{1}{2}(k_1 + k_2)$

The two lotteries are no longer necessarily equivalent, with the case of $k > 0$ implying preference for the more equitable solution, and the case of $k < 0$ implying preference for the solution offering compensation between attributes.

4.3.4. Generalizations for $m > 2$ attributes

Either of the aggregation formulae for $m = 2$ (i.e. (4.1) or (4.2)) can be generalized for $m > 2$ to suggest either a multilinear form of aggregation:

$$U(\mathbf{z}) = \sum_{i=1}^{p} k_i u_i(z_i) + \sum_{i=1}^{p} \sum_{i<j\leq p} k_{ij} u_i(z_i) u_j(z_j)$$
$$+ \ldots + k_{12\ldots p} u_1(z_1) u_2(z_2) \ldots u_p(z_p)$$

or multiplicative aggregation:

$$1 + kU(\mathbf{z}) = \prod_{i=1}^{p} [1 + kk_i u_i(z_i)]$$

which is in fact a special case of the multilinear form in which the coefficient corresponding to any product of q out of the p $u_i(z_i)$ terms is simply k^q times the product of the corresponding k_i terms. Both forms include additive aggregation as a special case.

In contrast to the situation for $m = 2$, the multilinear and multiplicative forms are no longer equivalent for $m > 2$, and thus correspond to

102 MULTIPLE CRITERIA DECISION ANALYSIS

somewhat different sets of assumptions for which we do not have space here to consider in detail. (See, for example, Keeney and Raiffa, 1976 for a more complete treatment.) It is worth, nevertheless, to look briefly at the assumptions underlying the multiplicative form, which has a much more parsimonious parameterization than the general multilinear form. To this end, let us define $\bar{z}_i$ as the set of all attributes except z_i. The critical assumption can thus be expressed as:

Assumption: Suppose that $\bar{z}_i$ is UI of z_i for each i, i.e. that preferences for lotteries on the $m-1$ vector $\bar{z}_i$ for a fixed value of z_i are independent of what this value is.

This can be shown to be equivalent to the seemingly stronger assumption of mutual utility independence, viz. that every subset of $z_1, z_2, \ldots, z_p$ is UI of its complement.

An important result is that the above assumption implies either multiplicative or additive aggregation, but no other form of aggregation. In other words, if mutual utility independence holds, and if both the partial and aggregated utility functions are standardized so that $U(\mathbf{z}^0) = 0$, $u_i(z_i^0) = 0$ for all i, $U(\mathbf{z}^1) = 1$ and $u_i(z_i^1) = 1$ for all i, then one of the following two relationships must hold:

$$U(\mathbf{z}) = \sum_{i=1}^{m} k_i u_i(z_i)$$

or:

$$1 + kU(\mathbf{z}) = \prod_{i=1}^{m} [1 + k k_i u_i(z_i)].$$

In the latter case, the parameter k is related to the other k_i according to the relationship:

$$1 + k = \prod_{i=1}^{m} [1 + k k_1].$$

The additive model corresponds in fact to the multiplicative model in the limiting case as k tends to zero, and thus only applies in a very special case, so that multiplicative aggregation should really be viewed as the standard consequence of utility independence.

Implementation of the multiplicative model requires not only the assessment of the m partial utility functions (each of which requires the decision maker to evaluate certainty equivalents for lotteries on the relevant criterion), but also estimation of the m parameters $k_1, k_2, \ldots, k_m$. The parameter estimation requires the establishment of indifferences between at least m pairs of multivariate lotteries. (For details, see Keeney and Raiffa, 1976.) As this involves rather difficult and demanding cognitive tasks for the decision maker, an important question is whether the gains in decision quality are worth the effort involved. What do we lose by simply using the expectations (if necessary) of an additive value function as defined in Section 4.2? We turn briefly to this question in the next subsection.

4.3.5.   Use of additive utility functions

The question at this stage relates to the degree to which preference orderings may be sensitive to the use of additive (rather than multiplicative) aggregation of partial utilities (in spite of the counter example raised previously). The question has important practical implications, as additive models are easier to understand and to construct, and only require the estimation of marginal probability distributions, rather than the full multivariate distribution. Reading of the literature suggests that in practice many applications of multiattribute utility theory do in fact tend to use additive models (probably for the reasons noted above).

In Stewart (1995) a number of simulation studies are reported in which the effects on preference orderings of using additive rather than multiplicative aggregation under conditions of mutual utility independence, are assessed. Details may be found in the cited reference, but in essence it appeared that the errors introduced by using the additive model were generally extremely small for realistic ranges of problem settings, and were in any case considerably smaller than those introduced by incorrect modeling of the partial utility functions (such as over-linearization of the partial functions). Related work (Stewart, 1996b) has demonstrated that violations of preferential independence are also a serious potential source of error.

Our overall conclusion is thus that although the multiattribute utility theory described above is of value in identifying pathological problems in which additivity may be inappropriate, in practice the use of additive models for decision making under uncertainty is likely to be more than adequate in the vast majority of settings. The imprecisions and uncertainties involved in constructing the partial utilities, which need to be addressed by careful sensitivity analysis, are likely to far outweigh any distinctions between the additive and multiplicative models. Since, as previously seen, the utility function is also a value function when applied to certain outcomes, it would appear that for most realistic problems it may be preferable simply to construct an additive value function by direct assessment, to be described in the nexttwo chapters, rather than to use the conceptually more difficult lottery questions.

4.4. SATISFICING AND ASPIRATION LEVELS

In this case, we operate directly on the partial preference functions $z_i(a)$ without further transformation. The assumption, however, is that the preference function values have cardinal meaning, i.e. are more than simply ordinal or categorical, and relate to operationally meaningful and measurable attributes.

The satisficing model suggested by Simon (e.g., Simon, 1976), but expressed in our MCDM terminology, was that decision makers focus initially on seeking improvements to what is perceived to be the most important criterion. In effect, available courses of action (alternatives) are systematically eliminated until, in the view of the decision maker, a satisfactory level of performance for this criterion has been ensured. At this point, attention shifts to the next most important criterion, and the search continues amongst the remaining alternatives for those which ensure satisfactory performance on this criterion. The process should not be viewed too literally in terms of a single pass through the criteria, as it would in most cases be dynamic in the sense that:

- If all but one alternative has been eliminated, but there still remain significant criteria which have not yet been taken into consideration, the decision maker might backtrack with a view to being less demanding when eliminating alternatives;

- If a number of alternatives remain once all criteria have reached satisfactory levels, the decision maker might cycle through the criteria again (not necessarily in the same order as before).

It is important to note that Simon did not intend this model to be normative in the sense of suggesting how decisions should be made. It was meant as a descriptive model of a heuristic which people use, as a result of bounded rationality, in making difficult decisions. In the Introduction to the third edition of his book on *Administrative Behavior*, Simon (1976) refers to: "... human beings who satisfice because they have not the wits to maximize". This needs to be emphasized as some views on goal programming (to be described in Chapter 7) seem to elevate the satisficing concept to an end in itself.

In spite of the fact that the satisficing model is not normative, there is nevertheless still merit in adopting the associated heuristic as a basis for designing methods of decision aid and support. If the heuristic has helped skilled decision makers to make at least good (if not "optimal") decisions, then it should also have the potential to provide systematic guidance in other complex, i.e. multicriteria, decision making problems.

This is certainly an important motivation for the goal programming techniques discussed in Chapter 7. As heuristics, these techniques are perhaps most valuable when used in situations in which the decision maker is unwilling or unable to provide the levels of inputs necessary for the in-depth preference modelling implied by the value and utility measurement approaches described earlier in this chapter. There are at least two contexts in which such situations might arise:

- *Preliminary stages of investigation:* At early stages of the decision making process, it is often necessary first to extract a "shortlist" of alternatives from the potentially vast (or even infinite) number of those available (cf. the case example described in Sections 2.2.3 and 3.5.3). It will typically not be feasible to undertake in-depth evaluations of all of the vast number, so that the extraction of the shortlist needs to be done quickly and efficiently without requiring intense debate on trade-offs and other preference information. Thereafter, more detailed evaluations can be applied to the elements in the shortlist. It is important, however, to have some assurance that potentially good alternatives are not excluded at the preliminary stage, and it is here that the satisficing heuristic may be very valuable. This is particularly true for strategic level decisions, and especially those in a strongly group decision making context, when the extraction of the shortlist may well be delegated to a small working group, who may not have complete understanding of the group preferences.

- *Routine decisions:* Some decisions may involve multiple criteria, but may be part of relatively routine management activities (for example, selecting new computer equipment). The extent of the impacts of the decisions may be quite limited, not really involving strategic goals of the organization. Under these circumstances, decision makers might not wish to spend much time in developing a full model of their preferences, so that the satisficing heuristic may be a valuable model, especially for more repetitive decisions in which decision makers have a very good idea *a priori* as to what are realistically achievable satisfactory performance levels.

The "goal programming" approach to MCDA, as discussed in Chapter 7, is essentially a simple operational implementation of the satisficing heuristic. The satisficing levels are generally assumed to be specified *a priori* as "goals", after which a mathematical programming algorithm is used to approach these goals as closely as possible. As we shall discuss at that time, there are a number of alternative modes of implementing the basic goal programming concepts. We shall argue there that some form of Tchebycheff, or min-max, measure of deviation from the goals appears

106 *MULTIPLE CRITERIA DECISION ANALYSIS*

to come closest to the dynamic nature of satisficing. The min-max approach is not widely adopted in the conventional goal programming literature, but is more-or-less standard in the "reference point" approach, or generalized goal programming, as introduced by Wierzbicki (1980, 1999).

The satisficing approach to decision analysis, as embodied in the goal programming and reference point approaches of Chapter 7, offers relatively simple and appealingly intuitive statements of preference from the decision maker. A still open research question, however, relates to the manner in which decision makers form perceptions of what constitutes a level of satisfactory performance. Observations of real world decision makers suggest quite divergent psychologies, with some decision makers setting goals or aspirations which are very much more ambitious than others. Wierzbicki's reference point approach (see Section 7.3) does perhaps recognize this more explicitly than standard goal programming approaches, by allowing the solution search to continue beyond the initially specified reference level, and recognizing the dynamically changing nature of satisficing levels. Real-world applications of goal programming and related techniques need to be accompanied by a substantial level of sensitivity analysis to assess the effects of changing the goals or aspirations.

4.5. OUTRANKING

As with the satisficing models, outranking models are applied directly to partial preference functions $z_i(a)$, which are assumed to have been defined for each criterion. These preference functions may correspond to natural attributes on a cardinal scale, or may be constructed in some way, typically as ordinal or ordered categorical scales. These preference functions do not need to satisfy all of the properties of the value functions introduced in Section 4.2, although ordinal preferential independence would still be necessary (i.e. it must be possible to rank order alternatives on one criterion independently of performances in terms of the other criteria).

A comprehensive treatment of the outranking concept is given by Roy and Bouyssou (1993) and Roy (1996). The general principle can be seen as a generalization of the concept of dominance described earlier in this chapter. We noted that if for two alternatives a and b, $z_i(a) \geq z_i(b)$ for all criteria i (with strict inequality $z_i(a) > z_i(b)$ for at least one criterion), then we can immediately conclude that a should be preferred to b (provided, of course, that the set of criteria is sufficiently complete). In this event, we could say that the evidence favouring the conclusion that alternative a is as good or better than alternative b is unarguable, and

Preference Modelling 107

a was said to *dominate b*. More generally, we shall say that *a* *outranks* alternative *b* if there is "sufficient" evidence to justify a conclusion that *a* is at least as good as *b*, taking all criteria into account. There are two key aspects of the outranking definition which distinguish it from the preference relationships corresponding to value functions. These are:

1 The emphasis is on strength of evidence for (or credibility of) the assertion that "*a* is at least as good as *b*", rather than on strength of preference *per se*. It is possible that strong preferences may ultimately be revealed, even though the evidence at an earlier stage of constructing preferences may be highly conflicting, leading to no definite conclusion, i.e. no "outranking" (for example, when alternatives differ widely on a large number of criteria). On the other hand, the evidence for preferring one alternative over another may be conclusive (one "outranks the other"), even though preferences are weak (for example, when one alternative dominates another, but none of the differences are particularly important). These concepts are illustrated in Example Panel 4.4.

2 It follows from the previous point that even when neither *a* nor *b* outranks the other, a state of indifference is not necessarily implied. In comparing two alternatives, therefore, four situations may arise (as opposed to the three implied by the strict adherence to a value function): a definite preference for one alternative over the other, indifference, or "incomparability" (i.e. a lack of decisive evidence, as in the case of comparing alternatives *a* and *b* in Example Panel 4.4).

As with value function or utility methods, outranking starts by consideration of the individual partial preference functions $z_i(a)$. The approach generally recognizes explicitly that the preference functions will usually be rather imprecise measures, so that alternative *b* can only be said to be conclusively preferred to *a* in terms of criterion *i* if the amount by which $z_i(b)$ exceeds $z_i(a)$ is above an "indifference threshold". In some applications, two thresholds may be sought so as to recognize a distinction between *weak* and *strict preference* respectively (with the sense being that "strict" preference for *b* over *a* on even just one criterion may make it much more difficult to reach a definite conclusion that *a* outranks *b*). As we have previously noted, it will not in general be true that equal increments in the preference function values will be of equal importance, so that the thresholds for indifference and strict preference may depend on the actual value of $z_i(a)$. Two threshold functions $p_i[z]$ and $q_i[z]$ are thus defined for each criterion *i* such that:

Consider a situation in which three alternatives are differentiated on four criteria, all of which are measured on a 9-point scale, with values as follows:

Alternative	z_1	z_2	z_3	z_4
a	9	9	2	2
b	1	1	9	9
c	8	8	2	2

The alternatives may also be represented by displaying their performance profiles as follows:

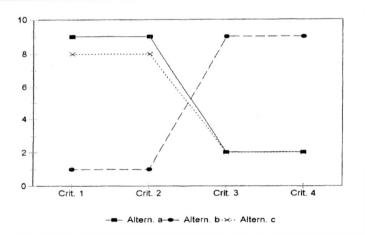

In view of the fact that alternatives a and b have such substantially different profiles, it may ultimately turn out that the decision maker very strongly prefers one or the other, so that *indifference* between a and b is not supported by the evidence. Without further information regarding decision maker values, however, we could not draw any definite conclusions concerning direction of preference, so that the alternatives a and b would have to be viewed as *incomparable* at this stage.

On the other hand, it is clear that a dominates c; but the differences are small, so that the preference for a over c might easily be reversed if another criterion were identified for which c was preferred to a. The preference for a over c could thus at best be described as weak, and the decision maker might ultimately be *indifferent* between the two if the differences on the first two criteria are perceived to below the threshold of indifference.

Example Panel 4.4: Illustration of differences between indifference and incomparability

- Alternative b is weakly preferred to alternative a in terms of criterion i if:

$$z_i(b) > z_i(a) + q_i[z_i(a)]$$

- Alternative b is strictly preferred to alternative a in terms of criterion i if:

$$z_i(b) > z_i(a) + p_i[z_i(a)]$$

(where, for consistency, we must have $p_i[z_i(a)] > q_i[z_i(a)]$). Note that if $z_i(b) \leq z_i(a) + q_i[z_i(a)]$ and $z_i(a) \leq z_i(b) + q_i[z_i(b)]$, then there is no evidence to support preference for either of the alternatives from the point of view of criterion i; in other words, in terms of criterion i there is indifference between a and b.

Once preferences in terms of each criterion have been modelled (by the partial preference functions and the indifference and preference thresholds), this evidence is aggregated across criteria in order to summarize the current state of information regarding discrimination between alternatives. The process is explicitly dynamic, and the situation will change as different "rules of evidence" are brought into play. Such rules of evidence are not uniquely defined, leading to the existence of many varieties of "outranking" methods for MCDA. There are, nevertheless, two recurring themes running through most of these methods:

- If a is demonstrably as good as or better than b according to a sufficiently large weight of criteria, then this is considered to be evidence in favour of a outranking b (the "concordance" principle);

- If b is very strongly preferred to a on one or more criteria, then this is considered to be evidence against a outranking b (the "discordance" principle).

The above two principles are rather imprecisely defined, and are certainly open to a number of interpretations. The concordance principle is usually made operational by some form of weighted pairwise voting procedure. Each criterion is allocated an importance "weight". Then, for any pair of alternatives a and b, each criterion in effect "votes" for or against the assertion that "a is at least as good as b". Outranking methods do differ according to how this vote is exercised, and the various implementations are discussed in detail in Chapter 8. In some sense, however, the proportion of the total votes which support the assertion that "a at least as good as b" provides a measure of concordance, i.e. of evidence supporting the assertion.

It is worth reflecting briefly on the meaning of the weights in this case. In contrast to the interpretation of weights in the value function, utility

110   *MULTIPLE CRITERIA DECISION ANALYSIS*

function and aspiration level models, the outranking weights do not represent tradeoffs or scaling factors introduced to ensure commensurability between criterion measures. The weights measure the influence which each criterion should have in building up the case for the assertion that one alternative is at least as good as another, and are uninfluenced by monotonic transformations of the preference functions. It is less clear to what extent these weights are or should be adapted to the range of outcomes available on the criteria (i.e. greater importance to be placed on criteria on which alternatives differ very widely), although on *prima facie* grounds it does seem that if there are substantial differences between alternatives in terms of one criterion, its influence *should* be greater than if there is less contrast between alternatives. We shall comment further on these issues in Section 4.7 and later in Section 9.9 (where we discuss the implementation of MCDA methods).

Discordance can be understood in terms of a "veto", in the sense that if performance represented by $z_i(a)$ is below a minimally acceptable level, or the difference $z_i(b) - z_i(a)$ is greater than some threshold, then no conclusion that "a is at least as good as b" is allowed, i.e. a cannot be said to outrank b. The implication is that the performance of a in terms of criterion i is so unacceptable, that it can never be compensated for by better performance on other criteria. An operational measure of the degree of discordance may be provided by the maximum (over all criteria) of expressions of the form $(z_i(b) - z_i(a))/S_i$, where S_i is some form of scaling factor (which may be generalized to depend upon the values of $z_i(a)$ and/or $z_i(b)$). Where the $z_i(a)$ are similarly scaled for all criteria, the same scaling factor may be used for all criteria. For example, the maximum of $|z_i(a) - z_i(b)|$ taken over all pairs of alternatives and over all criteria may be used as a common scaling factor. It should be noted that no importance weight is used in the definition of discordance, the point being that a sufficiently large difference in favour of b will "veto" a definite conclusion in favour of a, no matter how much more important the concordant criteria are.

The definitions of concordance and discordance do not yet define a full aggregation model, as the concordant and discordant evidence has yet to be aggregated. Two broad approaches can be identified, and may be termed "crisp" and "fuzzy" outranking relationships. In the "crisp" approach, a final conclusion is based on thresholds for the concordance and discordance measures respectively. Alternative a is asserted to outrank b if the concordance measure exceeds a minimum threshold level (for the evidence to be convincing), and if the discordance does not exceed the threshold (for the veto effect to be compelling). In the "fuzzy" approach, no definite conclusion is stated, but the strength of evidence is summarized by a value between 0 and 1 (a measure of credibility of the assertion), obtained by multiplying the concordance measure by a factor that tends to zero as discordance tends to its maximum value. Precise definitions of these thresholds, and of the fuzzy strength of evidence measures, will be presented for some specific outranking methods in Chapter 8. It should be noted here, however, that some outranking methods may use a number of different means of aggregating the outranking evidence in order to reach a recommendation for action or choice.

4.6. FUZZY AND ROUGH SETS

In this section we provide a brief introduction to the concepts of fuzzy sets and rough sets as used in preference modelling. We do not classify these concepts as separate schools of preference modelling, but rather as tools which may be applied in any of the models that have been discussed in this chapter. It is not our intention to provide a comprehensive review of fuzzy and rough set theories in this book, but it is useful to provide the reader with some appreciation of the basic concepts. Fuzzy set theory was introduced by Zadeh in 1965, and the first application to the multicriteria decision making context appears to be Bellman and Zadeh (1970), while a more recent exposition may be found in Lootsma (1997). A detailed review of the applications of fuzzy and rough set theory to decision making is given by Greco et al. (1999).

Traditional mathematics is based on "crisp" logic, in which any statement (for example, in our context, that an alternative a is preferred to b in terms of a specific criterion) is either true or false. This has expression in set theory, in which for any set A defined on a space of objects X, an object $x \in X$ either is or is not an element of A. For example, if we define P_a as the set of decision alternatives which are preferred to the alternative a, then for any other alternative b, either $b \in P_a$ or $b \notin P_a$.

A crisp set $A \subset X$, can be represented in terms of an *indicator function* $I_A(x)$, such that for any object $x \in X$, $I_A(x) = 1$ if $x \in A$, while $I_A(x) = 0$ otherwise, so that $I_A(x)$ indicates the truth of the statement $x \in A$. Such precise or crisp logic is sometimes found to be a limitation in modelling human perceptions and preferences, however, and for this reason Zadeh (1965) introduced the concept of a *fuzzy set*, by replacing the indicator function of a set by a *membership function* $\mu_A(x)$ such that $0 \leq \mu_A(x) \leq 1$. The value of $\mu_A(x)$ indicates the extent to which x "belongs" to the set A, or the "truth value" of the statement that $x \in A$. For example, if X represented all people alive today, and A were the set of tall people alive today, then A is not crisply defined: it is a fuzzy set. A particular person who is 2m tall presumably definitely belongs to A

112   MULTIPLE CRITERIA DECISION ANALYSIS

(membership in the set is 1), while someone who is 1.5m tall is definitely not in A (membership in the set is 0). However, someone who is 1.8m tall is not so clearly classifiable, and would have membership in A which is neither 0 nor 1, but rather some fractional value between 0 and 1.

The basic concept of fuzzy sets is very appealing in the context of MCDA, as it is undeniably true that the expression of human preferences is more often than not fuzzy in nature. It is not surprising, therefore, that concepts from fuzzy set theory find their way into many of the methods of MCDA. Ordinal preference statements, and the outranking concepts from Section 4.5, are frequently found to be expressed in fuzzy terms, associating a "truth value" with statements such as a is preferred to b, or a outranks b. The concepts of goals or aspiration levels discussed in Section 4.4 can also be expressed in a fuzzy manner, in the sense that a truth value can be associated with statements such as "this value of z_i satisfies our goal for criterion i". Once again, there will be values which definitely or do not satisfy goals or aspirations, but a range of intermediate values which are satisfactory to some extent (with membership in the set of satisfactory solutions being between 0 and 1). For example, in choosing a car we might specify a goal for fuel consumption of not more than 8 litres per 100km, but concede that a car having a consumption of 8.1 litres per 100km still has some non-zero level of membership in the set of cars satisfying the fuel consumption goal.

Even the superficially crisp and precise value measurement concepts can be made fuzzy, by expressing the assessed values (the $v_i(a)$) as "fuzzy numbers". In other words, when decision makers assert that the value of alternative a in terms of criterion i is at least 20 on a 0-100 scale, then this may not be interpreted crisply. Depending on the degree of precision associated with the assertion, the statement that $v_i(a) = 18$ may still be deemed to have some non-zero "truth value", while the truth value corresponding to $v_i(a) = 15$ may be essentially 0. Another way of expressing this is that 18 (but not as low as 15) may still have non-zero membership in the fuzzy set of outcomes with values expressed as "more-or-at least equal to 20".

While, as we have said, the fuzzy set concepts are appealing for representation of imprecision in MCDA, it must also be recognized that there are some practical problems in implementation which we should briefly mention.

Specification of the membership function: While the existence of a fuzzy membership function $\mu_A(x)$ appears to be quite plausible, it is difficult to envisage an operational procedure for eliciting the perceived "truth value" in anything more than an ordinal sense ("x has greater membership in A than y"). In the application of fuzzy set theory to MCDA, the shape of the membership function appears often to be chosen more-or-less arbitrarily for mathematical convenience (e.g. a linear function between two extreme values) rather than modelling decision maker preferences directly.

Definition of unions and intersections: In conventional (crisp) set theory:

- x belongs to the *union* of A and B (x belongs to A or B) if $I_A(x) = 1$ or $I_B(x) = 1$ (or both). In formal algebraic terms this can be expressed in a number of ways, such as $I_{A\cup B}(x) = \max\{I_A(x), I_B(x)\}$, or $I_{A\cup B}(x) = I_A(x) + I_B(x) - I_A(x)I_B(x)$.

- x belongs to the *intersection* of A and B (x belongs to A and B) if $I_A(x) = I_B(x) = 1$. Once again, in formal algebraic terms this can be expressed in a number of ways, such as $I_{A\cap B}(x) = \min\{I_A(x), I_B(x)\}$, or $I_{A\cap B}(x) = I_A(x)I_B(x)$.

For fuzzy sets, the concepts of union and intersection are less well-defined, and the alternative expressions above, when applied to membership functions, do not give the same answer. It is not so easy to confirm what people mean by the terms *and* and *or* in the fuzzy sense, and empirical work (Zimmermann and Zysno, 1979) has suggested that the true meaning may need to be modelled by much more complicated functions than those for crisp sets. In practical applications, however, users of fuzzy set concepts in MCDA seem to tend, without much discussion, to use one of the above simple algebraic expressions applying to crisp sets.

While fuzzy set theory attempts to model imprecision in defining sets (for example of tall people, or of alternatives which are satisfactory according to one specific criterion), the theory of rough sets deals more with imperfection in information regarding whether an object does or not belong to a particular set. In the absence of incomplete information, two objects (perhaps two decision alternatives) may be indistinguishable in the sense that no difference between them can be discerned at the current level of information or precision (reminiscent of the incomparability concept from outranking theory). They may nevertheless turn out to be distinctly different if analysis is taken to greater levels of detail. The basic idea in rough sets theory is to provide two approximations to any set, namely a *lower approximation* consisting of those elements definitely known to be in the set, and an *upper approximation* consisting of all elements which may belong to the set (i.e. not definitely excluded from the set).

114   *MULTIPLE CRITERIA DECISION ANALYSIS*

Rough set theory also makes use of the concept of a membership function to represent a degree of belief in the assertion that $x \in A$, but this is defined in a more precise manner, analogous to conditional probability concepts, and based on standard concepts of cardinality in (crisp) set theory. The theory is mathematically quite sophisticated and beyond the scope of the present text. The reader is referred Greco et al. (1999) for a more detailed review.

4.7. RELATIVE IMPORTANCE OF CRITERIA

Fundamental to any analysis of multiple criteria decision problems is the issue of the relative importance to be attached to each of the conflicting criteria, and thus the correct treatment of comparative importance is critical to the implementation of any of the models discussed in this chapter. Typically, some form of numerical weight parameter is used in most MCDA methods in order to model relative importance, and this we have recognized explicitly in the discussion of value measurement methods (Section 4.2) and of outranking (Section 4.5). Weight parameters are not always made explicit in goal programming models (Section 4.4), but are nevertheless implied in most applications.

The assessment and interpretation of importance weights has often been a matter of heated controversy, which regretfully often misses the point that the meaning of the numerical weight parameter will differ according to the particular preference model being used, and often also according to the range of alternatives under consideration. In discussing the various methodologies for MCDA in the next few chapter, we shall attempt to clarify the meanings of the weight parameters used in each. It is useful, however, include some general comments on the issue of weight parameters at this point.

Experience and other anecdotal evidence suggests that many people are quite happy to express opinions regarding relative importance in ratio terms, speaking easily of one criterion being "much more important than" (or even "three times as important as") another. The Analytic Hierarchy Process (AHP) approach (refer Section 5.7), for example, makes direct use of such intuitive statements, by allowing decision makers to give verbal descriptions of relative importance in terms such as "moderately", "strongly" or "absolutely" more important, which are converted into assumed ratios. It seems possible that one of the reasons for the popularity of AHP (and the associated Expert Choice software) is in fact the natural appeal of such semantic scales for purposes of expressing relative importance. It is by no means evident, however, that intuitive importance ratios expressed in this way correspond even approximately to the meaning of the weight parameter in a specific preference model.

Even if the intuitive weights are appropriate to one decision model, they cannot apply equally to *all* decision models, since (as we have noted) the meanings of weights differ between models. The readiness of people to express importance ratios intuitively and in the absence of context may thus well be a hindrance rather than a help to the implementation of MCDA.

Substantial evidence exists (for example, Mousseau, 1992, and von Nitzsch and Weber, 1993), to the effect that when people are asked to make direct assessments of relative importance, they are not only willing to do so, but also express the relative importance in a more-or-less context-free sense. In other words, people will express essentially the same ratio of importance for one criterion relative to another, irrespective of the context of the specific decision problem. For example, in evaluating new highway proposals, road safety might be stated to be three times as important as cost, irrespective of whether the options on the table differ by \$10m or \$100m in cost. We conjecture that such ratio judgements may be influenced by prior experience in related problems (for example, how often safety rather than cost seemed to be the critical issue). This might not be a problem if the current decision setting is similar in nature to that in which the prior experience was gained, but may create problems in eliciting weights if the current problem setting is substantially different to past experience (which is all too often the case in a rapidly changing world). In principle, one might be able to compensate for changes in context by careful construction of scenarios provided that the context of the prior experience were known explicitly, but this would seldom be the case.

When using the additive value function model described in Section 4.2, the algebraic meaning of the weight parameters is perhaps more clearly defined than for many of the other models. For any two criteria, say i and k, the ratio of weights (w_i/w_k) is simply the increment on the value scale for criterion k (i.e. the change in $v_k(a)$) which should just compensate for a unit loss on the value scale $v_i(a)$ for criterion i. Clearly, therefore, if the scaling (generally an arbitrary choice) used in the model for value assessments on criterion i is changed, then the ratio of w_i to w_k for all other criteria must change accordingly. As we shall see in Chapter 5, such scale changes may well occur in some value measurement applications when options are added or deleted to the set of alternatives under consideration. We shall return to consideration of this issue in Section 5.4.

In our discussion of goal programming concepts in Section 4.4, we have not at this stage made explicit mention of the issue of importance weights. In fact, some versions of goal programming ("preemptive goal

116 MULTIPLE CRITERIA DECISION ANALYSIS

programming") do not make use of weight parameters at all, representing relative importance purely in terms of a rank ordering of the criteria. Weights do, nevertheless, play an important role in other variations of goal programming, as part of the process of defining an overall measure of deviation from the specified goal or aspiration levels. In this case, the weights do again represent tradeoffs, in the sense of measuring the extent to which a closer approach to the goal on one criterion may compensate for greater deviations from goals on other criteria. We shall return to this point in Chapter 7.

The relative importance of criteria is modelled in two separate ways in outranking methods. The discordance principle allows a sufficiently poor performance on a particular criterion to "veto" acceptance of an alternative, irrespective of how well it performs on other criteria. Such a veto, by definition, does not allow any concept of trade-off between the criteria, and cannot be expressed as a weight. But the level of performance which would generate such a veto is, nevertheless, a measure of the importance of the criterion (in the sense that the more stringent the veto thresholds, the greater the impact of the criterion on the final results, and *vice versa*). In Section 4.5 we have explicitly expressed the concordance principle in terms of the "weight of criteria" favouring one alternative over another, usually defined operationally by a sum of weights for concordant criteria. As indicated there, and as stressed in the outranking literature, these weights do not represent trade-offs, but are rather a form of "voting power" associated with each criterion.

As part of the discussion of implementation of MCDA methods in Chapter 9, we shall review again the distinctions between the various methodologies as regards the meaning of the weight parameters, and comment on the implications for the practice of MCDA.

Apart from the model-dependent aspects of the meaning of weights, it is important also to record here, for completion, some of the behavioural issues which are known to affect the assessment of weights, irrespective of the underlying preference model being used. These have been summarized concisely by Weber and Borcherding (1993). Two issues in particular need to be stressed. The first relates to the assessment of weights in hierarchical value tree structures such as that illustrated in Figure 4.1, in which two key points emerge of importance to the applications of MCDA:

- *Effects of splitting attributes:* If a criterion is disaggregated into two or more sub-criteria, then the sum of the weights of the subcriteria should correspond to the effective weight of the original criterion (as will be discussed in some detail in Section 5.4.3 for the case of value function methods). A number of studies (for example, Weber et al.,

1988) have found, however, that the sum of sub-criterion weights obtained by disaggregated assessment is consistently larger (relative to weights on other criteria) than the weight obtained by direct evaluation of the parent criterion without disaggregation.

- *Effect of hierarchical level:* Different methods of problem structuring may result in different value tree structures. In particular, the hierarchical level at which a criterion is represented may be influenced by the procedures used, appearing, for example, either directly as a child of the overall objective, or grouped together with other criteria as components of a higher level composite criterion. For the same set of lowest level criteria (i.e. those not decomposed into further sub-criteria), the weight associated with a particular criterion tends to be higher when represented at a higher level in the hierarchical tree.

The second important issue discussed by Weber and Borcherding is that of the effect of reference points. As is known from the work of Kahneman and Tversky (1979), people react differently to stimuli perceived as losses to those perceived as gains, but the perception of what constitutes "gain" or "loss" is influenced by the *framing* of the problem through shifts of reference points. This framing problem carries over to the assessment of weights. If a particular criterion is framed in terms of losses relative to some reference point it is likely to be allocated a higher weight than if the same criterion (with the same set of outcomes) is framed in terms of gains relative to another reference point. Care thus needs to be taken in ensuring that decision makers understand and are satisfied with the implied reference points used in the model.

For these behavioural biases there is no one "correct" structuring and framing. The analyst must, however, remain consciously aware of the existence of these biases during both the problem structuring and weight assessment phases of the analysis, trying to convey to the client the effects that these may have, possibly making a deliberate attempt to get the client to look at other structures and framings. At very least, the potential effect of these biases needs to be evaluated at the sensitivity analysis stage.

4.8. FINAL COMMENTS

In this chapter we have attempted to give a brief overview of the paradigms and philosophical principles which underly different schools of thought in the modelling of decision maker preferences. As will be clear from the references given, we have only scratched the surface, as the concepts behind each school, and especially those of value theory and of outranking, have been developed in considerable detail (to which

118 MULTIPLE CRITERIA DECISION ANALYSIS

entire books have been devoted). It is true that we have perhaps devoted more space in this chapter to the value and utility schools of thought. This should not, however, be taken to imply that the value and utility approach has any special claim to being superior to the other approaches. The point rather is that the value / utility school is based on a somewhat more clearly defined set of axioms regarding rational preferences, which thus requires the more extensive explanation and discussion.

The basic modelling principles do not immediately define operational procedures for providing useful decision aid or support, and each school has spawned a rich variety of methodologies. The problem, often, is simply to know which method to use in a particular context. Over the next four chapters, we shall explore the methodologies emerging from each school in greater detail.

Chapter 5

VALUE FUNCTION METHODS: PRACTICAL BASICS

5.1. INTRODUCTION

In this chapter we focus on the use of value function methods, or *multiattribute value theory (MAVT)* for multi-criteria decision support, illustrating how the concepts of value measurement theory, as outlined in Section 4.2, are used in practice.

The problem structuring and model building activities described in detail in Chapter 3 lead to a description of the decision makers' objectives, often structured as a hierarchical value tree. If the value tree is to be used as the basis for assessment of a value function describing the decision makers' preferences then the implied preference structure should conform to the conditions of preferential independence outlined in Section 4.2.

Value function methods synthesise assessments of the performance of alternatives against individual criteria, together with inter-criteria information reflecting the relative importance of the different criteria, to give an overall evaluation of each alternative indicative of the decision makers' preferences. However, it should be stressed that the learning and understanding which results from engaging in the whole process of analysis is far more important than numerical results. The results should serve as a sounding board against which to test one's intuition. To further this learning, the evaluation should incorporate extensive sensitivity analysis and robustness analysis. Although we shall use the term "elicitation" with respect to decision makers' values and preferences, we are of the view that more often than not the process is a constructive one, helping decision makers to build a model of their values which will

120   *MULTIPLE CRITERIA DECISION ANALYSIS*

be useful in informing action, rather than seeking to make explicit a pre-existing preference function.

As discussed in Chapter 3, it is likely that in practice the process of evaluation will lead to some restructuring of the model. This may happen when it becomes apparent that it is not as easy as originally thought to assess performance with respect to a particular criterion, or when the analysis yields surprising results which prompt the decision makers to think of other important criteria.

The process of evaluation using a multiattribute value function (MAVF) is described in detail here, illustrated by reference to the Business Location case study introduced in Section 2.2.1. The value tree which resulted from the problem structuring process described in Chapter 3 was shown there as Figure 3.6; this will form the basis of our analysis. The problem structuring process also identified a selection of possible locations, embracing the range of possibilities which were put forward for initial evaluation. These are: Paris; Brussels; Amsterdam; Berlin; Warsaw; Milan; and London.

We begin with a brief review of the nature of the underlying model, go on to discuss in detail the process of eliciting information and values and then the use of the model to inform decision making.

The model

As discussed in chapter 4 the simplest, and most widely used form of value function is the additive model:

$$V(a) = \sum_{i=1}^{m} w_i v_i(a) \qquad (5.1)$$

where:

$V(a)$ is the overall value of alternative a

$v_i(a)$ is the value score reflecting alternative a's performance on criterion i

w_i is the weight assigned to reflect the importance of criterion i

We shall describe the construction of the value function in this additive form. Essentially the same principles will apply to the alternative multiplicative form mentioned in Section 4.2, except that elicitation of values will be in terms of ratios rather than the differences used in the additive model. More complicated value functions (i.e. not reducible to the additive form by transformation) are rarely, if ever, used in practice.

Remember that for the additive (or multiplicative) model to be appropriate, the criteria should satisfy the condition of preferential inde-

Value Function Methods: Practical Basics 121

pendence defined in Section 4.2. In essence, this states that the tradeoffs a decision maker is willing to accept between any two criteria should not be dependent on any other criteria.

Once an initial model structure and a set of alternatives for evaluation have been identified, then the next step is to elicit the information required by the model. There are two types of information, sometimes referred to as intra-criterion information and inter-criterion information, or alternatively as scores and weights. This process of elicitation is described in detail in the following sections.

5.2. ELICITING SCORES (INTRA-CRITERION INFORMATION)

5.2.1. Overview of the Scoring Process

Scoring is the process of assessing the value derived by the decision maker from the performance of alternatives against the relevant criteria. That is, the assessment of the partial value functions, $v_i(a)$ in the above model. If criteria are structured as a value tree then the alternatives must be scored against each of the bottom-level criteria (leaves) of the tree. As discussed in Section 4.2, these values need to be assessed on an interval scale of measurement, i.e. a scale on which the *difference* between points is the important factor. A ratio of values will only have meaning if the zero point on the scale is absolutely and unambiguously defined, which is seldom the case in practice. Thus, to construct a scale it is necessary to define two reference points and to allocate numerical values to these points. These are often taken to be the bottom and top of the scale, to which are assigned values such as 0 and 100 (which we shall use for illustration in this discussion), but other reference points (and other values) can be used. The minimum and maximum points on the scale can be defined in a number of ways, but it is useful to distinguish between a *local scale* and a *global scale*, as described below.

A local scale is defined by the set of alternatives under consideration. The alternative which does best on a particular criterion is assigned a score of 100 and the one which does least well is assigned a score of 0. All other alternatives will receive intermediate scores which reflect their performance relative to these two end points. The use of local scales permits a relatively quick assessment of values and can be very useful for an initial "roughing out" of a problem, or if operating under tight time constraints.

A global scale is defined by reference to the wider set of possibilities. The end points may be defined by the ideal and the worst conceivable performance on the particular criterion, or by the best

122   *MULTIPLE CRITERIA DECISION ANALYSIS*

and worst performance which could realistically occur. The definition of a global scale requires more work than a local scale. However, it has the advantages that it is more general than a local scale and that it can be defined before consideration of specific alternatives. This latter consideration also means that it is possible to define criteria weights before consideration of alternatives, as will be discussed in Section 5.4.

Another approach to the definition of a global scale is described by Bana e Costa and Vansnick (1999) who specify reference points describing "neutral" and "good" performance levels. The use of central rather than extreme reference points may guard against inaccuracies arising because of possible non-linearity in values occuring at extreme points, a factor which is particularly important in the assessment of weights.

Valid partial value functions can be based on either local or global scales. The important point is that all subsequent analysis, including assessment of the weights w_i, must be consistent with the chosen scaling. We shall return to this point later. The distinction between local and global scales for one of the criteria relevant to the business location case study is illustrated in Example Panel 5.1.

Suppose that in the office location problem, we use the number of direct flights per week to Washington DC on our preferred airlines as an appropriate attribute to represent level of achievement for the criterion *accessibility from US* (see Figure 3.6). It has been established, say, that these numbers of flights vary from 2 (for Warsaw and Milan) to 15 (for Amsterdam and London). A local scale could then be constructed, with $v_i(2) = 0$, and $v_i(15) = 100$ (where, of course, the argument now indicates values of the attribute flights per week).

The local scale might be quite satisfactory if no other cities are ever considered. But what if at a later stage we wish to consider other cities for which the numbers of flights fall outside of this range? In order to set up a global scale, we note firstly that the minimum possible number of flights is 0. At the other extreme, after some consideration we may reach the conclusion that there is unlikely to be any city having more than 4 flights per day on our preferred airlines, so that the effective maximum per week is 28. On this basis, we might construct a partial value function in such a way that $v_i(0) = 0$ and $v_i(28) = 100$.

Example Panel 5.1: Illustration of local and global scales

Once the reference points of the scale have been determined consideration must be given to how other scores are to be assessed. This can be done in the following three ways:

Definition of a partial value function. This relates value to performance in terms of a measurable attribute reflecting the criterion of interest (such as the flights per week in the above example).

Construction of a qualitative value scale. In this case, the performance of alternatives can be assessed by reference to descriptive pointers, or word models (to which appropriate values are assigned)

Direct rating of the alternatives. In this case, no attempt is made to define a scale which characterises performance independently of the alternatives being evaluated. The decision maker simply specifies a number, or identifies the position on a visual analogue scale, which reflects the value of an alternative in relation to the specified reference points.

5.2.2. Definition of a partial value function

The first step in defining a value function is to identify a measurable attribute scale which is closely related to the decision makers' values. For example, in assessing the accessibility of different locations from Washington, we had suggested that the number of direct flights per week by favoured airlines may serve as an appropriate indicator. If it is not possible to identify an appropriate quantitative scale, or if such scales as are available are only remotely related to the decision makers' values then it will be necessary to construct a value scale as described in the Section 5.2.3. The value function reflects the decision makers' preferences for different levels of achievement on the measurable scale. Such a function can be assessed directly or by using indirect questioning. Direct assessment will often utilise a visual representation.

Direct assessment of a value function

The decision maker should begin by determining whether:

- The value function is monotonically increasing against the natural scale – i.e. the highest value of the attribute is most preferred, the lowest least preferred, as is the case with the *accessibility to US* criterion represented by flights per week. Some possible functions are shown later in Figures 5.1 and 5.2.

- The value function is monotonically decreasing against the natural scale – i.e. the lowest value of the attribute is most preferred, the highest least preferred, as occurs with cost criteria.

- The value function is non-monotonic – i.e. an intermediate point on the scale defines the most preferred or least preferred point. For

124   *MULTIPLE CRITERIA DECISION ANALYSIS*

example, for a criterion such as distance from the railway station in choosing a house, a buyer might wish to be neither too close to the station (because of noise or other disruption), nor too far (to allow for convenient access).

Von Winterfeldt and Edwards suggest that if the problem has been well structured then the value functions should be regular in form – i.e. no discontinuities - as are those suggested in Figures 5.1 and 5.2. They go further to argue that all value functions should be linear or close to linear and suggest that the analyst should consider restructuring a value tree to replace non-monotonic value functions by one or more monotonic functions. Whilst we agree that an extremely non-linear value function, in particular a non-monotonic function, may indicate a need to revisit the definition of criteria, we caution against over-simplification of the problem by inappropriate use of linear value functions. Experimental simulations by Stewart (1993, 1996b) suggest that the results of analyses can be sensitive to such assumptions; thereby, the default assumption of linearity, which is often made, may generate misleading answers.

A non-monotonic value function is often an indication that the proposed measure actually reflects two conflicting values. For example, in the case of distance from the local railway station mentioned above, we could structure the house-buyer's concerns in terms of two separate criteria, namely attractiveness of the area and noise avoidance on the one hand, and reduction of travel time on the other. A stated ideal distance, such as 10 minutes walking time, is really an *a priori* assessment of what would represent a good compromise between these two conflicting criteria. But a more explicit examination of these two conflicting criteria might reveal different trade-offs and/or other alternatives. As discussed in Chapter 3, a value focused approach to model building should identify such underlying values, or objectives, more explicitly.

Indirect assessment methods

These methods assume that the value function is monotonically increasing or decreasing over the range of attribute measurement considered. As previously noted, the end points of the scale must be defined first. Thereafter, two methods of assessment are widely used, namely the *bisection* and the *difference* methods as described by von Winterfeldt and Edwards (1986) and Watson and Buede (1987).

Bisection Method. To illustrate this approach we make reference to the criterion, *Availability of Staff*, which we assess by information on the number of qualified applicants per post (obtained perhaps from recruitment agencies in each city). We begin by defining the end points

– the information that is available indicates that we should expect at least 4 qualified applicants for an advertised post and that it could be as high as 50. Value increases, that is, *availability of staff* is adjudged to be better satisfied, as the number of applicants per post increases.

Using the bisection approach the decision maker is asked to identify the point on the attribute scale which is halfway, in value terms, between the two end points. To help the decision maker identify the midpoint value it may be helpful to begin by considering the midpoint on the objective scale and posing the following question: *is the increase in number of applicants from 4 to 27 a greater or lesser increase in value than an increase from 27 to 50?* Suppose the decision maker responds that the increase from 4 to 27 represents a greater increase in value, the analyst might then ask how an increase from 4 to 20 compares with an increase from 20 to 50, continuing until the midpoint is identified. Suppose this is found to be 10 applicants. The next step would be to find the midpoints between 4 and 10 applicants and between 10 and 50 applicants - suppose these are found to be 8 and 20 respectively. It is generally accepted that 5 points (the 2 endpoints and 3 "midpoints") give sufficient information to enable the analyst to sketch in the value function, as illustrated in Figure 5.1. The aforementioned simulations (Stewart, 1993, 1996b) confirm the robustness of analyses to the use of 5 point estimates for value functions.

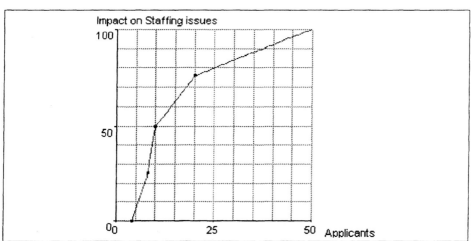

Figure 5.1: Value Function for Availability of Staff

Difference methods. These could be viewed as a collection of methods rather than a single one, but all require the decision maker to consider increments on the objectively measured scale and to relate these

126 *MULTIPLE CRITERIA DECISION ANALYSIS*

to differences in value. In the first approach (described by Watson and Buede, 1987), the attribute scale is divided into, say, 4 equal intervals, as shown in Table 5.1 for the criterion *Accessibility to US*, measured by the number of direct flights to Washington each week. The minimum number of direct flights is zero, the maximum 28 per week. Since preference is for more flights, an increase in the number results in an increase in value. The decision maker is asked to rank order the specified differences according to increase in associated value. For example, is the increase in value which occurs in going from 0 to 7 flights equal to, greater than, or less than that achieved in going from 7 to 14? Suppose the information elicited from the decision maker is as given below. The ranking gives an idea of the shape of the value function. In this example, the increase in value is greatest for lower numbers of flights, suggesting a concave, increasing value function. The curve could be sketched directly on the basis of this information, as illustrated in Figure 5.2, or may be further refined by asking the decision maker to assess the relative magnitude of value increases.

Increase in Number of Flights		Increase in Value
From	to	
0	7	1 = Greatest increase in value
7	14	2
14	21	3
21	28	4

Table 5.1. Intervals on "Accessibility from US" measured by the number of flights per week

A second approach (described by Von Winterfeldt and Edwards, 1986) is to begin by defining a unit level on the attribute scale (they suggest between one tenth and one fifth of the difference between the minimum and maximum points). To illustrate this approach, consider again the criterion *Accessibility from US*, measured as above. The minimum and maximum points on this scale are 0 and 28 flights, thus let the specified unit be equal to 3 flights per week (close to one tenth of the range). To assess the value function using this method we would first ask : *what is the number of flights, F, such that an increase from 3 to F flights results in the same increase in value as an increase from 0 to 3 flights?* Suppose the decision maker suggests that F should be 7. We next pose the question: *what is the value of F such that an increase from 7 to F flights is equal in value to the increase from 3 to 7?* The decision maker

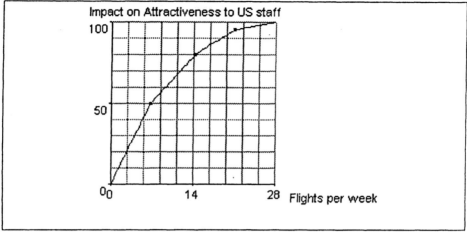

Figure 5.2: Value Function for Accessibility from US

responds that, having achieved the level of one flight per day (i.e. 7 per week), it would be necessary to double the frequency to 2 per day, 14 per week, to achieve the same increase in value. The additional value of an extra daily flight then diminishes further, the increase from 2 per day to 4 perhaps equating in value to the increase from 1 to 2, but beyond 4 flights per day extra flights do not add value. These responses give rise to a value function, specified in Table 5.2 for Accessibility which is very similar in shape to that defined using the previous method.

Number of flights	Value (per units defined above)	Value (0 to 100 scale)
0	0	0
3	1	25
7	2	50
14	3	75
28	4	100

Table 5.2. Value Function for Accessibility

Constructed measurement scales

The measurement scales used for the assessment of value functions for Availability of Staff and Accessibility arise naturally in the given context. Von Winterfeldt and Edwards (1986) comment that "A natural scale which is linear in value is obviously the most economical device for

128   *MULTIPLE CRITERIA DECISION ANALYSIS*

communicating value-relevant information". However, in some instances a simple natural scale may not exist and it becomes necessary to construct an appropriate measurement scale. For example, in evaluating *Business Potential* in the *public* and *private sectors* the decision makers wish to take into account volume (number and size of businesses) and accessibility. The principal concern is with business potential in the immediate, or near vicinity; most large cities in Europe are accessible to each other within a day by air so at that level the different locations are not distinguished. Thus it may be suggested that an index for *Private Business Potential* be constructed to take into account the number of large, medium and small company head offices (the most likely source of business) located within the city centre and within the city suburbs. For *Public Sector Business Potential*, a similar suggestion may be that the index take into account the number of government offices and hospitals within the same area. Hypothetical figures for purposes of the example are displayed in Table 5.4 (at the end of Section 5.2). Value functions can be assessed in relation to these proposed indices, precisely as for the more natural scales of numbers of flights and of job applicants above, provided that the constructed scale is operationally meaningful to all concerned.

5.2.3.   Constructing a Qualitative Value Scale

Often it is not possible to find a measurable attribute which captures a criterion. In such circumstances it is necessary to construct an appropriate qualitative scale. As discussed above, it is necessary to define at least two points on the scale (often taken as the end points). Intermediate points may also be defined. An example of such a scale in regular use is the well known Beaufort scale for measuring the strength of wind, shown in Table 5.3. Points on the scale are defined descriptively and draw on multiple concepts in the definition (Note: the points on the scale are also defined in terms of actual wind speed). An alternative approach to defining a scale could be to associate specific alternatives, with which the decision makers are familiar, with points on the scale.

Qualitative scales should have the following characteristics:

- Operational: allow the decision makers to rate alternatives not used in the definition of the scale.

- Reliable: two independent ratings of an alternative should lead to the same score

- Value relevant: relates to the decision makers' objective

Force	Conditions
0	Calm, sea like a mirror.
1	Light air, ripples only.
2	Light breeze, small wavelets (0.2m). Crests have a glassy appearance.
3	Gentle breeze, large wavelets (0.6m), crests begin to break.
4	Moderate breeze, small waves (1m), some white horses.
5	Fresh breeze, moderate waves (1.8m), many white horses
6	Strong breeze, large waves (3m), probably some spray
7	Near gale, mounting sea (4m) with foam blown in streaks downwind.
8	Gale, moderately high waves (5.5m), crests break into spindrift
9	Strong gale, high waves (7m), dense foam, visibility affected
10	Storm, very high waves (9m), heavy sea roll, visibility impaired. Surface generally white
11	Violent storm, exceptionally high waves (11m), visibility poor
12	Hurricane, 14m waves, air filled with foam and spray, visibility bad

Table 5.3. Beaufort scale

- Justifiable: an independent observer could be convinced that the scale is reasonable.

The process of constructing a qualitative value scale has many parallels with what Roy (1996) describes as building a criterion. An illustration of the process is presented in Example Panel 5.2

The approach described above and in Example Panel 5.2 directly assigns values to the qualitative statements; MACBETH, which is described further in Section 6.2, can be used to build a value scale from a category scale by a process of pairwise comparisons requesting ordinal judgements about preference differences. The output of MACBETH is a range of values associated with each category, consistent with the judgements input to the analysis. The decision maker may choose to work with the midpoints of these intervals as the corresponding value scale, or may wish to further refine the input judgements to arrive at a tighter definition of values. It is possible that the initial judgements are ordinally inconsistent, in which case the method highlights inconsistencies and suggests revisions which would move towards consistency.

5.2.4. Direct Rating of Alternatives

Direct rating can be viewed as the construction of a value scale, but defining only the end points of the scale. A local or a global scale may be used, the former creating minimal work for the decision makers. If using a local scale, the alternative which performs best of those under consid-

130 *MULTIPLE CRITERIA DECISION ANALYSIS*

In the value tree drawn up for the Business Location example the criteria which may be most appropriately measured by qualitative scales are "quality of life" and "ease of set up and operations". Quality of life is seen as a complex mix of issues encompassing culture, climate, language, standard of living, etc. Defining a scale which is accepted by all actors in the decision process could be quite difficult, as they will have very different perceptions of the concept. One way of handling this may be to further decompose the criterion. However, we have decided to work with this relatively high-level concept and begin by suggesting the 10 point qualitative value scale described below. This quality of life scale is determined by the following factors, each of which is rated as: unfavourable, acceptable or favourable:

Climate	A favourable climate is one which is generally warm with no extremes of temperature or rainfall. An unfavourable climate is one which suffers extremes of temperature (too hot or too cold) for a significant part of the year or one which is persistently cool or wet.
Standard of living	A favourable standard of living means affordable, good quality housing and schooling. Good facilities for shopping and leisure activities.
Ease of adapting to the culture	In a favourable situation English would be widely spoken, it would not be difficult to adapt to day to day living.
Quality of social / cultural life	In a favourable situation music, theatre, art, would be widely accessible.
Quality of the environment (pollution, noise, etc.)	A favourable environment would have a low pollution count, good public transport facilities
Safety considerations (crime level, etc.)	In a favourable environment crime rate is low, anxiety about personal safety is low, etc.

These considerations are taken together in building the value scale as follows:

Value	Description
10	All factors are favourable
	Balance of factors is better than all acceptable
5	All factors are acceptable or at most one unfavourable factor may be balanced by a favourable factor.
	Balance of factors is worse than all acceptable
0	No factors are favourable and three or more factors are unfavourable

---

Example Panel 5.2: Development of a qualitative scale

Value Function Methods: Practical Basics **131**

eration is given the highest score, usually 100 or 10, and the alternative which performs least well (not necessarily badly in any absolute sense) is given a score of 0. All other alternatives are positioned directly on the scale to reflect their performance relative to the two reference points. Although no attempt is made to relate performance to a measurable scale, the positioning of alternatives can generate extensive discussion, yielding rich information on the decision makers' values. Ideally this information should be recorded for future reference. A disadvantage of using a local scale is that if new alternatives are introduced into the evaluation this may necessitate the redefinition of scales, something which has consequences for the weighting of criteria, as discussed in Section 5.4.

A good example may be to use direct rating for assessing the criterion *ease of set up and operations* in the office location problem. There are a number of factors which impact on this, as discussed in Sections 2.2 and 3.5, for example, whether or not there are partnership arrangements which would facilitate setting up, prior experience in the systems, the nature of the tax and legal systems, availability of office space, etc. It might be decided to use a global scale to allow for the easy incorporation of other locations at a later stage. The reference points may then be taken as opening up a new office on the East coast of the US, a procedure which is already familiar to the company managers and would involve minimal hassle, and setting up an office in Japan or China, countries whose languages and customs are relatively unknown. These two points may be assigned values of 10 and 0 respectively, if the decision maker is satisfied that this allows for sufficient discrimination. All seven locations being evaluated are considered to lie between these two extremes. A conversation between the facilitator (F) and decision maker (DM) may go as illustrated in Example Panel 5.3, continuing further until both are satisfied with the outcome. Often the decision maker is very willing to give scores and an important role of the facilitator is to reflect these back, seek justification and check consistency across judgements. The process is aided by a visual representation of the scale as illustrated in Figure 5.3; this may be simply a sketch on a flipchart or it may be an interactive computer implementation.

Sometimes the decision maker may have difficulty in allocating scores. This may be because the criterion is not well defined, or because it is too high level a concept and the decision maker is having difficulty taking account of all it comprises. In this latter situation it may be appropriate to extend the value tree to incorporate further detail: such iterations between problem structuring and later stages of modelling should be expected.

132 MULTIPLE CRITERIA DECISION ANALYSIS

> *F:* As a first step, can you rank the options against this criterion?.
>
> *DM:* Well, the best would be London - there would be no language difficulties and having worked there for a time I have a sense of how things operate. The worst of the 7 would be Warsaw, we have no experience of working in Eastern European countries, no one knows the language and I just have a sense that things are still changing rapidly there - whilst that might be an advantage in some respects, I think it could cause problems with operational aspects. Taking everything into account, I would rank the others in the order Berlin, Milan, Amsterdam, Paris, and Brussels. The local region has partnership agreements with the regions in which Berlin and Milan are located - I know from the experiences of other companies that this can be a great help, both practically and financially.
>
> *F:* OK. In relation to your reference points are these 7 locations spread across the scale, or are they clustered to the top or bottom?
>
> *DM:* I think they are reasonably spread - perhaps there's more of a tendency towards the top.
>
> *F:* Let's try to position the bottom one first. Think about the increase in benefit - that is, in ease of set up and operations - that would come from locating an office in Warsaw rather than the Far East. Is this more or less than half the difference between the Far East and the US?
>
> *DM:* Definitely less than half - I'd say it was about 3.
>
> *F:* What about London, which you said would be the best of the 7; where would you position that between the Far East and US?
>
> *DM:* Pretty close to the top - about 9. And then I'd put Berlin and Milan together on 8.
>
> *F:* So the difference between Milan and London is about the same as the difference between London and the US?
>
> *DM:* .... Let me think ... yes, I guess so. Even though there is the partnership support in Milan the language issue makes quite a difference and the system is less familiar. Yes, I'm comfortable with those values ..

Example Panel 5.3: Illustration of a direct rating process

5.3. DIRECT RATING OF ALTERNATIVES BY PAIRWISE COMPARISONS

The use of pairwise comparisons is implicit in all scoring procedures as scores are assessed relative to reference points rather than in an absolute sense. Furthermore, in order to check consistency of judgements a facilitator may incorporate questioning procedures which make explicit pairwise comparisons between alternatives, as illustrated above. However, even if explicit, such comparisons tend to be ad-hoc and do not consider all possible comparisons. A systematic pairwise comparison

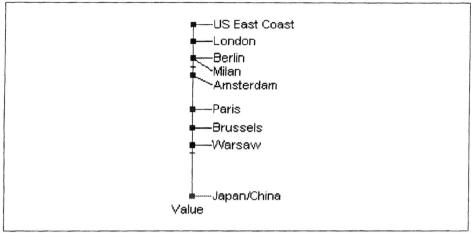

Figure 5.3: Visual representation of scale

approach is one of the cornerstones of the Analytic Hierarchy Process (AHP) described in more detail in Section 5.7. The AHP employs a procedure for direct rating which requires the decision maker to consider all possible pairs of alternatives with respect to each criterion in turn, to determine which of the pair is preferred and to specify the strength of preference according to a semantic scale (or associated numeric 1-9 scale). However, the AHP treats responses as ratio judgements of preferences, which is not consistent with the value function approach. The underlying mathematics is easily modifiable to be consistent with difference measurement. The MACBETH approach mentioned earlier, which is founded on difference measurement and also based on pairwise comparisons, can be used to derive direct ratings.

One of the potential drawbacks of pairwise comparison methods is the large number of judgements required of the decision maker ($n(n-1)/2$ for each criterion, where n is the number of alternatives). Nevertheless, the approach is a powerful one which can be effectively utilised if decision makers find the direct rating procedure difficult. With some pairwise comparison approaches it is not necessary to compare all possible pairs and considerable work has been done to derive appropriate sampling procedures (e.g., Islei and Lockett, 1988; Millet and Harker, 1990; Salo and Hämäläinen, 1992).

5.3.1. Scores for the Location Decision Example

In illustrating the methodologies for assessing partial value functions, we have suggested means by which the seven potential locations might be evaluated in terms of each of the bottom level criteria in the value

134　*MULTIPLE CRITERIA DECISION ANALYSIS*

tree. For ease of reference in later use of the same example, we record here in Table 5.4 the resulting assessments for each criterion. For the first, second, fourth and fifth criteria, the values in the table refer to the relevant attributes (either natural or constructed). For use in the value function formulation, these attribute values need still to be converted into partial value scores, by making use of value functions such as those illustrated for the first two attributes in Figures 5.1 and 5.2. As indicated above, the scores shown for the third and sixth attributes are derived from a qualitative scale and direct rating of alternatives, respectively.

	Staffing Issues			Business Potential		Ease of set
	Avail. of staff	Access. from US	Qual. of life	Public sector	Private sector	up and operations
Paris	10	10	7	65	80	5
Brussels	6	5	5	85	60	4
Amsterdam	4	15	5	75	90	7
Berlin	15	3	3	50	70	8
Warsaw	20	2	2	45	55	3
Milan	18	2	8	55	70	8
London	12	15	6	50	65	9

Table 5.4.　Ratings of alternatives in the business location problem

5.4.　ELICITING WEIGHTS (INTER-CRITERION INFORMATION)

5.4.1.　What is meant by relative importance?

It is clear that in any evaluation not all criteria carry the same weight, thus it is desirable to incorporate an assessment of the relative importance of criteria. This aspect of analysis has been the focus of extensive debate. It is clear that decision makers are able and willing to respond to questions such as: "What is more important to you in choosing a car, safety or comfort?". Furthermore, they are able and willing to respond to questions asking them to rate the relative importance of safety and comfort against a numerical or verbal scale. The Analytic Hierarchy Process is founded on such questions. However, it has been argued by many that the responses to such questions are essentially meaningless. The questions are open to many different interpretations, people do not respond to them in any consistent manner and responses do not relate to the way in which weights are used in the synthesis of information. The weights which are used to reflect the relative importance of criteria in

Value Function Methods: Practical Basics 135

a multi-attribute value function are, however, well defined. The weight assigned to a criterion is essentially a scaling factor which relates scores on that criterion to scores on all other criteria. Thus if criterion A has a weight which is twice that of criterion B this should be interpreted that the decision maker values 10 value points on criterion A the same as 20 value points on criterion B and would be willing to trade one for the other. These weights are often referred to as *swing weights* to distinguish them from less well defined concept of *importance weights*. Thus the notion of swing weights captures both the psychological concept of "importance" and the extent to which the measurement scale adopted in practice discriminates between alternatives. One of the commonest errors in naive scoring models is to assume that weights are independent of the measurement scales used; it is clear from the algebraic structure of (5.1), however, that the effect of the weight parameter w_i is directly connected to the scaling used for $v_i(a)$, so that the two are intimately connected.

5.4.2. Assessing Weights

The swing weight method

The "swing" which is usually considered is that from the worst value to the best value on each criterion. If the value tree is small, then the decision maker may be asked to consider all bottom-level criteria simultaneously and to assess which swing gives the greatest increase in overall value; this criterion will have the highest weight. The process is repeated on the remaining set of criteria, and so on, until the order of benefit resulting from a swing from worst to best on each criterion has been determined, thereby defining a ranking of the criteria weights. To assign values to the weights the decision maker must assess the relative value of the swings. For example, if a swing from worst to best on the most highly weighted criterion is assigned a value of 100, what is the relative value of a swing from worst to best on the second ranked criterion? It is important to remember that these weights are dependent on the scales being used for scoring as well as the intrinsic importance of the criteria. This means that it is not possible to assign swing weights until the scales for each criterion have been defined. If an intrinsically important criterion does not differentiate much between the options - that is, if the minimum and maximum points on the value scale correspond to similar levels of performance - then that criterion may be ranked quite low.

Note that, although it has been customary to derive swing weights by reference to swings over the whole range of value measurement, it is

136   *MULTIPLE CRITERIA DECISION ANALYSIS*

perfectly valid to use any two reference points on the criteria scales. As mentioned earlier, Bana e Costa and Vansnick (1999) have used definitions of "neutral" and "good" as reference points. Decision makers may feel more at ease comparing swings standardised in this way rather than swings between extreme points, particularly if the degree of differentiation differs substantially across criteria (for example, on one criterion the "worst" and "best" reference points may both represent very acceptable performance, discriminating little between the alternatives under consideration, whereas on another "worst" may be truly awful, whereas "best" is extremely good).

There are many ways of eliciting the values of swing weights. Example Panel 5.4 contains a hypothetical conversation between facilitator (F) and decision maker (DM) which gives an idea of how questioning might proceed in the context of the office location problem.

F: The aim of this stage of the process is to assess the relative contribution of individual criteria to the overall evaluation of alternatives. It is essentially a process of determining how much 100 points on one criterion is worth in comparison with 100 points on another. To begin with, consider a hypothetical option which scores zero on all criteria - i.e. worst on all scales which we have considered. Now imagine that you are allowed to increase just one criterion to its maximum level on the relevant scale. Which would you choose?

DM: Well, business potential is very important, particularly in the public sector. The difference between 0 and 100 on that scale is quite substantial, so I think that would be the one I would choose.

F: OK. So business potential in the public sector is the most highly rated criterion. Now imagine that you are in a situation where the maximum possible score of 100 is achieved on this criterion, but all others remain at zero. You can now select a second criterion to be raised to the maximum level. What would it be this time?

DM: Business issues are still the most important - it would be business potential in the private sector.

F: I'm sure you know what I'm going to ask next ... if both of these criteria were raised to the maximum score, what would be the next most important swing?

DM: The next most important would be availability of staff, followed by accessibility then ease of set up and operations, and finally quality of life.

Example Panel 5.4: Conversation between facilitator (F) and decision maker (DM) while assessing swing weights

Having established a rank order for the criteria weights, the next step is to assign values to them. Once again, there are a number of ways of doing this. The decision maker could be asked directly to compare each of the criteria in turn with the most highly ranked criterion. For each criterion the decision maker is asked to assess the increase in *overall value* resulting from an increase from a score of 0 to a score of 100 on the selected criterion as a percentage of the increase in overall value resulting in an increase from a score of 0 to 100 on the most highly ranked criterion.

Decision makers are generally comfortable working with visual analogue and may be willing to assess the relative magnitude of the swing weights directly using this means, as illustrated in Figure 5.4. These provide a means for communicating a good sense of the magnitude of judgements whilst removing the need for numerical precision. However, it is important that this degree of imprecision is not forgotten when information is aggregated.

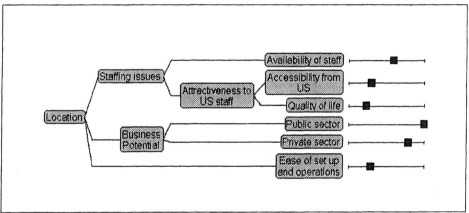

Figure 5.4: Swing weights - visual analogue scale

Normalisation

The weights implied by the visual representation in Figure 5.4 may be translated into numerical values, as shown in Table 5.5. The second column of the table lists the weights as they are displayed in Figure 5.4, i.e. standardized with the largest weight set to 1. It is usual, although not essential, to normalise weights to sum to 1 or 100, as shown in the third column of Table 5.5. Such normalization allows decision makers to interpret for example the weight of *availability of staff* in Table 5.5 as constituting 19% of the total importance weight. This seems often to be a useful interpretation. However, in specific cases decision makers may find it more intuitive to specify a reference criterion whose units are weighted at 1 and against which all other criteria are compared, as

138 MULTIPLE CRITERIA DECISION ANALYSIS

shown be the original weights with *public sector business potential* as the reference criterion.

Criterion	Original weights	Normalized weights
Availability of staff	0.6	0.19
Accessibility from US	0.3	0.09
Quality of life	0.225	0.07
Business potential – public sector	1	0.31
Business potential – private sector	0.8	0.25
Ease of set up & operations	0.3	0.09

Table 5.5. Swing weights - original and normalised values

An alternative explanation of swing weights

Consider a problem in which alternatives are assessed according to 5 criteria, C_1 to C_5, with the scores for any alternative being represented by the set of values $(c_1, c_2, c_3, c_4, c_5)$. The scores on each criterion range from 0 to 100. In assessing the rank order for criteria weights the decision makers are asked to imagine that the status quo is an alternative rated $(0,0,0,0,0)$ and they are being offered the opportunity to exchange this for one of the following 5 alternatives:

$$A = (100, 0, 0, 0, 0) \quad B = (0, 100, 0, 0, 0) \quad C = (0, 0, 100, 0, 0)$$
$$D = (0, 0, 0, 100, 0) \quad E = (0, 0, 0, 0, 100)$$

Which would they choose? Suppose alternative A is chosen; the implication then is that the increase from 0 to 100 on criterion C_1 is valued more highly than the increase on any other of the criteria and consequently the criterion C_1 will have the highest weight.

The decision maker is then offered the choice between:

$$B' = (100, 100, 0, 0, 0) \quad C' = (100, 0, 100, 0, 0) \quad D' = (100, 0, 0, 100, 0)$$
$$E' = (100, 0, 0, 0, 100)$$

This determines the criterion with the second highest weight, and so on. Once the rank order of criteria weights is determined each criterion can be compared with C_1. Essentially the decision makers are asked to answer questions such as: *How does the increase in overall value resulting in a change from the status quo to B compare with a change from the status quo to A?* If the response is that the first increase is worth 60% of the second, then the value of weight assigned to criterion 2 should be 0.6 of the weight assigned to criterion 1.

Value Function Methods: Practical Basics   139

An alternative approach to the elicitation of swing weights, again assuming that the first criterion is the most highly ranked, requires the decision makers to specify values for X such that an alternative defined by $(X,0,0,0,0)$ is valued equally to each of B, C and D respectively. Suppose $(50,0,0,0,0)$ is considered to be of equal value to alternative C, then: $W_1 \times 50 = W_3 \times 100$, that is, $W_3 = 0.5 \times W_1$.

5.4.3.   Weights within Value Trees

When the problem is structured as a multi-level value tree consideration has to be given to weights at different levels of the tree. It is useful to define *relative* weights and *cumulative* weights. Relative weights are assessed within families of criteria - i.e. criteria sharing the same parent - the weights within each family being normalised to sum to 1 (or 100). The cumulative weight of a criterion is the product of its relative weight in comparison with its siblings and the relative weights of its parent, parent's parent, and so on to the top of the tree.

By definition, the cumulative weights of all bottom-level criteria (leaves of the tree) sum to 1 (or 100) – thus the normalised weights shown in Table 5.5 are cumulative weights. The cumulative weight of a parent criterion is the total of the cumulative weights of its descendants.

As illustrated for the example problem, if the value tree does not have too many leaves, then weights can be assessed by directly comparing all bottom-level criteria to give the cumulative weights. Weights at higher levels of the tree are then be determined by adding the cumulative weights of all members of a family to give the cumulative weight of the parent. Relative weights are determined by normalising the cumulative weights of family members to sum to 1. Relative and cumulative weights for the example problem are illustrated in Figure 5.5.

For larger models it is easier to begin by assessing relative weights within families of criteria. Weights at higher levels of the value tree can be assessed top-down or bottom-up. The top-down approach would assess relative weights within families of criteria by working from the top of the tree downwards. However, the analyst must be aware of the difficulty of interpreting weights at higher levels of the value tree - the weight of a higher level criterion is the sum of the cumulative weights of all its subcriteria. Thus, in comparing two higher level criteria the decision maker should be thinking in terms of a swing from 0 to 100 on all subcriteria of the two higher level criteria. For example, if comparing *Availability of Staff* with *Attractiveness to US Staff*, the decision maker should compare a swing from worst to best on *Availability of Staff* (which has no subcriteria) with a swing from worst to best **simultaneously on both** *Accessibility from US* and *Quality of Life*. If the top-down approach is

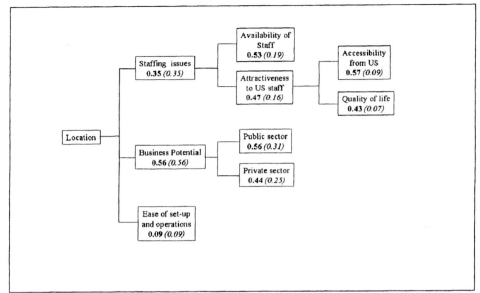

Figure 5.5: Relative Weights (in boldface) and Cumulative Weights (in italics)

used it is important to carry out cross family checks on the cumulative weights of bottom level criteria. For example, working with the example problem, having assessed relative weights within each of the criteria families (*Staffing Issues, Attractiveness to US Staff* and *Business Potential*) we can calculate the implied cumulative weights for all bottom-level criteria and check that the trade-offs between, say, *Business potential - public sector* and *Accessibility from US*, which these imply are acceptable to the decision makers.

The bottom-up approach begins by assessing relative weights within families which contain only bottom-level criteria and then carrying out cross family comparisons using one criterion from each family (perhaps the most highly weighted criterion in each family) and comparisons with any unitary bottom-level criteria. This process would eventually give the cumulative weights of bottom-level criteria which can be aggregated to higher levels as described above. In the example problem first we would assess the relative weights within the family *Attractiveness to US staff*. Suppose this allocates values of 1.0 and 0.75 to *Accessibility from US* and *Quality of Life* respectively (seen in the top part of column A of Table 5.6). We might then consider *Accessibility from US* in comparison with *Availability of staff* - all the time thinking in terms of swing weights as described above. Suppose the response is that the weight of

Availability should be twice that of *Accessibility* (recorded in column B). Independently of these criteria we consider the family of criteria labelled *Business Potential* and determine that the weight of *Private Sector* is considered to be 80% of *Public Sector* (seen in the bottom part of column A). Taking the most highly weighted criterion from each family, the final comparison is between *Availability of staff*, *Business potential in the public sector* and *Ease of set-up and operations*: the decision maker responds that these should be in the ratio 6:10:3 (column C). We can combine all of these responses (column D) to give values for the cumulative weights shown in the right hand column of Table 5.6.

	A	B	C	D	Cumulative Weights
Staffing Issues					
Availability of staff		1	0.6	0.6	0.19
Attractiveness to US staff					
Accessibility from US	1	0.5		0.3	0.09
Quality of life	0.75			0.23	0.07
Business Potential					
Public sector	1		1	1	0.31
Private sector	0.8			0.8	0.25
Ease of set up and operations			0.3	0.3	0.09
SUM:				3.23	1

Table 5.6. Relative and cumulative weights for the example problem

5.4.4. Consistency Checks

It is good practice to carry out more than the minimum number of comparisons necessary to specify the set of criteria weights, thus building in a check on the consistency of the decision makers' judgements. As with scores, the assessment of weights is also implicitly a process of pairwise comparison. This may be formalised by specifying a reference criterion against which all others are compared (requiring the minimal number of comparisons), or each criterion may be compared with every other one giving full specification (requiring $m(m-1)/2$ comparisons) as in the AHP or MACBETH approaches. Alternatively, something between these two extremes may be sought by judicious choice of pairs of criteria to be compared.

5.4.5. Working with Weaker Information

The process of determining values for criteria weights calls for a lot of hard thinking on the part of the decision maker. Questions such as those

142 MULTIPLE CRITERIA DECISION ANALYSIS

posed above are difficult to answer. Depending on the circumstances of the decision, an alternative way of proceeding might be to use the rank order of to give crude initial estimates of criteria weights and then to use this as a starting point for extensive sensitivity analysis. This may show that the preferred alternative is insensitive to changes in weights which preserve the rank order - in which case it would not be necessary to specify more precise values. Or it may indicate that attention should be focused on the weight assigned to a specific criterion. We shall illustrate later in this chapter, in the context of the business location case study, how even such weak ordinal information may lead to the elimination of all but a few decision alternatives. In the next chapter, we shall discuss in greater detail how ordinal information can be used systematically to identify all "potentially optimal" alternatives, i.e. those which can achieve the largest value for $V(a)$ amongst all alternatives for some set of weights satisfying the stated rank orders.

Some writers (for example Edwards and Barron, 1994, discussing their "SMARTER" approach) suggest that, in the presence of only ordinal information, an initial analysis be carried out using weights which are in some sense most central in the region defined by $w_1 > w_2 > w_3 > \cdots > w_m > 0$. One possibility is to estimate weights by the centroid, i.e. the arithmetical average of the extreme points of the region. When normalizing the weights to sum to 1, it is easily confirmed that the m extreme points are: $(1, 0, \ldots, 0)$, $(1/2, 1/2, 0, \ldots, 0)$, $(1/3, 1/3, 1/3, 0, \ldots, 0)$, $\ldots$, $(1/m, 1/m, \ldots 1/m)$. In our view, the use of the centroid to generate a set of weights is not entirely satisfactory, as it leads to rather extreme values. For example, with $m = 3$ criteria, the centroid weights are 0.611, 0.278 and 0.111 respectively, so that the ratio of w_1 to w_3 is 5.5, so that the third criterion will only have a very marginal influence on outcomes, contrary to what appears to be meant by the inclusion of all three criteria in the analysis. The situation becomes more extreme as m increases; with $m = 5$, the centroid weights are 0.457, 0.257, 0.157, 0.090 and 0.040. The ratio of w_1 to w_5 is now over 11:1, so that the fifth criterion has an almost vanishingly small influence. Experience seems to suggest that a weight ratio of 5 or more indicates an almost absolute dominance of one criterion over another, so that larger ratios should not really be expected amongst criteria which have been retained in the value tree.

Two possibilities for ameliorating the extreme effects of using the centroid may be:

- Inclusion of a further constraint to the effect that no criterion i will have been included in the model if the ratio of w_1 (the largest weight) to w_i exceeds some factor R: For example with $R = 9$, the extreme points of the feasible weight region for $m = 3$ will be:

$(9/11, 1/11, 1/11)$, $(9/19, 9/19, 1/19)$ and $(1/3, 1/3, 1/3)$. The centroid weights based on these extremes are 0.541, 0.299 and 0.159.

- Assume a geometrically decreasing set of weights, with each w_i being a constant proportion r of the next most important weight w_{i-1}: The centroid weights do in fact decrease in an approximately geometric fashion, but with a high rate of decrease (with a value of r around 0.4–0.45 for $m = 3$, and around 0.5–0.55 for $m = 5$). Some anecdotal experience suggests a rather less dramatic rate of reduction in weights, certainly well above 0.5. For example, even with r increased up to 0.6, the estimated weights for $m = 3$ are much more moderate, namely: 0.510, 0.306 and 0.184.

5.5. SYNTHESISING INFORMATION

5.5.1. Overall Evaluations

The overall evaluation of an alternative is determined by first multiplying its value score on each bottom-level criterion by the cumulative weight of that criterion and then adding the resultant values. If the values relating to individual criteria have been assessed on a 0 to 100 scale and the weights are normalised to sum to 1 then the overall values will lie on a 0 to 100 scale. If the criteria are structured as a value tree then it is also informative to determine scores at intermediate levels of the tree. In the business location problem this allows the alternatives to be compared, for example, on *Attractiveness to US staff*, *Staffing Issues and Business Potential*. Overall and intermediate scores for hypothetical assessments in this example problem are presented in Table 5.7.

The overall values can be displayed visually using many simple graphical techniques, such as a linear thermometer scale, or a set of bar graphs, both of which are illustrated in Figure 5.6.

However, the determination of an overall value should by no means be viewed as the end of the analysis, but simply another step in furthering understanding and promoting discussion about the problem. Although the underlying model is simple and static that should not be a limitation in its use. It provides a powerful vehicle for reflecting back to decision makers the information they have provided, the judgements they have made, and an initial attempt at synthesising these. The extent to which the model will be a successful catalyst for discussion of the problem and for learning about ones own and other's values, depends on the effectiveness with which feedback can be provided. Simple, static visual displays are an effective means of reflecting back information provided and well designed visual interactive interfaces provide a powerful vehicle for exploring the implications of uncertainty about values.

144   *MULTIPLE CRITERIA DECISION ANALYSIS*

	Staffing Issues			Business Potential		Ease of set up & oper.	Aggregated Scores			
	Avail. of staff	Attractiveness to US staff		Public sector	Private sector		Attract. to US staff	Staffing issues	Business potential	Overall
		Access. from US	Quality of life							
Paris	50	64	70	78	91	50	67	58	84	72
Brussels	12	38	50	94	74	40	43	27	85	61
Amsterdam	0	82	50	87	96	70	68	32	91	68
Berlin	63	23	30	21	83	80	26	46	48	50
Warsaw	75	15	20	16	42	30	17	48	27	35
Milan	70	15	80	42	83	80	43	57	60	61
London	55	82	60	21	79	90	73	63	47	56
Cum. weights	0.19	0.09	0.07	0.31	0.25	0.09				
Rel. weights		0.57	0.43							
Rel. weights				0.56	0.44					
Rel. weights	0.53						0.47			
Rel. weights						0.09		0.35	0.56	

Table 5.7.   Synthesis of information for the business location case study

Value Function Methods: Practical Basics   145

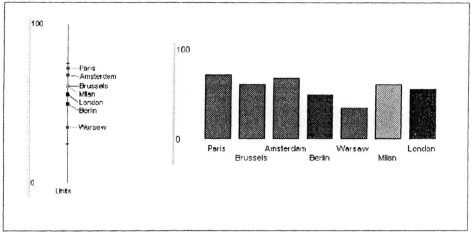

Figure 5.6: Overall evaluation of alternatives – visual displays

In exploring the model decision makers should test the overall evaluation and partial aggregations of information against their intuitive judgement. Are the results in keeping with intuition? If not, why not? Could some values have been wrongly assessed? Is there an aspect of performance which is not captured in the model? Is the additive model inappropriate? Or does the model cause the decision maker to revise their intuitive judgement? The aim of the analysis should be to arrive at a convergence between the results of the model and the decision makers' intuition.

Decision makers should look not only at the overall evaluation of alternatives, but at their profiles. How is an alternative's overall value made up? Is it a good "all rounder" or does it have certain strengths and weaknesses? Alternatives with similar overall scores can have very different profiles. Are there any dominating, or dominated alternatives? One option dominates another if it does at least as well on all criteria relevant to the decision. In simple terms, if there is an option which dominates all others it should be preferred, or if an option is dominated by another it should not be a candidate for choice. However, rather than acting as rigid guidelines, these concepts should be used as catalysts for further thought and learning about the problem situation.

The performance of the alternatives in the example on the three top-level criteria can be seen in the profile graph shown in Figure 5.7. This presentation highlights a number of points. Although Paris has the highest overall score it is not the highest rated alternative in any of three key areas. However, it performs well with respect to both *Staffing Issues* and *Business Issues*, whereas those locations which outperform Paris on

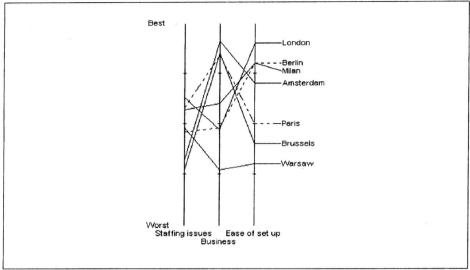

Figure 5.7: Profile graph - top-level criteria

one factor (London, Berlin, Amsterdam) perform significantly worse on the other factor. It is interesting to note that, at this level of aggregation, Amsterdam dominates Brussels, Milan dominates Berlin, while Paris and London dominate Warsaw. However, if we look at profiles of performance across the six bottom-level criteria, as shown in Figure 5.8, then it can be seen that no location is completely dominated by another. By scaling this graph to indicate the criteria weights we are able to appreciate better which are the stronger alternatives, as seen in Figure 5.9. This presentation emphasises the superiority of Brussels, Paris and Amsterdam on the highly weighted criteria.

The investigations described above should all be facilitated by the software tool used to support the analysis, which may be a standard spreadsheet or customised software. In practice, the nature of the analysis is dictated by the software tools available, particularly if working interactively with decision makers. Thus, it is important that the tools are flexible and easy to use as well as providing for appropriate display facilities and analyses, an issue which will be discussed in more detail in Chapter 9.

Value Function Methods: Practical Basics 147

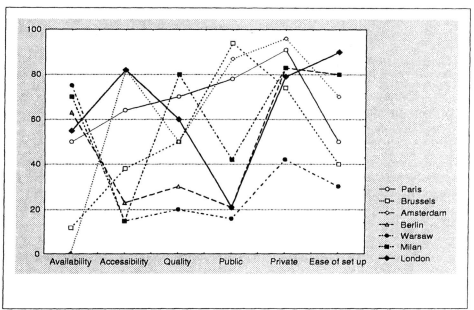

Figure 5.8: Profile graph – bottom-level criteria

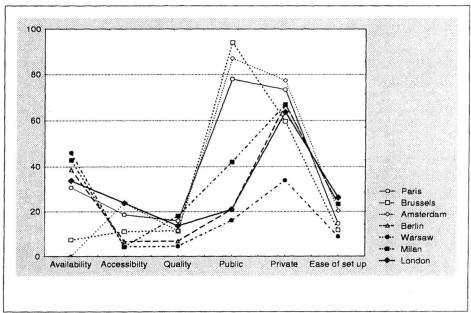

Figure 5.9: Weighted profile graph - bottom-level criteria

148   *MULTIPLE CRITERIA DECISION ANALYSIS*

5.6.   SENSITIVITY AND ROBUSTNESS ANALYSIS

5.6.1.   Perspectives on Sensitivity Analysis

Analysis should be carried out to investigate whether preliminary conclusions are robust or if they are sensitive to changes in aspects of the model. Changes may be made to investigate the significance of missing information, to explore the effect of a decision maker's uncertainty about their values and priorities or to offer a different perspective on the problem. On the other hand, there may be no practical or psychological motivation for changing values; the exploration may be driven simply by a wish to test the robustness of results. Sensitivity analysis can be viewed from the following three perspectives.

Technical perspective: From a technical perspective sensitivity analysis is the objective examination of the effect on the output of a model of changes in input parameters of the model. The input parameters are the value functions, scores and weights as determined by the decision makers. The output is any synthesis of this information - the overall evaluation of alternatives or the aggregation of values to any intermediate level of the value tree. A technical sensitivity analysis will determine which, if any, of the input parameters have a critical influence on the overall evaluation - that is, where a small change in a criterion weight or an alternative's score can affect the overall preference order.

Individual perspective: The function of sensitivity analysis from an individual's perspective is to provide the sounding board against which they can test their intuition and understanding of the problem. Do they feel comfortable with the results of the model? If not, why not? Have important criteria been overlooked in the analysis?

Group perspective: The function of sensitivity analysis within the group context is to allow the exploration of alternative perspectives on the problem, often captured by different sets of criteria weights. For example, if the problem were one of determining future energy policy one might look at the decision from the perspective of an economist, an environmentalist, different industry representatives.

Figures 5.8 and 5.9 highlight the fact that Paris performs relatively poorly with respect to availability of staff and suggests the importance of carrying out sensitivity analysis with respect to that parameter. Some of the softwares which are available facilitate interactive sensitivity analysis, allowing the user to investigate the effect of specified changes. A display such as that presented in Figure 5.10 is also widely used. This line graph shows how the alternatives' scores on Location (i.e. the overall scores) change as the relative weight allocated to *Availability of Staff* is varied. The current weight of 0.53 (relative to the parent criterion of *staffing issues*) is indicated by the dashed vertical line. We can see that, despite the weak performance of Paris on this criterion, it remains the preferred location overall when the allocated weight is increased. This may seem counter-intuitive initially, but careful consideration reveals that the other locations which have high overall scores perform even less well than Paris on this criterion. The graph shows that a reduction in weight to below about 0.33 would lead to Amsterdam being the preferred alternative; however, the decision makers are concerned about the current weight being too low rather than too high.

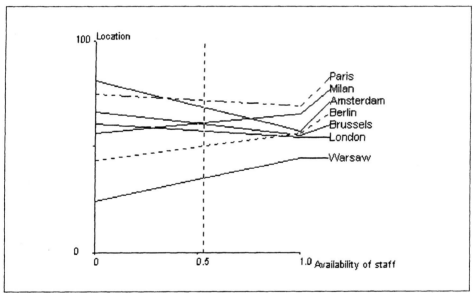

Figure 5.10: Sensitivity analysis: effect of varying the weight on Availability of Staff

The underlying linear model means that it is possible to carry out extensive sensitivity analyses using linear programming. These analyses are essentially an exploration of multi-dimensional weight space which can be divided into sub-regions, each corresponding to the set of weights for which a particular alternative is preferred. They are similar in form, and can be viewed as convergent with analyses which work with weaker data, described in detail in the next chapter. We conclude this section by describing briefly two examples of such analyses.

Preference Regions

When working with three criteria, information about potentially optimal alternatives can simply and clearly be displayed in a 2-dimensional projection of weight space. Each point in the triangular region illustrated in Figure 5.11 represents a set of weights defined by the perpendicular distances from each baseline. Thus each vertex (labelled in the figure by the three top level criteria in our example) represents a situation in which all the weight is allocated to one criterion. The triangle can then be divided into "preference regions", i.e. areas of weight space in which a particular alternative would be the preferred option. The same display can be used to investigate the robustness of preferred alternatives to changes in any three criteria weights. We use the display of Figure 5.11 to explore the weights allocated to the three top-level criteria in the example problem. The current weight allocation is indicated by the black square. As we can see, the weight space is divided more or less evenly between the three locations, Paris, Amsterdam and London, with a tiny space in the centre (not easily visible in the diagram) in which Milan is preferred. The current weight, as we already know, indicates Paris as the preferred alternative. This presentation illustrates that given a stronger emphasis on *Business Potential* and *Staffing Issues* rather than *Ease of set up and operations*, the preferred position is in the lower part of the triangle and Paris represents a robust compromise between these first two factors.

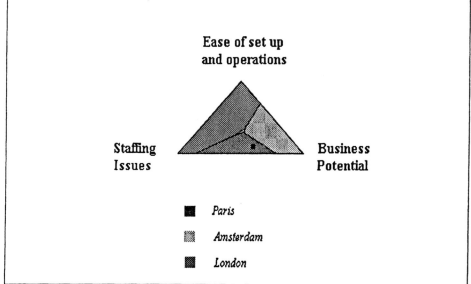

Figure 5.11: Working with ordinal information on criteria weights

As discussed earlier, the region of weight space consistent with $w_1 > w_2 > w_3 > \cdots > w_m$ is defined by the m points $(1, 0, \ldots, 0)$, $(1/2, 1/2, 0, \ldots, 0)$, $(1/3, 1/3, 1/3, 0, \ldots, 0)$, $\ldots$, $(1/m, 1/m, \ldots, 1/m)$. Thus, if we are confident that the weight to be assigned to *Business Potential* is greater than *Staffing Issues* which in turn is greater than *Ease of set up and operations*, we can focus our attention on that part of weight space indicated by the area marked by vertical lines in Figure 5.12, i.e. the triangle with the 3 extreme points as vertices. As illustrated in the figure, this restriction on the weights should focus attention on Paris and Amsterdam as preferred locations.

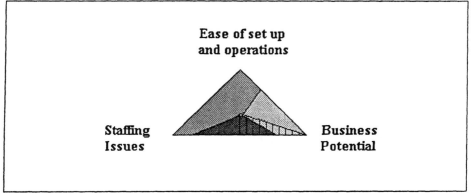

Figure 5.12: Implication of weight restrictions

Potentially Optimal Alternatives

For any selected alternative, it is possible to determine whether or not we can specify weight values which make it the preferred alternative. It may also be useful to know how far it would be necessary to move from the current set of weights. Analysis of the example problem reveals that all locations except Berlin are *potentially optimal*, i.e. best for some set of weights. Figure 5.13 shows the minimum change in weights which would be necessary to make Warsaw the preferred location; it is easily seen that a substantial change would be required. Refer to Rios Insua (1990) for a more detailed discussion of potential optimality.

5.7. THE ANALYTIC HIERARCHY PROCESS

The Analytic Hierarchy Process (AHP), a method for MCDA developed by Saaty (1980), has in its implementation many similarities with the multi-attribute value function (MAVF) approach. Both approaches are based on evaluating alternatives in terms of an additive preference function such as that given by (5.1). In this sense, AHP can be viewed as

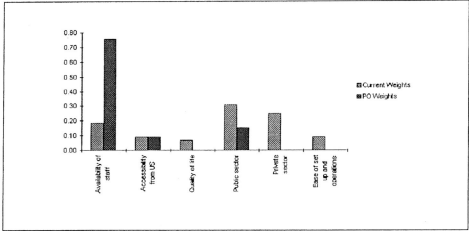

Figure 5.13: Comparison of current weights and potentially optimal weights for Warsaw

an alternative means of eliciting a value function, although it rests on different assumptions about value measurement as will be discussed below. However, it was developed independently of decision theory, and some AHP proponents insist that it is not a value function method. It is interesting to note, nevertheless, that the similarity of the AHP and MAVF approaches is evidenced by the convergence of supporting software; a number of available packages support both elicitation approaches.

As with the MAVF approach, the initial steps in using the AHP are to develop a hierarchy of criteria (value tree) and to identify alternatives. The major factors which differentiate the AHP from the MAVF approach from a practical viewpoint are the use of pairwise comparisons in comparing alternatives with respect to criteria (scoring) and criteria within families (weighting), and the use of ratio scales for all judgements. In the standard AHP procedure, alternatives are not differentiated from criteria, but are treated as the bottom level of the hierarchy (the ultimate means to the end) and all comparisons follow the same procedure. Rather than constructing a value function or an explicit qualitative scale against which the performance of alternatives is assessed, the decision maker is required to respond to a series of pairwise comparison questions which leads to an implied numerical evaluation of the alternatives according to each criterion.

Before we describe and illustrate this procedure further, however, we need to note that Saaty (1990b), apparently largely in response to the "rank reversal" problem to be discussed below, has suggested that in place of pairwise comparison of alternatives, a so-called "absolute mea-

Value Function Methods: Practical Basics   153

surement mode" be used. In this approach, a number of "absolute" levels of performance on each criterion are defined, and it is these levels rather than the alternatives which are compared pairwise, to generate numerical scores for each level of performance. Values for each alternative are derived from those of the absolute performance levels for each criterion to which it most closely corresponds. These performance levels form, in effect, a qualitative scale as discussed in Section 5.2.3, and the approach is thus very close in spirit to the use of partial value functions in MAVF. An advantage of the absolute measurement mode is that the resultant scaling of the scores for each criterion is independent of the alternatives, and may be relatively similar across criteria, which may in principle facilitate the interpretation of weights. However, reported applications of AHP seem seldom to refer to use of absolute measurement, and for this reason we restrict further attention to the standard form of AHP (i.e. direct pairwise comparison of alternatives).

Let us consider, then, the application of AHP to the office location problem discussed earlier. For example, in comparing alternatives with respect to *quality of life*, decision makers would be asked: "Thinking only about the *quality of life* of the locations, which of the two alternatives p or q do you prefer?" If no clear preference for one over the other is expressed, they move on to the next pair. If (say) p is selected as the more preferred of the two options from the point of view of quality of life, then the decision makers are asked to indicate the strength of their preference for p over q on the following scale (Saaty, 1980):

1	*Equally preferred*
3	*Weak preference*
5	*Strong preference*
7	*Demonstrated preference*
9	*Absolute preference*

Intermediate values may be used when decision makers hesitate between two of the descriptions. Decision makers may be shown just the verbal (or semantic) scale, or both the verbal and numeric scales.

Once all pairs of alternatives have been compared in this way, the numeric values corresponding to the judgements made are entered into a pairwise comparison matrix. All diagonal entries are by definition equal to 1. The method interprets the above numerical scale of strengths of preference in a ratio sense. Thus if alternative p is preferred to alternative q, with strength of preference given by $a_{pq} = S$ (where a_{pq} is the entry in the p-th row and q-th column of the comparison matrix), then the comparison of q with p is the reciprocal of that value, i.e. $a_{qp} = 1/S$. Table 5.8 illustrates a set of pairwise comparisons of the 7 alternatives

154   MULTIPLE CRITERIA DECISION ANALYSIS

	Paris	Brussels	Amsterdam	Berlin	Warsaw	Milan	London
Paris	1	3	4	5	7	1/3	3
Brussels	1/3	1	1	5	7	1/5	1/2
Amsterdam	1/4	1	1	3	5	1/5	1/2
Berlin	1/5	1/5	1/3	1	5	1/7	1/5
Warsaw	1/7	1/7	1/5	1/5	1	1/9	1/7
Milan	3	5	5	7	9	1	3
London	1/3	2	2	5	7	1/3	1

Table 5.8.   Comparison matrix for *quality of life*

with respect to *quality of life* (the values being based roughly on the scores given in Table 5.4).

To determine a set of relative priorities amongst n alternatives, only $n-1$ judgements are in principle needed. By asking for a complete set of pairwise comparisons ($n(n-1)/2$ in total), more information than necessary is provided, hopefully reducing the impact of a poor, or erroneous response, and introducing the possibility of assessing the consistency of judgements. The first step in synthesising these judgements is to reduce the pairwise comparison matrix to a comparison vector, i.e. a set of scores (or partial values) representing the relative performance of each alternative. The values in the pairwise comparison matrix are interpreted as ratios of these underlying scores. For example, the value of 3 in the first row and second column of Table 5.8 is interpreted as an assessment that the score for Paris on this criterion (say v_1) is approximately 3 times as large as the corresponding score for Brussels (say v_2). We note briefly that ratios of scores are only meaningful if there exists an unambiguous zero point against which all other outcomes are assessed. For example, a temperature of 40 degrees Celsius cannot meaningfully be interpreted as being "twice as hot as" 20 degrees Celsius, since the zero point (the freezing point of water under standard conditions) is essentially arbitrary. Someone brought up on the Fahrenheit scale would view these temperatures as 104 and 68 degrees respectively, giving a ratio of 1.53. (Physicists do work in terms of degrees Kelvin relative to an absolute zero, but this seldom enters into regular conversation.) We shall return to discussion of this point later.

Within this context, the aim in AHP is to find the set of values $v_1, \dots, v_n$, such that the matrix values a_{pq} are approximated as closely as possible by the corresponding ratios v_p/v_q. The standard AHP method of doing this is to extract the eigenvector corresponding to the maximum eigenvalue of the pairwise comparison matrix, i.e. the solution to the m simultaneous equations of the form $\sum_{q=1}^{n} a_{pq}v_q = \lambda v_p$ giving the largest value of λ. Readers unfamiliar with eigenvalue analysis of matrices need not concern themselves with details here. It suffices to know that the procedure is iterative, and not easily performed by hand, but that software for extracting the relevant values is easily available. For a fully consistent matrix, i.e. one in which there exist scores $v_1, \ldots, v_n$, such that $a_{pq} = v_p/v_q$ for all p and q (also implying that $a_{pr} = a_{pq}a_{qr}$ for all p,q and r), the maximum eigenvalue will equal the size of the matrix (n), and thus discrepancies between the maximum eigenvalue and n is a measure of inconsistency (see below). The elements of the vector of scores are normalised to sum to unity. Using the nomenclature of (5.1) this implies that for each criterion i, the scores $v_i(a)$ are standardized such that $\sum_a v_i(a) = 1$, so that $v_i(a)$ represents in some sense the proportion of total available value on criterion i (within the set of alternatives) which is contributed by alternative a. The algebraic implication is that the introduction of a new alternative, or the deletion of a current alternative, alters the scale of measurement of performance.

For the pairwise comparison matrix given by Table 5.8, the vector of relative preference scores with respect to *quality of life*, normalized to unit sum, turn out to be as follows:

Paris	0.23
Brussels	0.10
Amsterdam	0.08
Berlin	0.04
Warsaw	0.02
Milan	0.38
London	0.14

Consistency Index

The eigenvalue (say λ_{max}) in the above example turns out to 7.58, which is somewhat larger than the value of 7 applying to a fully consistent matrix. In order to provide a measure of the severity of this deviation, Saaty defined a measure of consistency, or consistency index (CI) by:

$$ \mathrm{CI} = \frac{\text{Principal eigenvalue - size of matrix}}{\text{size of matrix - 1}} = \frac{\lambda_{max} - n}{n - 1}. $$

The consistency index is compared to a value derived by generating random reciprocal matrices of the same size, to give a consistency ratio (CR) which is meant to have the same interpretation no matter what the size of the matrix. The comparative values from random matrices are as follows for $3 \leq n \leq 10$ (e.g. Saaty, 1996, p. 52):

Size of matrix	3	4	5	6	7	8	9	10
Comparative value:	0.52	0.89	1.11	1.25	1.35	1.40	1.45	1.49

In the above example, the consistency index is thus $(7.58-7)/6=0.097$. Since $n = 7$, the consistency ratio is $0.097/1.35=0.07$. A consistency ratio of 0.1 or less is generally stated to be acceptable, so that the assessments in Table 5.8 would be deemed to be acceptably consistent.

Aggregating preferences

A vector of relative preferences (analogous to the scores in the multi-attribute value function approach) is determined, as described above, by comparing the alternatives with respect to each of the criteria at the next level of the hierarchy (i.e. the lowest level criteria in the value tree as described in the previous sections). The next step is to compare all criteria which share the same parent using the same pairwise comparison procedure, deriving a vector indicating the relative contribution of the criteria to the parent (analogous to the weights in the MAVF approach: however, as discussed below, care should be exercised in interpreting these weights). The decision maker is asked questions similar to that above, but now comparing criteria, for example: "Thinking about *staffing issues*, which is more important, *availability of staff* or *accessibility from the US*?" If the response is that *availability of staff* is more important, then the decision maker is asked by how much, using the same 9-point verbal scale presented above. We shall return shortly to problems of interpreting precisely what this ratio of importance actually means.

The judgements are aggregated by working upwards from the bottom of the hierarchy, as with the multi-attribute value function. The strength of the approach once again will depend on the way it is used to facilitate understanding, learning and discussion, which depend on interaction with the decision makers and the effectiveness of information displays.

Sensitivity Analysis

It is impractical, in part because of the number of judgements involved, but also as a consequence of the use of the eigenvector method to derive preference vectors, to investigate easily the sensitivity of results to individual judgements (pairwise comparisons). Sensitivity analysis which takes as its starting point the vectors of preferences (analogous to the scores and weights in a multi-attribute value function) can be carried out as described for the MAVF approach.

5.7.1. Debate about the AHP

There has been extensive debate about the AHP, centring on a number of issues with both practical and theoretical significance. Some of these issues will be discussed in more general terms in Chapter 9. An early series of correspondence was published in *Omega* (Belton and Gear, 1982, 1985; Watson and Freeling, 1982, 1983; Saaty et al., 1983; Saaty and Vargas, 1984; and Vargas, 1985). This was followed more recently in *Management Science* (Dyer, 1990a, 1990b; Saaty, 1990a; and Harker and Vargas, 1990). It would take too long for our purposes here to cover all the issues raised in detail, but we briefly summarize the main points of concern and debate in the following.

The interpretation of criteria weights. It is clear from the algebraic structure of the additive model used in both MAVT and AHP that the weight parameters w_i define the desirable levels of tradeoffs between performances on the different criteria, when the measures of performance are given by the scores $v_i(a)$. In other words, if alternatives a and b differ only on criteria i and k, with $v_i(a) > v_i(b)$ but $v_k(b) > v_k(a)$, then the two alternatives will adjudged (by the model) to be equally preferred if and only if:

$$\frac{w_i}{w_k} = \frac{v_k(b) - v_k(a)}{v_i(a) - v_i(b)}.$$

In the case of multi-attribute value theory, this tradeoff implication is utilised directly by scaling all partial value functions to a fixed range (typically 0-100), so that the weight parameter for each criterion becomes the relative worth of the swing between the two reference points or levels of performance defining the 0 and 100 points on the scale function, as described in Section 5.4.

Because of the scaling of the partial scores to sum to 1, the implied meaning of weight in the standard AHP procedure is the relative worth, not of the swings, but of the "total" or "average" score on different criteria (Belton, 1986) – thus it can only be defined by all the alternatives under consideration and is, in our view, much more difficult to conceptualise. It is not at all evident that when decision makers express relative weight ratios, they have this interpretation in mind. The AHP can be adapted with relative ease, by changing the normalisation procedures adopted, to work with other interpretations of weight (Belton and Gear, 1982; or Schoner and Wedley, 1989), but this adaptation appears to be resisted by the "purists".

The assumption of a ratio scale of preference. The AHP assumes that all comparisons can be made on a ratio scale - this

158 MULTIPLE CRITERIA DECISION ANALYSIS

implies the existence of a natural zero, the natural reference point. This means that if comparing A and B we could state our preference for one over the other as a ratio - for example, that a is twice as good as b. In doing this we are implicitly considering the "distance" of a and b from the natural reference point - the zero - and saying that a lies twice as far away from it as does b. This makes good sense if we are dealing with something like distance, or area, which are natural ratio scales, but not if we are dealing with something like comfort, image, or quality of life, for which no clear reference level exists. Proponents of the MAVF approach, suggest that an interval scale should be used to measure preferences. This calls for the definition of two reference points, as discussed in Section 5.2, the difference between these points defining the unit of measure. In effect, the partial values are then a measure of distance from the lower reference point, acting as an explicit "worst case" (locally or globally). In principle, it is not strictly necessary that the worst case reference point be made explicit, as long as it remains stable during all pairwise assessments. Behavioural research such as that of Kahnemann and Tversky (1979) demonstrates quite clearly, however, that reference points are strongly influenced by the framing of problems, while the framing will almost inevitably change from one pairwise comparison to another, so that stable reference points cannot be expected to occur in general.

Numerical interpretation of the semantic scale. Some concern has been expressed about the appropriateness of the conversion from the semantic to the numeric scale used by Saaty as a measure of strength of preference, and various alternatives have been suggested (for example, Belton, 1986; or Lootsma, 1997, Chapter 5). The general view, supported by experimental work, seems to be that the extreme point of the scale defined semantically as "absolute preference" is more consistent with a numeric ratio of 1:3 or 1:5 than the 1:9 used in AHP.

The eigenvector method of estimation. A number of analysts have suggested that in place of the eigenvector approach for reconciling inconsistencies in pairwise comparisons, one could use logarithmic least squares. Formally, we could then select the $v_1, \ldots, v_n$ so as to minimize:

$$\sum_{p=1}^{n} \sum_{q<p} (\log a_{pq} - \log v_p + \log v_q)^2.$$

Value Function Methods: Practical Basics   159

There is a sound statistical basis for the use of least squares, derived from a minimisation of errors argument. From a practical perspective, the least squares approach is easier to work with as no iteration is required to derive the solution. Furthermore, the process is easier to explain to mathematically less sophisticated users, and does not require that all $n(n-1)/2$ pairwise comparisons be carried out. Saaty strongly defends the eigenvector approach as an ideal, on the basis that it preserves certain mathematical properties which he finds compelling, but nevertheless concedes that the least squares solution may still be used as approximation to the maximal eigenvector.

The first of the above issues has led to a debate about "Rank Reversal" extensively referred to in the *Omega* and *Management Science* papers and correspondence. Rank reversal refers to the fact that in certain situations, the introduction of a new alternative which does not change the range of outcomes on any criterion may lead to a change in the ranking of the other alternatives as determined by AHP. We feel that most of this debate is irrelevant to the issue raised originally when the observation was made by Belton and Gear (1982), namely the interpretation of the relative importance of criteria as operationalised in the AHP. At an algebraic level, rank reversal occurs because the new alternative changes the scaling of the scores differently for each criterion. In view of the interpretation of AHP weights in terms of the importance of total or average scores, the weights *should* change with the addition or deletion of alternatives, in a manner which would compensate for the changes in scaling. It seems, however, that most decision makers do not see any reason to change their weight assessments in the light of a new alternative which does not introduce new levels of performance. Much of the resulting debate has ignored this fundamental issue of the interpretation of relative importance, and has focused on justifying rank reversal and/or seeking *ad hoc* means of avoiding it. Whilst the impact of the ambiguity surrounding this concept may in practice be of little consequence, we feel uneasy using concepts which we are unable to define and would recommend decision makers to work with the well defined notion of swing weight as used in the MAVF approach.

The extent of practical implications of the second and third issues is open to debate. Experimental work has shown that the fourth is of little practical significance.

5.8.   CONCLUDING COMMENTS

In this chapter we have looked at the basics of building and using a value function. We have focused on the use of the simple additive model,

160 MULTIPLE CRITERIA DECISION ANALYSIS

introduced in Chapter 4, as this is the approach that seems to have been most widely adopted in practice (see, for example, Belton 1985, 1993; Butterworth, 1989; Buede and Choisser, 1992). We have also described the Analytic Hierarchy Process, a structurally similar additive model, although founded on different assumptions about preference measurement. The AHP has also seen widespread practical application (see Saaty's book, *Decision Making for Leaders*, 1982, for examples; and Zahedi, 1986, for a bibliography of applications). We feel that the popularity of these methods derives from their transparency, the intuitive appeal of their simplicity and their ease of use. These attributes are at the same time a reason for caution – are the assumptions underlying the use of this simple model valid? – have decision makers understood the questions posed, particularly in relation to the importance of criteria? The reader is referred back to Chapter 4 for a discussion of these issues and of the fundamentals of preference modelling and measurement on which the current chapter is premised.

Although the main focus of the discussion in this chapter is on the technical and practical aspects of deriving a value function and using it for decision support, we have emphasized throughout the importance of this as a learning process. Decision makers are helped to make sense of an issue, to better understand the organizational context, to explore their own values and priorities and to appreciate the perspectives of other parties to the decision. Once again we would like to stress that we see the model as a tool for learning, a sounding board against which decision makers can test their intuition, not as a means of providing an "answer" which is in some way "objective" or "right".

In Chapter 6 we shall go on to look in detail at other approaches that accept a value function as a reasonable representation of preferences. These can be classified as indirect and interactive methods. The former also seek to elicit a value function, but using less direct and less demanding questioning procedures than those described in the present chapter. In contrast, interactive methods do not seek to make the value function explicit, instead moving progressively towards a preferred solution, guided by a process of interactive questioning. The indirect and interactive methods make use of somewhat more complicated mathematical models than have been used up to this point, in order to extrapolate from the weaker preference information provided by the decision maker. For this reason, their description will need to include a corresponding greater level of mathematical detail, so that less technically oriented readers may wish to skim through Chapter 6 without attempting to master all the mathematical details.

Later in the book, in Chapter 9, we shall return to the practical issues which need to be considered when working with a group of decision makers using these, or any other, multicriteria approaches. In Chapters 10 and 11 we shall present the case for considering use in integration with other multicriteria approaches and other tools of OR/MS, as well as the need to appreciate the relevance and importance of a broad range of work on the psychology and sociology of decision making in organizations.

Chapter 6

VALUE FUNCTION METHODS: INDIRECT AND INTERACTIVE

6.1. INTRODUCTION

In the previous chapter, we discussed different methods for assessing value functions expressed in the additive form:

$$V(a) = \sum_{i=1}^{m} w_i v_i(a) \qquad (6.1)$$

on the assumption that the relevant preferential independence axioms hold. For the purposes of this chapter, it will be convenient to reformulate the above into the form:

$$V(a) = \sum_{i=1}^{m} u_i(a) \qquad (6.2)$$

where $u_i(a) = w_i v_i(a)$. In other words, the partial value functions are now scaled in proportion to their importance weight.

The above formulation is expressed in what we have previously termed the extensive form (Section 4.1) of the value function, i.e. represented in terms of the lowest level criteria of the value tree. For ease of presentation we will adopt this form to describe the methods introduced in this chapter. As we saw in Chapter 4, a value function can, without loss of generality, be expressed in extensive form even if the problem has been structured as a value tree with more than one hierarchical level. It should be clear to the reader that the ideas described in this chapter can in principle be extended to multi-level value trees, although practical implications may mitigate against doing so in some cases.

The methods discussed in Chapter 5 require decision makers to assess relative strengths of preference between different outcomes either by

164   *MULTIPLE CRITERIA DECISION ANALYSIS*

direct quantification or by means of semantic scales interpreted quantitatively. While such assessments in principle completely identify the value function, they can be very demanding on those having to provide the assessments. Although we would always recommend extensive sensitivity analysis to alleviate the difficulties of assessment, there may arise situations in which decision makers may be unwilling or unable to provide the necessary inputs. This may, for example, be true at earlier stages of analysis in which the aim is primarily to establish a shortlist of alternatives for more detailed analysis (before there is time for comprehensive value assessments in decision workshops, for example). This phase of the analysis may be delegated to a working group, whose members may not feel able to represent decision makers' values to the required level of precision, and may well be carried out in "do-it-yourself" mode, i.e. without the assistance of an expert facilitator. Similar situations may arise in more routine operational decision problems, in which there are still substantial conflicts between criteria, but in which the magnitude of the consequences might not warrant complete decision workshops. Decision makers might then seek a more "quick-and-dirty" MCDM aid.

For situations as described in the previous paragraph, it is useful to develop procedures for estimating a value function from rather weaker or less precise inputs than those described in Chapter 5. The results obtained would still need to be subject to relevant sensitivity analyses and could be modified by more direct assessment (as in Chapter 5) at a later stage. In the next section, we discuss the situation in which clients (typically representatives or agents of the true decision makers) express preferences in ordinal or categorical terms only, i.e.:

Ordinal: Simple comparative statements such as that one criterion is more important than another, or that the difference between a pair of performance levels on one criterion is more important than the difference between another pair;

Categorical: The extension of ordinal statements into categories of importance, typically expressed as a semantic scale (e.g. very much more important, moderately more important, etc.).

Also in the next section, we consider imprecise numerical judgements (e.g. that the importance weight for one criterion is between 2 and 4 times as great as that for another).

In Section 6.3 we examine methods which require only that the client provide holistic statements of preference between a small number of specific alternatives (which may be a subset of the actual set of decision alternatives, or may be sets of hypothetical outcomes). These holistic judgements are used to infer something of the structure of the underlying value functions, and hence extend the preferences to the full set of alternatives.

The treatment of imprecise and/or holistic judgements leads naturally to consideration of the so-called "interactive methods" for MCDA, to be described in Section 6.4. Interactive methods, sometimes termed "progressive articulation of preferences", lead decision makers or their agents through a sequence of simple choices which gradually restrict the range of alternatives which need to be considered until the point is reached at which the decision maker is happy to make a final selection. In Section 6.4, we consider only interactive methods constructed within the framework of value function theory; other interactive methods are based on reference point or goal programming principles, and are discussed in Section 7.3.

The reader should be warned that portions of this chapter are more technical (mathematically) in nature than the previous, and are perhaps more directed towards the developers of *decision support systems* for MCDM problems than at the typical "decision maker". In order to facilitate reading of the chapter, some of the mathematics has been placed in boxes with a reduced font, as a signal to the reader that these parts can be skipped at a first reading. For ease of reference by more technical readers, the main notational conventions used are grouped together in tabular form at the end of the chapter.

The implementation of many of the procedures described in this chapter do require some familiarity with basic concepts of linear programming. The discussion is included here, as facilitators of workshops in which the resultant software support is used, as well as more technically inclined decision makers, will wish to have some understanding of the principles of the methods discussed in the next few sections, and perhaps even to implement these in spreadsheet form. For this reason we shall illustrate their implementation by means of practical examples, based on the land use planning case example (Sections 2.2.3 and 3.5.3), with particular reference to the problem of selecting a small subset of the large number of feasible policy scenarios, and on the game reserve stocking problem (Sections 2.2.2 and 3.5.2).

6.2. USE OF ORDINAL AND IMPRECISE PREFERENCE INFORMATION

In this section, we consider the identification of a value function, or a family of value functions, which is in some sense most consistent with imprecise preference information, typically expressed in one or more of the following forms:

166   *MULTIPLE CRITERIA DECISION ANALYSIS*

- Ordinal statements, i.e. to the effect that certain outcomes (combinations of attribute values) are preferred to others;

- Classification of outcomes into semantic categories (such as "weakly preferred", "strongly preferred", etc.);

- Interval assessments of magnitudes, for example that the ratio of importance weights for two criteria lie between stated lower and upper bounds.

The issue of exploiting imprecise information has been addressed in a variety of ways by a number of authors (e.g. Cook and Kress 1991, Salo and Hämäläinen 1992, and Bana e Costa and Vansnick 1994), all of whom make use of some form of linear programming to obtain a consistent estimate of the value function. For the purposes of the discussion here, we shall present the different approaches in an integrated manner, rather than to describe the details of each individually.

As discussed in Section 5.2, the partial values $v_i(a)$ used in (6.1) (and, equivalently, the scaled partial values $u_i(a)$ used in (6.2)) can be constructed in two different ways, depending upon the context of the problem:

Explicit function of measurable attributes (cf. Subsection 5.2.2): Performance in terms of the criterion is assumed to be represented by the surrogate attribute z_i, defined in such a way that increasing values are preferred. The aim is to construct a function of the form $u_i(z_i)$, which can then applied to any particular outcome. The imprecision in the inputs implies that the constructed function will be an approximation. Simulation studies (Stewart 1993, 1996b) have shown that piecewise linear approximations for each partial value function are adequate for many purposes.

An approximation to $u_i(z_i)$ based on ν_i piecewise linear segments is defined by the values of $u_i(z_i)$ at each of $\nu_i + 1$ "breakpoints", say $z_{i0} < z_{i1} < \cdots < z_{i\nu_i}$, where z_{i0} and $z_{i\nu_i}$ are the endpoints of the scale which may be defined locally or globally as discussed in Section 5.2. It is then only necessary to obtain estimates for $u_{ij} = u_i(z_{ij})$ at each breakpoint. By definition, $u_{i0} = 0$, while $u_{i\nu_i}$ will be proportional to the importance weight of criterion i. The estimate for $u_i(z_i)$ at any other value of z_i is obtained by linear interpolation.

The construction of such a piecewise linear structure is illustrated in Example Panel 6.1, and by Figure 6.1.

Qualitative or categorical scale (cf. Subsection 5.2.3): A number of categories of performance in terms of the criterion are defined ver-

Value Function Methods: Indirect and Interactive   167

> In the land use planning case example (Section 3.5.3), one of the measurable attributes was the lowest flow levels expected in the dry season. Suppose that the range of possible outcomes for plausible scenarios is from 0 to 1.4 cubic metres per second. If the value function is approximated by $\nu_i = 4$ piecewise linear segments, then $z_{i0} = 0$ and $z_{i4} = 1.4$. Possible choices for the intermediate breakpoints may be $z_{i1} = 0.4$, $z_{i2} = 0.7$ and $z_{i3} = 1.0$, so that we only need estimates of $u_{i0} = u_i(0)$, $u_{i1} = u_i(0.4)$, ..., $u_{i4} = u_i(1.4)$.
>
> Note that the segments need not be of equal lengths. In a situation such as this, values below 0.4 m^3/sec may all be considered to be highly unsatisfactory, while values above 1 m^3/sec may offer relatively low further marginal value. The partial value function will thus vary most rapidly between 0.4 and 1, so that the piecewise segments are made shorter in this region, in order to give better discrimination. This is illustrated by the piecewise linear function shown in Figure 6.1

Example Panel 6.1: Illustration of breakpoints

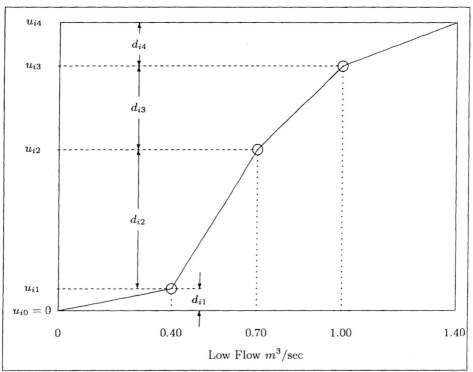

Figure 6.1: Piecewise linear value function for low flow

168   MULTIPLE CRITERIA DECISION ANALYSIS

bally (either in simple terms, such as "excellent", "good", etc., or as more comprehensively described scenarios). The evaluation of any alternative in terms of this criterion will thus consist of allocation to one of the categories. It turns out to be algebraically convenient to index the categories from 0 (the category corresponding to the least desired outcome) to ν_i say (most desired outcome), where once again "most" and "least" desired can refer to a local or to a global scale (see Section 5.2). Note that the number of categories is thus $\nu_i + 1$. We then define u_{ij} as the value of $u_i(a)$ for an alternative whose performance in terms of criterion i is assessed as being in category j (for $j = 0, 1, \ldots, \nu_i$). See Example Panel 6.2 for an illustration.

One of the other attributes suggested for the land use planning example was a composite assessment of water quality on a categorical scale. A five point scale is often convenient, perhaps expressed as "Excellent" (Close to pristine quality), "Good" (Good quality improvement), "Status quo", "Poor" (Substantially degraded relative to status quo) and "Very bad" (Seriously polluted). In our terminology, the scale is represented as follows:

Category	Index (j)	Partial Value (u_{ij})
Very bad	0	$u_{i0} = 0$
Poor	1	u_{i1}
Status quo	2	u_{i2}
Good	3	u_{i3}
Excellent	4	u_{i4}

Note that the descriptions might well refer to specific situations experienced elsewhere, so that there is no implication that the categories are evenly spaced as regards desirability. This is illustrated by representation of the categories on a thermometer scale in Figure 6.2.

Example Panel 6.2: Illustration of categories

It is worth noting here that if the number of alternatives is small, then it may be possible to rank order all alternatives in terms of the criterion under consideration. Each rank position might then in this case be represented as a "category" (1st ranking, 2nd ranking, etc.), and the estimation of values corresponding to each category becomes in effect a direct rating of alternatives (cf. previous chapter).

The approximated value function is fully defined in both cases by the parameters u_{ij}, for $i = 1, 2, \ldots, m$ (i.e. for all criteria, whether based on measurable attributes or on categories) and for $j = 0, 1, \ldots, \nu_i$. By definition, we have that $u_{i0} \le u_{i1} \le \cdots \le u_{i\nu_i}$.

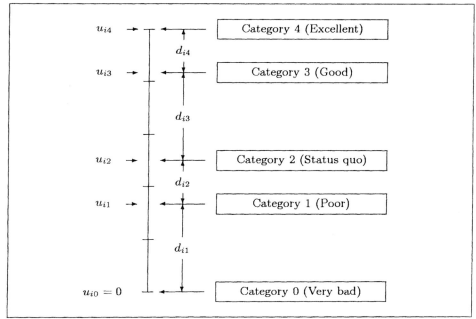

Figure 6.2: Categorical scale values for water quality

Irrespective of how the parameters u_{ij} are ultimately estimated, the partial value functions themselves must retain the interval scale property previously discussed. This implies *inter alia* that only differences of the form $V(a) - V(b)$, or $u_i(a) - u_i(b)$, have any relative meaning. Thus without loss of generality we can set $u_{i0} = 0$ for all i (as we have done in the above examples), in which case $u_{i\nu_i}$ provides a measure of the relative importance of criterion i. If a hierarchical structure is explicitly used (cf. Section 4.2), then $\sum_{i \in L_I} u_{i\nu_i}$ will be a relative measure of importance of higher level criterion I.

It is convenient to define the value function in terms of the *differences* $d_{ij} = u_{ij} - u_{i,j-1} \geq 0$ for $j = 1, 2, \ldots, \nu_i$. The value function model can then be represented entirely in terms of the d_{ij} parameters, since for $q > 0$:

$$u_{iq} = \sum_{j=1}^{q} d_{ij}.$$

Note also that the value difference in terms of criterion i between the p-th and q-th ordered outcomes (i.e. categories or breakpoints) for $q > p$

170 MULTIPLE CRITERIA DECISION ANALYSIS

is given by:

$$u_{iq} - u_{ip} = \sum_{j=p+1}^{q} d_{ij}.$$

The differences d_{ij}, and their relationships to the scaled partial values u_{ij}, are indicated in Figures 6.1 and 6.2, for the piecewise linear value functions and categorical scales respectively. Once again we emphasize the importance of focussing on the *differences*. For example, for a categorical scale such as water quality as illustrated in Figure 6.2, it needs to be ensured that the gaps or intervals between successive categories represent the relative magnitudes of perceived gains in moving from one category to the next. Thus the thermometer scale in Figure 6.2 indicates that the gain in moving from "status quo" to "good" is about twice as important as the gain in moving from "poor" to "status quo".

The partial values $u_i(a)$ can thus be expressed entirely in terms of linear functions of the d_{ij} parameters. For criteria defined in terms of categories, we need only the values at each of the categories. As indicated above and illustrated in Figure 6.2, these values are given by:

$$u_{iq} = \sum_{j=1}^{q} d_{ij}.$$

In the case of the piecewise linear form, $u_i(z_i)$ will in general be obtained by interpolation between the values at two of the breakpoints. For example, if for the function illustrated in Figure 6.1 we need to find $u_i(z_i)$ for a value of z_i in the range $0.7 < z_i < 1$, then:

$$u_i(z_i) = d_{i1} + d_{i2} + \frac{z_i - 0.7}{0.3} d_{i3}.$$

For ease of notation, it is useful to represent the partial values for an arbitrary alternative a directly as a linear function of the d_{ij} as follows:

$$u_i(a) = \sum_{j=1}^{\nu_i} \kappa_{ij}^a d_{ij}. \tag{6.3}$$

When values are defined directly on ordered categories, $\kappa_{ij}^a = 1$ if alternative a is placed in category j or higher (starting from the least preferred as $j = 0$) in terms of criterion i, while $\kappa_{ij}^a = 0$ otherwise. When expressing performance in terms of the surrogate attribute z_i, $\kappa_{ij}^a = c_{ij}(z_i^a)$, where $c_{ij}(z_i) = 0$ for any j such that $z_i \leq z_{i,j-1}$, $c_{ij}(z_i) = 1$ for any j such that $z_i \geq z_{ij}$, and:

$$c_{ij}(z_i) = \frac{z_i - z_{i,j-1}}{z_{ij} - z_{i,j-1}} \tag{6.4}$$

for the segment j (unique if it exists) for which $z_{i,j-1} < z_i < z_{ij}$.

> For example, if we wished to evaluate the partial value for an alternative a having a low flow of $z_i^a = 0.9 \ m^3/\text{sec}$ (based on the approximation illustrated in Figure 6.1), we would have $c_{i0} = c_{i1} = 1$, $c_{i2} = (0.9 - 0.7)/0.3 = 0.667$, and $c_{i3} = c_{i4} = 0$, so that $u_i(a) = d_{i1} + d_{i2} + 0.667 d_{i3}$.

Assessment of the d_{ij} parameters involves consideration of two rather different issues, namely the evaluation of the partial value functions for each individual criterion, and the evaluation of importance weights (i.e. the relative scalings for the $u_i(a)$). We examine each of these issues in turn, giving consideration to the types of preference information which might be provided by the *user* of the system (the decision maker, or an expert working group).

1. *Evaluation of the partial value functions for a specific criterion*: This involves assessment of the magnitudes of $\sum_{j=1}^{p} d_{ij}$ for $p = 1, \ldots, \nu_i - 1$ relative to $\sum_{j=1}^{\nu_i} d_{ij}$ for each criterion i. Recall that by definition we have $u_{i0} \le u_{i1} \le \cdots \le u_{i\nu_i}$, which is ensured by the non-negativity of the d_{ij}. In most cases, however, further information regarding these relative magnitudes can be supplied at various levels of precision:

    (a) *Ordering of "gaps" between categories:* When performance on a criterion is represented in terms of categories, the user may be able to state that the value difference between a pair of consecutive categories (categories $p - 1$ and p, say), is greater than the corresponding difference between another pair (categories $q - 1$ and q, say). Such a statement implies the inequality:

    $$d_{ip} > d_{iq}.$$

    In the situation illustrated in Figure 6.2, it may for example have been stated that the gain in moving from category 2 (status quo) to category 3 (good) is greater than the gains in moving either from category 1 (poor) to category 2 (status quo), or from category 3 (good) to category 4 (excellent), as illustrated by the positioning of these categories on the thermometer scale in Figure 6.2. This implies $d_{i3} > d_{i2}$ and $d_{i3} > d_{i4}$

    (b) *Specification of functional shape:* A similar ordering may be possible for the "gaps" between breakpoints when constructing a value function on a measurable attribute z_i. The more natural assessment in this case may, however, be to assess changes in the *gradient* (slope) of the function, i.e. its shape. The gradient of the value function between breakpoints $z_{i,j-1}$ and z_{ij} is given by

    $$\frac{d_{ij}}{z_{ij} - z_{i,j-1}} = h_{ij} d_{ij} \ \text{(say)}$$

172   *MULTIPLE CRITERIA DECISION ANALYSIS*

where we define $h_{ij} = 1/(z_{ij} - z_{i,j-1})$. Judgements regarding the shape of the function can thus be expressed in terms of inequalities involving $h_{ij}d_{ij}$ terms.

In the piecewise linear function illustrated by Figure 6.1, the user might have specified a larger gradient between 0.4 and 0.7 than between 0 and 0.4. These gradients are respectively $d_{i2}/0.3$ and $d_{i1}/0.4$, so that a larger value for the former implies that $3.33d_{i2} > 2.5d_{i1}$.

Constraints on relative slopes often arise if the function is known to be either concave (decreasing slope with increasing attribute values, or "decreasing marginal returns to scale") or convex (increasing slope with increasing attribute values). Thus for a concave function we would require that $h_{ij}d_{ij} > h_{i,j+1}d_{i,j+1}$ for $j = 1, \ldots, \nu_i - 1$ (and *vice versa* for a convex function). For arbitrary shapes it can get quite complicated to specify all the relevant inequalities. It is worth noting, however, that the sigmoidal, or "S-shaped", functional form (which seems often to arise in practice) can be enforced in the case of $\nu_i = 3$ or 4 (which is usually sufficient accuracy for most purposes) by simply introducing the two inequalities $h_{i1}d_{i1} < h_{i2}d_{i2}$ (so that the slope increases after the first breakpoint) and $h_{i\nu_i-1}d_{i\nu_i-1} > h_{i\nu_i}d_{i\nu_i}$ (so that the slope decreases for the final segment).

(c) *Classification of differences into importance classes:* This is the basis of the "MACBETH" approach (**M**easuring **A**ttractiveness by a **C**ategorical **B**ased **E**valuation **T**ec**H**nique, developed by Bana e Costa and Vansnick, 1994). Instead of directly rank ordering the d_{ij}, or the gradients $h_{ij}d_{ij}$, the value "gaps" in moving from performance level (i.e. category or breakpoint) p to level q, i.e. the differences $u_{iq} - u_{ip}$, are classified according the perceived strength of preference for the value increments they represent, expressed on a semantic scale. The questions posed are of the form: "How strong is the increase in attractiveness realized by moving from level p to level q on criterion i?" The magnitude of the increase is then expressed in terms of a predefined set of importance classes. Bana e Costa and Vansnick suggest the use of the following six classes:

C1	Between negligible and weak
C2	Weak
C3	Between weak and strong
C4	Strong
C5	Between strong and extreme
C6	Extreme

Value Function Methods: Indirect and Interactive   173

We use the notation $(p, q; i) \in C_r$ to denote that the increase in value from level p to q is in importance class C_r .

As in AHP, the standard MACBETH approach suggests that the evaluations be carried out for all pairs of categories or breakpoints, although in view of the potential tedium of the process this could be restricted, for example, only to cases in which $q = p + 1$ or $p + 2$, as illustrated in Example Panel 6.3.

Consider the categorical water quality criterion as illustrated in Figure 6.2. The user would be asked to classify the increase in attractiveness of moving from category p to category q into one of the six predefined importance classes, for each of a number of pairs of categories such that $p < q$.

The classifications are conveniently summarized in a table such as that below, in which the entries represent the importance class associated with the difference between the level indicated by the row and that indicated by the column. An 'X' entry indicates that no classification has been made. The rows and columns have been ordered from most to least preferred, so that only entries in the upper triangular portion of the matrix are relevant (i.e. such that the row level is preferred to the column level).

	3	2	1	0
4	C_2	C_4	X	X
3		C_3	C_5	X
2			C_2	C_5
1				C_4

The preferences recorded in the above table are for example such that:

- the gain in moving from the lowest water quality category to category 2 is between "strong" and "extreme" (the entry in the row indexed 2 and column indexed 0); while

- the gain in moving from water quality category 3 to category 4 is only "weak" (the entry in the row indexed 4 and column indexed 3).

A similar tabular assessment can be constructed for comparisons between breakpoints rather than categories

---

Example Panel 6.3: Pairwise comparisons in MACBETH

Although different users would be inclined to interpret the semantic scales differently, it would nevertheless remain true that if the gain from (category or breakpoint) p to q is classified into a higher importance class than the gain from r to s, then $u_{iq} - u_{ip} > u_{is} - u_{ir}$, i.e. $\sum_{j=p+1}^{q} d_{ij} > \sum_{j=r+1}^{s} d_{ij}$. In illustration of this, note that for the judgements recorded in Example Panel 6.3:

174   *MULTIPLE CRITERIA DECISION ANALYSIS*

- The gain in moving from category 2 to category 3 is in C_3, while that in moving from category 1 to category 2 is in C_2. Thus $u_{i3} - u_{i2} > u_{i2} - u_{i1}$, so that $d_{i3} > d_{i2}$.
- The gain in moving from category 1 to category 3 is in C_5, while that in moving from category 0 to category 1 is in C_4. Thus $u_{i3} - u_{i1} > u_{i1} - u_{i0}$, so that $d_{i2} + d_{i3} > d_{i1}$.

Since all preferences allocated to one importance class are (by assumption) strictly stronger than those allocated to a lower importance class, the above inequalities should be strict. Strict inequalities are ensured in the MACBETH approach through the introduction of additional variables $s_2, \ldots, s_6$ which act as separators between the categories in the sense described in the following box.

Two strictly positive constants, δ and ϵ, are chosen, and the following constraints are imposed on the value differences:

$$
\begin{aligned}
s_2 &\geq \delta \\
s_k &\geq s_{k-1} + \delta && \text{for } k = 3, 4, 5, 6 \\
\epsilon &\leq u_{iq} - u_{ip} \leq s_2 - \epsilon && \text{for all } (p, q; i) \in C_1 \\
s_k + \epsilon &\leq u_{iq} - u_{ip} \leq s_{k+1} - \epsilon && \text{for } k = 2, \ldots, 5 \text{ and} \\
& && \text{all } (p, q; i) \in C_k \\
s_6 + \epsilon &\leq u_{iq} - u_{ip} && \text{for all } (p, q; i) \in C_6.
\end{aligned}
\tag{6.5}
$$

MACBETH seeks to find consistent values for the u_{ij} and s_k satisfying all the constraints. It does this by solving three separate linear programming problems similar to that described later in this section, aimed at testing consistency of the information, identifying sources of inconsistency, and generating ranges of consistent parameter values. Software is available implementing the MACBETH form of assessment directly. We shall not describe the specific linear programming formulation in detail here, as we shall incorporate this type of preference information into a broader framework which includes the other forms of imprecise assessment discussed in this section. Within this broader framework, the associated linear programming model both tests for consistency and achieves the strict inequality (if possible) by maximizing a minimum slack on the desired inequalities.

(d) *Interval assessments of $v_i(a)$ or $v_i(z_{ij})$:* For certain criteria, it may be possible, perhaps with the use of graphical aids, for the user to give direct quantitative assessments of the magnitude of the partial value function on a standardized scale (e.g. 0-100) at certain categories or breakpoints. As these assessments are unlikely to be precise in the context being discussed here, they would in general be represented as upper and lower bounds on the standardized value function, at relevant categories or breakpoints. If, for example, the partial value function is normalized to a 0-100 scale, and the decision maker states that the value at category or breakpoint r lies between 100α and 100β on this scale, then $\alpha < u_{ir}/u_{i\nu_i} < \beta$, no matter how the function may subsequently be scaled. For example in the case of the low flow levels illustrated in Figure 6.1, the user might state that if 100 points were to be allocated to a low flow of 1.4 m^3/sec, then the value of achieving a low flow of 0.7 m^3/sec would be between 55 and 75 points, so that $0.55 < u_{i2}/u_{i4} < 0.75$.

Assessments such as those of the previous paragraph generate two linear inequalities in the d_{ij}:

$$\sum_{j=1}^{r} d_{ij} - \alpha \sum_{j=1}^{\nu_i} d_{ij} > 0$$

and

$$\sum_{j=1}^{r} d_{ij} - \beta \sum_{j=1}^{\nu_i} d_{ij} < 0.$$

Interval assessments are also utilized in the direct value function approach discussed in the previous chapter. The point of introducing the concept here again, is to demonstrate that quantitative interval judgements for some, but not necessarily all, criteria can be combined with other imprecise judgements such as functional shape.

Note that all of the above conditions are represented as linear inequality constraints on the d_{ij} parameters for a fixed i.

2 *Evaluation of importance weights*: In essence, this requires the estimation of the relative scaling of the partial value functions. Assuming evaluation in the extensive form, this requires that every pair of criteria be compared (implicitly or explicitly) in terms of overall importance. In view of the algebraic meaning of weights in additive value functions, the comparisons need to be made of the basis of some convenient but clearly defined "swing", i.e. a range of values which needs to be selected for purposes of comparison, as discussed in Section 5.4. This range would be represented by a specified pair of categories or breakpoints for each criterion, say categories or breakpoints p_i and q_i ($p_i < q_i$) for criterion i. The user is thus asked to

176 MULTIPLE CRITERIA DECISION ANALYSIS

assess whether the gain with respect to achievement of criterion i by moving from level p_i to q_i is greater or less than the corresponding gains for each other criterion. In some cases, it may be convenient to consider the entire range of outcomes (i.e. to select $p_i = 0$ and $q_i = \nu_i$), but particularly when dealing with attributes defined on a measurable scale, users may feel more comfortable with comparing tradeoffs over less extreme ranges of values (for example, $p_i = 1$ and $q_i = 3$ when $\nu_i = 4$), in order to avoid difficult comparisons of extremes. Thus, in Example Panel 6.3, in considering the criteria of water quality and low flows, the user might be asked to compare the importance of improving from "poor" to "good" on quality ($p_i = 1$, $q_i = 3$), relative to that of improving low flows from 0.7 to 1.4 ($p_i = 2$, $q_i = 4$).

In either case, the comparison of these "swings" for two criteria (say k and ℓ) is equivalent to a statement about the relative magnitudes of $\sum_{j=p_k+1}^{q_k} d_{kj}$ and $\sum_{j=p_\ell+1}^{q_\ell} d_{\ell j}$.

As with the evaluation of partial value functions, the above comparisons may be made ordinally (the swing on one criterion is "more important" than the swing on another), categorically (the swing on a particular criterion is classified into one of the importance classes C_1–C_6 defined earlier), or in terms of interval estimates for weight ratios. The ordinal and categorical evaluations induce inequalities of the form:

$$\sum_{j=p_k+1}^{q_k} d_{kj} > \sum_{j=p_\ell+1}^{q_\ell} d_{\ell j}$$

while bounds on relative importances of two criteria would generate the pair of inequalities:

$$\sum_{j=p_k+1}^{q_k} d_{kj} - \alpha \sum_{j=r_\ell+1}^{s_\ell} d_{\ell j} > 0$$

and

$$\sum_{j=r_\ell+1}^{s_\ell} d_{\ell j} - \beta \sum_{j=p_k+1}^{q_k} d_{kj} < 0.$$

Comparisons of higher level criteria can in principle be used to generate similar inequalities relating terms of the form $\sum_{i \in L_I} \sum_{j=p_k+1}^{q_k} d_{kj}$ (cf. Section 4.2). It is more difficult to define "swings" in a meaningful manner in this case, however, and considerable care needs to be taken in the interpretation. For ease of presentation, thus, we shall not include higher level comparisons in our formulations here.

The inequalities which may be generated by consideration of the two criteria for the land use planning problem discussed in the above examples, are illustrated in Example Panel 6.4. We note here in passing, that the inequalities generated in Example Panel 6.4 are mutually consistent. This need not necessarily always be the case with such imprecise judgements, however, a point to which we shall return shortly.

The important point to note at this stage is that, as clearly illustrated in Example Panel 6.4, all of the value judgments discussed above translate into positivity of the d_{ij} parameters, and a number of inequalities involving linear functions of the d_{ij} that can be expressed in the general form:

$$\sum_{i=1}^{m}\sum_{j=1}^{\nu_i} \phi_{ijt}d_{ij} > 0 \qquad (6.6)$$

for $t = 1, 2, \ldots, T$, where T is the number of inequalities generated by the comparisons and value judgments, and t simply indexes each such inequality. For practical purposes, it is appropriate to re-scale the ϕ_{ijt} coefficients in such as way that the deviations to be defined in the next paragraph are comparable in magnitude. A convenient approach is to scale the ϕ_{ijt}'s to unit Euclidean norm, i.e. such that:

$$\sum_{i=1}^{m}\sum_{j=1}^{\nu_i} \phi_{ijt}^2 = 1$$

for each t. For example, consider the first inequality ($t = 1$) in Example Panel 6.4, i.e. $-2.5d_{11} + 3.33d_{12} > 0$. The sum of squares of the coefficients is $2.5^2 + 3.33^2 = 17.36$. The square root of 17.36 is 4.167, and dividing through by this number gives the appropriately scaled inequality: $-0.6d_{11} + 0.8d_{12} > 0$. There are $T = 11$ inequalities in Example Panel 6.4, and the corresponding normalized values (to two significant digits) for the ϕ_{ijt} are given in Example Panel 6.5.

In principle, we should now attempt to find values for the d_{ij} parameters which are maximally consistent with the inequalities (6.6). Two situations need to be addressed. The inequalities may be inconsistent, so that there is no set of d_{ij} values which simultaneously satisfy all. Alternatively, the set of consistent solutions to (6.6) may be large, leaving substantial freedom in choosing the parameter values. In the first situation, we would want to find parameter values which lead to the smallest violations of the desired inequalities. In the second, it would be useful to establish parameter values which satisfy the inequalities most easily (the so-called "middle-most" solution suggested by Zionts and Wallenius, 1983). Both situations can be addressed by reformulating (6.6)

178 *MULTIPLE CRITERIA DECISION ANALYSIS*

We examine the land use planning problem described in Section 3.5.3, but with only the two attributes used in this chapter, namely lowest flow levels (on a measurable scale between 0 to 1.4 m^3/sec), and water quality (represented on a 5-point categorical scale).

As previously illustrated, we represent values for the first criterion by a piecewise linear approximation, between the breakpoints 0, 0.4, 0.7, 1.0 and 1.4. The slopes of the four segments are given by $2.5d_{11}$ ($=d_{11}/0.4$), $3.33d_{12}$, $3.33d_{13}$ and $2.5d_{14}$ respectively. Suppose now that the user provides the following information for this criterion:

- The value function is "S-shaped", implying the constraints:

$$-2.5d_{11} + 3.33d_{12} \quad > \quad 0$$
$$3.33d_{13} - 2.5d_{14} \quad > \quad 0.$$

- Relative to the worst case of no flow, the value gain in achieving a low flow of 0.7 m^3/sec is not less than 55%, nor more than 75% of the value gain in achieving 1.4 m^3/sec. In other words:

$$0.55 \leq \frac{d_{11} + d_{12}}{d_{11} + d_{12} + d_{13} + d_{14}} \leq 0.75$$

which generates the two linear inequalities:

$$-0.25d_{11} - 0.25d_{12} + 0.75d_{13} + 0.75d_{14} \quad > \quad 0$$
$$0.45d_{11} + 0.45d_{12} - 0.55d_{13} - 0.55d_{14} \quad > \quad 0.$$

For the second criterion, suppose that the user has classified strengths of preferences as given in Example Panel 6.3. This enables us to draw a number of inferences concerning the relative magnitudes of the d_{2j}. For example:

- Since the increase in value in moving from category 0 to category 2 (measured by $d_{21} + d_{22}$) is placed in importance class C_5, while the increase in value in moving from category 2 to category 4 (measured by $d_{23} + d_{24}$) is placed in importance class C_4, it follows that:

$$d_{21} + d_{22} > d_{23} + d_{24};$$

- Since the increase in value in moving from category 2 to category 4 (measured by $d_{23} + d_{24}$) is placed in importance class C_4, while the increase in value in moving from category 1 to category 2 (measured by d_{22}) is placed in importance class C_2, it follows that:

$$d_{23} + d_{24} > d_{22}.$$

Example Panel 6.4: Illustration of inequalities generated from imprecise inputs

Value Function Methods: Indirect and Interactive   179

After elimination of redundant and trivial constraints, the classifications into importance classes given in Example Panel 6.3 can be seen to lead to the following five inequalities:

$$d_{21} + d_{22} - d_{23} - d_{24} \; > \; 0$$
$$-d_{21} + d_{22} + d_{23} \; > \; 0$$
$$d_{22} - d_{24} \; > \; 0$$
$$d_{21} - d_{23} \; > \; 0$$
$$-d_{22} + d_{23} \; > \; 0$$

One of the constraints derived earlier, namely $d_{23} + d_{24} > d_{22}$ does not appear in the above, as it is implied by $d_{23} > d_{22}$ and $d_{24} \geq 0$.

Finally the user might state that the importance of the "swing" from *poor* (category 1) to *excellent* (category 4) on quality is at least as important, but not more than twice as important as the "swing" from the first ($z_{11} = 0.4$) to the third ($z_{13} = 1.0$) breakpoints for low flow. This implies:

$$1 < \frac{d_{22} + d_{23} + d_{24}}{d_{12} + d_{13}} < 2$$

which generates the two linear inequalities:

$$d_{22} + d_{23} + d_{24} - d_{12} - d_{13} \; > \; 0$$
$$2d_{12} + 2d_{13} - d_{22} - d_{23} - d_{24} \; > \; 0.$$

Example Panel 6.4: Illustration of inequalities generated from imprecise inputs (Continued)

t	ϕ_{11t}	ϕ_{12t}	ϕ_{13t}	ϕ_{14t}	ϕ_{21t}	ϕ_{22t}	ϕ_{23t}	ϕ_{24t}
1	-0.60	0.80	0	0	0	0	0	0
2	0	0	0.80	-0.60	0	0	0	0
3	-0.22	-0.22	0.67	0.67	0	0	0	0
4	0.45	0.45	-0.55	-0.55	0	0	0	0
5	0	0	0	0	0.5	0.5	-0.5	-0.5
6	0	0	0	0	-0.58	0.58	0.58	0
7	0	0	0	0	0	0.71	0	-0.71
8	0	0	0	0	0.71	0	-0.71	0
9	0	0	0	0	0	-0.71	0.71	0
10	0	-0.45	-0.45	0	0	0.45	0.45	0.45
11	0	0.60	0.60	0	0	-0.30	-0.30	-0.30

Example Panel 6.5: The "ϕ_{ijt}" coefficients for the inequality constraints generated in Example Panel 6.4

180   *MULTIPLE CRITERIA DECISION ANALYSIS*

as:

$$\sum_{i=1}^{m}\sum_{j=1}^{\nu_i}\phi_{ijt}d_{ij} - \sigma_t = 0 \qquad (6.7)$$

where σ_t is unconstrained in sign. If $\sigma_t < 0$, then the model defined by the d_{ij}-values is inconsistent with the t-th inequality, and $|\sigma_t|$ is a measure of the extent of the inconsistency. On the other hand, a strictly positive value for σ_t provides a measure of the extent to which the inequality is satisfied. The basic principle underlying the linear programming formulations below is that if we choose the parameter values d_{ij} to maximize the minimum value of σ_t (taken over all of the inequalities), then: (a) if this value is negative, there are no consistent solutions (and sources of inconsistency are identified); while (b) a positive value implies a most consistent solution with strict inequality for every t. The assumption here is that the σ_t are comparable in magnitude, and it is for this reason that the coefficients in (6.6) needed to be normalized as described above.

Before describing the linear programming formulation, however, some points are worth noting, with some additional notational conventions:

- For comparisons related to assessment of the partial value function for a specific criterion k, $\phi_{ijt} = 0$ for all $i \neq k$. For convenience, let us denote by τ_i the number of such comparisons which are relevant only to criterion i. We suppose that the constraints of the form (6.6) are arranged such that the first τ_1 constraints refer to criterion 1, the next τ_2 constraints to criterion 2, etc. (Thus in Example Panel 6.4, $\tau_1 = 4$ and $\tau_2 = 6$.) If we further define $T_k = \sum_{i=1}^{k}\tau_i$ for $k = 1, 2, \ldots, m$ (and formally, $T_0 = 0$), then the constraints relevant to criterion k will be indexed from $t = T_{k-1} + 1$ to $t = T_k$. The total number of such within-criterion comparisons is then given by T_m ($=9$ in Example Panel 6.4 for $m = 2$).

- For comparisons between criteria, $\phi_{ijt} \neq 0$ for two or more criteria (e.g. the last two rows of Example Panel 6.5). Such comparisons will be represented by the constraints indexed by $t = T_m + 1, \ldots, T$ ($t = T_{11}$ and T_{12} in Example Panel 6.4).

We thus need to establish a set of parameters d_{ij} which are consistent (or at least minimally inconsistent) with all T constraints (6.6). In practice there are sometimes advantages, however, in splitting the analysis into two phases, in the first of which we only consider the parameters relevant to each individual criterion in turn. Thereafter, the inter-criterion issues can be addressed. Although (as we shall see), the two-phase approach may result in some minor loss of information, there are two important advantages in the two-phase approach:

Computational: The two-phase approach involves the solution of typically very small LP problems for each criterion in turn. These can be solved rapidly in interactive mode.

Interpretability: Estimation of the form of the partial value functions for each criterion is separated out from inter-criteria comparisons. This allows the decision maker to evaluate the partial value functions judgmentally, and to make direct adjustments to each, before proceeding to issues of inter-criterion importance.

The various linear programming formulations will now be discussed in the context of this two-phase approach.

6.2.1. Assessments for a single criterion

Consistent parameter estimates: The first requirement is to establish whether the information given for each criterion is internally consistent, i.e., for each criterion k to establish whether there exist d_{kj} for $j = 1, \ldots, \nu_k$, satisfying the constraints. In principle, we should include all T constraints, as even the weight bounds could conceivably place some indirect restrictions on the shapes of individual partial value functions. This effect is, however, likely to be small relative to the τ_k direct restrictions placed on the shape of the partial value function for criterion k, as given by constraints T_{k-1} to T_k. The idea of the two-phase approach is thus at this stage only to seek the set of d_{kj} values which most closely satisfy the τ_k constraints directly relevant to this criterion (k). The effects of the additional restrictions that may be imposed by the remaining $T - \tau_k$ constraints will in practice be small relative to the imprecisions in the inputs. We shall describe the estimation procedures in terms of the direct constraints only. It is, of course, a simple matter to reformulate the linear programming structures below in terms of all T constraints rather than just the τ_k direct constraints.

We require that the parameters d_{kj}, together with the variables σ_t, should satisfy the constraints (6.7) for $t = T_{k-1} + 1, \ldots, T_k$. At this stage, as we are examining each criterion independently of the others, it is sufficient to select an arbitrary scaling of the partial value function, for example such that:

$$\sum_{j=1}^{\nu_k} d_{kj} = 100. \tag{6.8}$$

182 MULTIPLE CRITERIA DECISION ANALYSIS

When we come to inter-criteria comparisons we will address the issue of scaling in proportion to importance weight.

As indicated earlier, our aim will be to maximize the minimum of the σ_t variables, which is achieved by maximizing a new variable Δ which satisfies:

$$\Delta - \sigma_t \leq 0 \qquad (6.9)$$

for $t = T_{k-1}+1, \ldots, T_k$. The variable Δ is also unconstrained in sign, with negative values corresponding to inconsistent solutions (in which case $|\Delta|$ is the min-max deviation from the desired inequality), and positive solutions corresponding to consistent solutions (in which case Δ measures the minimum level of slack). The maximization problem can be solved using standard linear programming (LP) software. In this way we obtain point estimates defining the value function for each criterion k at its specified breakpoints or categories, as well as an indication of whether the constraints (6.6) are consistent for this criterion taken on its own.

The LP formulation for testing consistency of the inputs regarding the low flow criterion in the land use planning problem, using the illustrative data from Example Panels 6.4 and 6.5, is recorded in Example Panel 6.6.

Max Δ
s.t.
$$-0.60d_{11} + 0.80d_{12} - \sigma_1 \qquad = \quad 0$$
$$0.80d_{13} - 0.60d_{14} - \sigma_2 \qquad = \quad 0$$
$$-0.22d_{11} - 0.22d_{12} + 0.67d_{13} + 0.67d_{14} - \sigma_3 \qquad = \quad 0$$
$$0.45d_{11} + 0.45d_{12} - 0.55d_{13} - 0.55d_{14} - \sigma_4 \qquad = \quad 0$$
$$d_{11} + d_{12} + d_{13} + d_{14} \qquad = \quad 100$$
$$\Delta - \sigma_1 \qquad \leq \quad 0$$
$$\Delta - \sigma_2 \qquad \leq \quad 0$$
$$\Delta - \sigma_3 \qquad \leq \quad 0$$
$$\Delta - \sigma_4 \qquad \leq \quad 0$$

The above LP is in fact easily solved on a spreadsheet optimizer, yielding the solution $\Delta = 9.55 > 0$, so that the information provided is fully consistent.

Example Panel 6.6: LP Formulation for consistency checking for the low flow criterion in the land use planning example

Bounding the partial value functions: If the above LP yields a strictly positive solution, then there exists a range of consistent partial value functions, and it is useful to characterize the full family of consistent solutions in some manner. Such a characterization is obtainable from the solutions to a number of further LP problems, in all of which the equality constraints (6.7) are replaced by the original inequality forms (6.6), but with strict inequality replaced by $\geq$ (forcing solutions to be consistent with the inequalities, or at worst on the boundary of consistency), while the variable Δ and its defining constraints (6.9) are dropped. The idea is to characterize the upper and lower trajectories of functions consistent with the constraints by respectively maximizing and minimizing values at each of the breakpoints or categories, subject to the constraints. One possible method of achieving this is described in the following box.

In principle, the aim is to identify upper and lower bounds for the marginal value function for criterion k, at each of breakpoints or ordered outcomes, i.e. upper and lower bounds on $\sum_{j=1}^{p} d_{kj}$ for each $p = 1, 2, \ldots, \nu_k - 1$, subject to the relevant constraints from (6.6). (Recall that $\sum_{j=1}^{\nu_k} d_{kj}$ has a fixed value from (6.8).) Separate optimizations for each p could, however, lead to incompatible solutions (as there may not be feasible sets of d_{kj} values which maximize or minimize $\sum_{j=1}^{p} d_{kj}$ for all p simultaneously). In some cases, it may be sufficient simply to maximize $\sum_{j=1}^{p} d_{kj}$ for $p = \nu_k - 1$ in order to obtain compatible upper values for all p, i.e. an upper value function trajectory, while minimizing d_{k1} might generate the lowest value function trajectory. But in the event that degenerate solutions arise, it may become necessary to include maximization or minimization of $\sum_{j=1}^{p} d_{kj}$ for other values of p, but with reduced weight. Expressing the corresponding LP objective in the form $\sum_{p=1}^{\nu_k - 1} \omega_p \sum_{j=1}^{p} d_{kj}$ is clearly equivalent to maximizing or minimizing functions of the form:

$$\sum_{j=1}^{\nu_k - 1} \gamma_j d_{kj} \tag{6.10}$$

where $\omega_{\nu_k - 1} = \gamma_{\nu_k - 1}$ and $\omega_p = \gamma_p - \gamma_{p+1}$ for $1 \leq p \leq \nu_{k-2}$.

In seeking the upper trajectory, we require $\omega_{\nu_k - 1} > \omega_{\nu_k - 2} > \ldots > \omega_1$. Geometrically decreasing weights with a factor of 2 are achieved by setting $\gamma_{\nu_k - 1} = 1$, $\gamma_{\nu_k - 2} = 1.5$, $\gamma_{\nu_k - 3} = 1.75$, etc. Similarly, for the lower trajectory, we require $\omega_1 > \omega_2 > \ldots > \omega_{\nu_k - 1}$, and geometrically decreasing weights with a factor of 2 are achieved by setting $\gamma_{\nu_k - 1} = 1$, $\gamma_{\nu_k - 2} = 3 \ (= 1 + 2)$, $\gamma_{\nu_k - 3} = 7 \ (= 1 + 2 + 4)$, etc.

For later reference, let us define λ_{kj} and μ_{kj} as the values of d_{kj} obtained when identifying the lower and upper value function trajectories respectively. Note that partial value functions defined by either the λ_{kj}'s or the μ_{kj}'s will be consistent with the ordinal and categorical information provided by the decision maker for criterion k. The resulting lower and upper trajectories for the value functions,

184   *MULTIPLE CRITERIA DECISION ANALYSIS*

given by $\sum_{j=1}^{p} \lambda_{kj}$ and $\sum_{j=1}^{p} \mu_{kj}$ respectively for $p = 1, 2, \ldots, \nu_k$, can be presented graphically to the user. At this point the user might well wish to adjust the solutions to better represent his or her preference judgments, before proceeding to the steps involving inter-criteria assessments.

The LP solution for obtaining bounds on the partial value function for the low flow criterion in the land-use planning problem is illustrated in Example Panel 6.7, in which the resulting bounds are also displayed graphically.

6.2.2.   Inter-criteria assessments

As we now work with all criteria simultaneously, it is useful to re-standardize the value functions so that the maximum aggregated value is fixed at a convenient level, such as 100. We thus replace (6.8) by:

$$\sum_{i=1}^{m} \sum_{j=1}^{\nu_i} d_{ij} = 100. \tag{6.11}$$

This re-normalization does not change the basic shape of each partial value function, but does imply for each criterion i that $\sum_{j=1}^{\nu_i} d_{ij}$ represents the effective importance weight of this criterion relative to the others.

Global consistency: We can repeat a consistency check similar to that applied to the partial value functions for each individual criterion, but now for global consistency (i.e. including all relative weight assessments), by maximizing a variable Δ subject to constraints (6.7) and (6.9) applied for all $t = 1, \ldots, T$, and to the global normalization (6.11).

Alternative rank orders: Suppose that globally consistent solutions exist, i.e. that the optimum value of Δ in the previous step is positive. It is then no longer necessary to retain the Δ and σ_t variables; we need only examine those solutions satisfying (6.6) (in the $\leq$ form) for all $t = 1, \ldots, T$.

An important consideration now may be to establish what rank orders of the alternatives are consistent with the available information. One important reason for this may be to identify a sub-set of alternatives which consistently rank highly for many sets of feasible parameter values. In the first phase of the land use planning problem, for example, this may be used to select a shortlist for more detailed evaluation

The first four constraints identified in Example Panels 6.4 and 6.5 are relevant to this criterion, for which $\nu_1 = 4$. The lower trajectory of consistent value functions is thus obtained by solving:

$$\begin{array}{rl} \text{Min} & 7d_{11} + 3d_{12} + d_{13} \\ \text{s.t.} & -0.60d_{11} + 0.80d_{12} \geq 0 \\ & 0.80d_{13} - 0.60d_{14} \geq 0 \\ & -0.22d_{11} - 0.22d_{12} + 0.67d_{13} + 0.67d_{14} \geq 0 \\ & 0.45d_{11} + 0.45d_{12} - 0.55d_{13} - 0.55d_{14} \geq 0 \\ & d_{11} + d_{12} + d_{13} + d_{14} = 100 \end{array}$$

The solution gives $d_{11} = 0$; $d_{12} = 55$; $d_{13} = 19.3$; and $d_{14} = 25.7$, which are thus also the values for λ_{1j}. The lower bounds on the value function at the breakpoints are given by the cumulative sums, i.e. 0; 0; 55; 74.3 and 100. Similarly, maximizing $1.875d_{11} + 1.75d_{12} + 1.5d_{13} + d_{14}$ subject to the constraints gives: $\mu_{11} = 43.0$; $\mu_{12} = 32.3$; $\mu_{13} = 24.7$ and $\mu_{14} = 0$, giving upper bounds at the breakpoints of 0; 43.0; 75.3; 100; 100. These lower and upper trajectories are illustrated in the following figure.

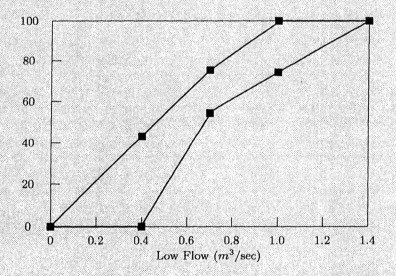

For completion we record here the λ_{2j} and μ_{2j} defining the bounds on the values at each category for the second criterion:

Increment (j)	1	2	3	4
λ_{2j}	25	25	25	25
μ_{2j}	50	25	25	0

Example Panel 6.7: LP Formulation to determine bounds on the partial value function for the low flow criterion

186 MULTIPLE CRITERIA DECISION ANALYSIS

in the second phase. Another important benefit derived from generating the consistent rank orders even applied to a relatively small number of options, may simply be to present political decision makers with a clear indication of what are or are not potentially optimal solutions according to the available information.

An alternative a is potentially optimal (i.e. optimal for some value function consistent with available preference information) if there exist parameter values satisfying the constraints, which are such that:

$$\sum_{i=1}^{m}\sum_{j=1}^{\nu_i} \kappa_{ij}^a d_{ij} \geq \sum_{i=1}^{m}\sum_{j=1}^{\nu_i} \kappa_{ij}^b d_{ij}$$

for all alternatives $b \neq a$, where κ_{ij}^a is as defined in (6.3). This can be established by maximizing a re-defined variable Δ_a (unconstrained in sign), subject to:

$$\Delta_a \leq \sum_{i=1}^{m}\sum_{j=1}^{\nu_i} (\kappa_{ij}^a - \kappa_{ij}^b) d_{ij}$$

for all $b \neq a$, to the constraints (6.6) in $\leq$ form for $t = 1, \ldots, T$, and to (6.11). The alternative a is potentially optimal if the solution is non-negative.

Maximizing Δ_a for each alternative in turn, we obtain n different sets of values for the alternatives with associated rank orders. While this does not necessarily identify all possible rank orders of the alternatives, it does provide a wide range of rank orders, which give a direct measure of the effects of the imprecisions in the inputs. If very similar rank orders arise in each case, then there may be no need for any further precision in the inputs. On the other hand, very divergent rank orders suggest that greater effort is needed in providing more precise inputs. The decision as to whether the differences in rank orders are such that more precision is needed must ultimately be left to the user.

Simplified form of inter-criteria assessments

The linear programming problems described above for the inter-criteria assessments can become quite large if there are many criteria, so that interactive implementation may be slow. We now describe a simplified form of assessment which leads to substantially smaller LPs, without serious loss of information.

Recall that in the assessments for the individual partial value functions, we had identified upper and lower trajectories of consistent partial value functions in terms of the sets of parameters λ_{ij} and μ_{ij} for each criterion i. We now restrict attention to the families of partial value functions in which for each criterion the d_{ij} can be expressed in terms of two new non-negative variables x_i and y_i as follows:

$$d_{ij} = \lambda_{ij} x_i + \mu_{ij} y_i. \tag{6.12}$$

The resultant partial value function for each i is simply a weighted average of the upper and lower trajectories, rescaled so that the maximum value $u_{i\nu_i} = \sum_{j=1}^{\nu_i} d_{ij} = 100(x_i + y_i)$ (since by definition $\sum_{j=1}^{\nu_i} \lambda_{ij} = \sum_{j=1}^{\nu_i} \mu_{ij} = 100$ by (6.8)).

The restriction of the d_{ij} to those satisfying (6.12) for some x_i and y_i does limit the range of partial value functions shapes, but these still satisfy all the specified constraints. The advantage is in the substantial reduction in size of the LPs. The number of variables is reduced from $\sum_{i=1}^{m} \nu_i$ to $2m$. But the important reduction is in the number of constraints. By definition, if either $d_{ij} = \lambda_{ij}$ for all i and j, or $d_{ij} = \mu_{ij}$ for all i and j, then (6.6) is satisfied for all $t = 1, \ldots, T_m$. This ensures that the d_{ij} defined by (6.12) also satisfy (6.6) for all $t = 1, \ldots, T_m$, so that these constraints (or those in the equality form (6.7)) need only explicitly to be included in the LPs for $t = T_m + 1, \ldots, T$.

For the record, we include here the relevant LP formulations obtained by substituting (6.12) into the previous formulations:

- The global consistency test is achieved by maximizing Δ subject to:

$$\sum_{i=1}^{m} \sum_{j=1}^{\nu_i} \phi_{ijt} [\lambda_{ij} x_i + \mu_{ij} y_i] - \sigma_t = 0$$

as well as to (6.9) for $t = T_m + 1, \ldots, T$, and the normalization:

$$\sum_{i=1}^{m} [x_i + y_i] = 1.$$

- The test for potential optimality of alternative a is achieved by maximizing Δ_a subject to:

$$\sum_{i=1}^{m} \sum_{j=1}^{\nu_i} \phi_{ijt} [\lambda_{ij} x_i + \mu_{ij} y_i] \geq 0$$

for $t = T_m + 1, \ldots, T$, to:

$$\Delta_a \leq \sum_{i=1}^{m} \sum_{j=1}^{\nu_i} [(\kappa_{ij}^a - \kappa_{ij}^b) \lambda_{ij} x_i + (\kappa_{ij}^a - \kappa_{ij}^b) \mu_{ij} y_i]$$

for all $b \neq a$, and the normalization:

$$\sum_{i=1}^{m} [x_i + y_i] = 1.$$

188 *MULTIPLE CRITERIA DECISION ANALYSIS*

6.3. HOLISTIC ASSESSMENTS AND INVERSE PREFERENCES

Suppose now that the decision maker (or user of the decision support system) is able to express a definite preference for an alternative a over another alternative b. By assumption, this implies that $V(a) > V(b)$ which in turn imposes constraints on the parameters of the value function model. Such constraints can replace or be added to those developed in the previous section, for purposes of estimating the models by linear programming. The more holistic comparisons there are, the more precise the resulting estimates will tend to be.

Sets of holistic choices between alternatives can arise in at least two ways. One possibility is that the decision maker is familiar from past experience with a subset of the available alternatives, and is able to express preferences between these. This information can then be used to provide some partial or tentative ranking orderings of the remaining unfamiliar alternatives. Alternatively, the analyst may construct a number of hypothetical alternatives which may be easier for the decision maker or user to rank order than the real alternatives (typically because the hypothetical alternatives may be designed in such a way that any pair differ only a small subset of the criteria). We now discuss each of these possibilities.

Real Alternatives

In this case, a subset (or "reference set") of the actual set of alternatives is selected for holistic evaluation by the decision maker. These may, as indicated earlier, be alternatives with which the decision maker is familiar, or may simply be a subset of alternatives selected for detailed evaluation. Typically, the decision maker would provide a rank ordering of alternatives in the reference set, which implies a sequential set of preferences. The linear programming models described in the previous section can then be used to identify potentially optimal solutions in the full set of alternatives, and a number of possible rank orders on this full set consistent with the information, based on the constraints implied by the holistic preferences (with or without any additional imprecise information). In effect, we extrapolate preferences on the reference set to a preference ordering (possibly partial or tentative) on the full set. Clearly this approach only makes sense if the actual number of alternatives (n) is relatively large, as it would typically require the rank ordering of at least 8-10 alternatives before the linear programming solutions are able to produce usefully precise outputs. The approach is thus well suited to a situation in which a shortlist of alternatives has to be extracted from a large set (possibly even a set defined implicitly in mathematical programming form).

A version of the approach described in the previous paragraph was suggested by Srinivasan and Shocker (1973), and these ideas have been developed further in the "UTA" (Utilité Additive) and related methods, sometimes termed "disaggregation" methods (Siskos, 1980; Jacquet-Lagrèze and Siskos, 2001) Software implementing UTA and related approaches include PREFCALC (Jacquet-Lagrèze, 1990), MIIDAS and MINORA (Siskos et al., 1999). The usual approach adopted in UTA is to minimize a sum of deviations rather than the maximum deviation recommended in Section 6.2, but the principle is otherwise the same.

An issue which does not appear to be much discussed in the related literature is that of choice of the reference set. In some cases, as we have indicated, there is no real choice, as the reference set is determined by the alternatives with which the decision maker is most familiar. In situations such as those described in the land use study (Section 3.5.3), the analyst might have considerable discretion as to which alternatives are selected for detailed evaluation. Certain choices may lead to much tighter constraints on the parameters, and will thus be much more informative, than others. As an extreme example of this, if the reference set consists of a sequence of alternatives in which each one dominates the next, then this will add no useful preference information. A suggestion made in Stewart (1986) was to generate the reference set sequentially, taking pairs of alternatives at a time. At each step, the pair of alternatives is selected which differ by the least amount according to the current best estimate of the value function. Once the decision maker has expressed a preference between these two, the LP is solved again, to get an updated estimate of the value function, before the next pair is chosen. This places the procedure within the context of "interactive" methods defined in the next section, and in fact the aim of Stewart (1986) was to combine some of the concepts of UTA with those of the Zionts-Wallenius approach discussed in Section 6.4.2.

Hypothetical Alternatives ("Conjoint Scaling")

In conjoint scaling, sets of hypothetical outcomes are generated which differ in terms of two criteria at a time only. The basic idea is that it may be far easier to compare alternatives differing on a small number of criteria, than to make holistic comparisons of real alternatives differing in general on all criteria. The methodology was originally developed as a tool for market research, in an attempt to identify a product profile which best satisfies consumer needs. The principles are easily adapted to MCDA, however. A useful by-product of the approach is a partial

190 *MULTIPLE CRITERIA DECISION ANALYSIS*

test of preferential independence of the criteria. If the decision maker has problems in rank ordering such hypothetical alternatives without referring to the other criteria, then this would indicate violation of preferential independence.

The following steps are necessary in order to apply conjoint scaling effectively in MCDA.

1 Select three or (at most) four representative attribute levels (categories, or possibly some of the "breakpoints" for the piecewise linear approximations) for each criterion: These should cover most of the range of values represented by the alternatives, and serve as a basis for constructing hypothetical outcomes. Supposing (for purposes of explanation) that three levels are chosen for each criterion, we shall denote these by $z_i^1 < z_i^2 < z_i^3$.

2 Select pairs of criteria to be compared with each other: Each criterion must be compared directly with at least one other, and any two criteria must be linked to each other by a chain of comparisons.

3 For each pair of criteria to be compared, say criteria k and ℓ, generate all 9 pairs of outcomes, and present these to a the decision maker for rank ordering. This may conveniently be done in tabular form, for example as follows:

Levels of Criterion k	Levels of Criterion ℓ		
	z_ℓ^1	z_ℓ^2	z_ℓ^3
z_k^1	9	–	–
z_k^2	–	–	–
z_k^3	–	–	1

Supposing (for purposes of illustration) that increasing values of each attribute are preferred, the pairs given by (z_k^3, z_ℓ^3) and (z_k^1, z_ℓ^1) must rank first and last (9th) respectively, as indicated in the table above. The user would be required to fill in the rank orders for all other pairs.

The conjoint scaling procedure is illustrated in Example Panel 6.8.

Model estimation

We continue to model the value functions as in Section 6.2. Suppose that the decision maker (or user of the decision support system) expresses a definite preference for an alternative r over s (which may be real or

Value Function Methods: Indirect and Interactive 191

> We continue with the land use planning case example, as used in the examples of Section 6.2. Suppose that conjoint scaling is to be applied to comparison of the low flow and water quality criteria defined in Section 6.2, with the intermediate breakpoints and categories used in each case as representative performance levels, viz. flows of 0.4, 0.7 and 1.0 m^3/sec, and the "poor", "status quo" and "good" water quality categories.
>
> Suppose then that the user provides the rank orderings given in the following table:
>
> | Water Quality | Low Flow Criterion (m^3/sec) | | |
> |:---:|:---:|:---:|:---:|
> | Criterion | 0.4 | 0.7 | 1.0 |
> | Poor | 9 | 8 | 5 |
> | Status Quo | 7 | 4 | 3 |
> | Good | 6 | 2 | 1 |
>
> The rank ordering is fully consistent with dominance properties. Some of the rank orderings (for example, the pairs ranked 3rd and 4th) are in fact implied by dominance considerations and will therefore not add new information to the model estimation. The pairs of successive rank orders which contribute non-trivial preference information are those which relate to outcome pairs in which one does not dominate the other, namely: the 2nd and 3rd; 4th and 5th; 5th and 6th; and 7th and 8th.

Example Panel 6.8: Conjoint scaling comparison of two criteria

hypothetical). With the notational convention introduced in (6.3), this implies the following constraint on the parameters d_{ij}:

$$\sum_{i=1}^{m}\sum_{j=1}^{\nu_i}[\kappa_{ij}^r - \kappa_{ij}^s]d_{ij} > 0 \qquad (6.13)$$

This constraint is of precisely the same form as (6.6), with $\phi_{ijt} = \kappa_{ij}^r - \kappa_{ij}^s$.

In comparing real alternatives, non-zero coefficients for the d_{ij} will typically be generated for all i and j. In the context of conjoint scaling, however, non-zero coefficients will only occur for the two criteria being compared. For example, the 4th and 5th ranked pairs in Example Panel 6.8 state that the status quo quality and a low flow of 0.7 m^3/sec is preferred to poor water quality and a flow of 1.0 m^3/sec. The constraint implied by (6.13) (letting $i = 1$ represent low flow, and $i = 2$ represent water quality) gives $d_{11} + d_{12} + d_{21} + d_{22} > d_{11} + d_{12} + d_{13} + d_{21}$, i.e. $d_{22} - d_{13} > 0$.

The estimation procedures discussed in the previous section apply equally well to constraints generated by (6.13), with or without those

192   *MULTIPLE CRITERIA DECISION ANALYSIS*

provided by the imprecise preference information described in Section 6.2. As before, these can provide point or interval estimates for the weights and partial value functions, and/or potentially optimal rank orderings on the full set of alternatives. For estimates based entirely on holistic comparisons, however, the constraints given by (6.13) will always involve more than one criterion at a time, so that there would be no point in adopting a two-phase form of solution (within and between criteria), or to express the d_{ij} in a form such as (6.12). The relevant LPs would thus be as defined in the following.

Check for consistent solutions: Maximize Δ subject to:

$$\sum_{i=1}^{m}\sum_{j=1}^{\nu_i}[\kappa_{ij}^r - \kappa_{ij}^s]d_{ij} - \sigma_{(r,s)} = 0$$

and

$$\Delta - \sigma_{(r,s)} \leq 0$$

for all pairs (r, s) of alternatives (real or hypothetical) such that r has been identified as preferred to s, and to the normalization constraint:

$$\sum_{i=1}^{m}\sum_{j=1}^{\nu_i}d_{ij} = 100.$$

As before, the d_{ij} variables are non-negative, but Δ and the $\sigma_{(r,s)}$ are unconstrained in sign.

Alternative rank orders: If a consistent solution exists (i.e. the value of Δ in the previous step is positive), then we can proceed to identify potentially optimal solutions by maximizing Δ_a for each real alternative a, subject to:

$$\Delta_a \leq \sum_{i=1}^{m}\sum_{j=1}^{\nu_i}(\kappa_{ij}^a - \kappa_{ij}^b)d_{ij}$$

for all $b \neq a$,

$$\sum_{i=1}^{m}\sum_{j=1}^{\nu_i}[\kappa_{ij}^r - \kappa_{ij}^s]d_{ij} \geq 0$$

for all pairs (r, s) of alternatives for which r is preferred to s, and to:

$$\sum_{i=1}^{m}\sum_{j=1}^{\nu_i}d_{ij} = 100.$$

There is, of course, no reason why the ordinal and other imprecise forms of information (from Section 6.2) and holistic information cannot be combined during part of the same analysis. The resulting linear programming formulations will then include constraints of both forms (6.6) and (6.13). In this case, it may be preferable to return to the two-phase analysis of Section 6.2, with the holistic comparisons (6.13) included with (6.6) for $t > T_m$. This concept has been incorporated as an option into the DEFINITE for Windows multicriteria decision support software.

6.4. INTERACTIVE METHODS BASED ON VALUE FUNCTION MODELS

At first sight, it may appear strange to find a specific section devoted to "interactive" methods, as all MCDM, and all applications of value functions methods, must involve interaction with the decision maker. Nevertheless, a number of MCDM approaches have come to be termed "interactive methods" (or sometimes "progressive articulation of preferences"), in that they involve the following characteristic steps (cf. also Gardiner and Steuer, 1994):

1 A feasible (and usually efficient) solution, or small number of solutions, is generated according to some specified procedure and presented to the decision maker.

2 If the decision maker is satisfied with the solution (or one of those) generated, then the process stops. Otherwise, the decision maker is requested to provide some local preference information in the vicinity of the solution(s) presented, such as direct comparisons between (actual or hypothetical) solutions or tradeoffs.

3 In the light of the local information provided, preference models are updated and/or parts of the decision space are eliminated, and the process returns to the first step.

Interactive methods in this sense are perhaps best suited to situations in which a single individual, or a small homogeneous group, needs quite quickly to find one or a few satisfactory alternatives to a decision problem, and when the criteria are well-represented in terms of quantitative attributes. The context may be that of taking a decision directly, especially in a relatively familiar setting in which similar decision problems arise frequently, or that of a first level of screening of alternatives in which a shortlist is to be generated for later more detailed evaluation (as in the first phase of the land use planning problems discussed in Section 3.5.3). The exploration of the efficient frontier which is facilitated by the interactive approach can provide considerable insight into the options and tradeoffs which are available. Thus the most important benefit may not be in the generation of an optimal solution *per se*, but rather in the insights and understanding generated for the decision maker in exploring the decision space in a systematic and coherent manner.

194   *MULTIPLE CRITERIA DECISION ANALYSIS*

There are certain advantages and disadvantages associated with the use of interactive methods for MCDM, and these have an influence on the types of MCDM problems for which they are or are not suited. The primary advantage of the interactive approach is that the value judgements which have to be made by the decision maker are set in a realistic context. This is in contrast to the situation with some MCDM methodologies in which global judgements (such as importance weights of criteria) have to be made, often involving the need to compare quite extreme combinations of outcomes not encountered in practice. The disadvantage of interactive methods is that the process is relatively unstructured, with the result that it may be difficult to motivate clearly the solution obtained.

From the above comments, it may be clear that interactive methods are not well suited to use in group decision making contexts in which there are substantial conflicts, in which many criteria are qualitative, or in which a clearly defensible justification for the solution obtained needs to be established (as is often the case in strategic public sector policy decisions). Nevertheless, even in such contexts individual members of the group may well find that the interactive methods are useful in exploring the options for themselves prior to the group sessions.

Shin and Ravindran (1991) provide a comprehensive review of interactive methods published up to about 1990. Their review covers interactive methods based both on goal programming and value function approaches to MCDM. In this section, we shall deal only with interactive methods based on the value function approach; we shall return to interactive goal programming or aspiration-based methods in Section 7.3.

6.4.1.   Methods using trade-off information

We start again with the assumption of the existence of a value function, say $V(\mathbf{z})$, expressed as a function of the vector $\mathbf{z}$ of attribute values. Interactive methods do not strive to fully specify or estimate this value function, and relatively mild assumptions, such as that of pseudo-concavity, are made concerning the functional form of $V(\mathbf{z})$. Stronger assumptions such as additivity (implying the need for preferential independence of criteria) may not necessarily be required, so that interactive value-based methods are potentially usable in a "do-it-yourself" mode (i.e. not necessarily under the direct guidance of a skilled MCDM facilitator), without the need for careful tests for assumptions such as preferential independence. Furthermore, the assumptions which are made tend only to be applied in a local sense, i.e. to provide sufficient guidance to the DM concerning potentially better regions of the decision space which need to be explored. The underlying philosophy is that preferences tend to evolve and to develop as greater understanding of the problem is attained, with the result that the value function itself may change during the process.

Perhaps the earliest interactive procedure of this type is that described by Geoffrion, Dyer and Feinberg (1972) for the case in which alternatives are associated with points $\mathbf{x}$ in a compact, convex set $\mathbb{X} \subset \mathbb{R}^n$ (for example the set of feasible solutions to a multiple objective linear programming problem). Attribute values are then assumed to be expressible as functions of $\mathbf{x}$, i.e. $z_i = f_i(\mathbf{x})$. In order to keep notation compact, the attribute vector corresponding to $\mathbf{x}$ will be denoted by $\mathbf{z} = \mathbf{f}(\mathbf{x})$. The procedure of Geoffrion et al. can then viewed as maximizing $V(\mathbf{f}(\mathbf{x}))$ subject to $\mathbf{x} \in \mathbb{X}$ using the Franke-Wolfe algorithm for non-linear programming, but using the decision maker, in effect, as a function and gradient evaluator. Details of the algorithm are given in the original reference as well as in, for example, Steuer (1986) or Stewart (1999a). A brief outline is provided in Figure 6.3

---

1 Generate an arbitrary feasible point, say $\mathbf{x}^1$. Set $k = 1$.

2 Calculate the corresponding attribute vector $\mathbf{z}^k = \mathbf{f}(\mathbf{x}^k)$. This is presented to the decision maker for evaluation.

3 One attribute is arbitrarily selected as a reference. Without loss in generality, we denote the reference attribute as z_1. Set the "weight" for this attribute as $w_1^k = 1$. For every other attribute i, w_i^k is set to the tradeoff between attributes 1 and i as assessed by the decision maker when considering the performance levels represented by $\mathbf{z}^k$, i.e. the amount of attribute 1 (the reference) which the decision maker is prepared to sacrifice at this point, in order to obtain a unit gain in attribute i.

4 The partial derivative of the value function with respect to x_j (the j-th element of $\mathbf{x}$), evaluated at the point $\mathbf{x}^k \in \mathbb{X}$, is estimated by:

$$\nabla V_j = \sum_{i=1}^{m} w_i^k \frac{\partial f_i(\mathbf{x}^k)}{\partial x_j}.$$

5 Obtain a direction of improvement by maximizing $\sum_{i=1}^{n} \nabla V_j y_j$, subject to $\mathbf{y} \in \mathbb{X}$ (where $\mathbf{y}$ is the n-dimensional vector with elements y_j). Let the solution be $\mathbf{y}^k$.

6 The decision maker is presented with the attribute vectors corresponding to a sequence of feasible solutions along the line between $\mathbf{x}^k$ and $\mathbf{y}^k$ (i.e. solutions of the form: $(1 - t)\mathbf{x}^k + t\mathbf{y}^k$ for $0 \leq t \leq 1$. If the decision maker sees no benefit in moving from the solution $\mathbf{x}^k$ then the process stops. Otherwise, the solution which the decision maker most prefers becomes the starting point for a new iteration, viz. x^{k+1}. The process repeats again from step 2 with $k = k + 1$.

Figure 6.3: Outline of the Geoffrion-Dyer-Feinberg algorithm

196   *MULTIPLE CRITERIA DECISION ANALYSIS*

The optimization steps in the algorithm are based on rather imprecise information, while relatively few iterations (possibly 6 or 8) can be executed in practice, in comparison with perhaps hundreds of iterations in a numerical optimization procedure. Geoffrion et al. do, nevertheless, provide some practical evidence that useful solutions (in terms of satisfying decision maker goals) can be obtained by use of this method.

One difficulty in principle with the approach is, however, the fact that preference information from one iteration is discarded before the next iteration. This is very inefficient use of information, the assessment of which is both a difficult task for the DM and imprecise. While local tradeoffs may be expected to change as the DM moves through the decision space, it is also true that in most cases the changes would be gradual. More effective use can be made of the tradeoff information if the assumption is made that the value function is *"pseudoconcave"*, i.e. such that a move in a direction of initially decreasing preference can never eventually lead to point of increased preference. It can be shown that in this case:

$$\sum_{i=1}^{m} w_i(z_i^b - z_i^a) \le 0 \qquad (6.14)$$

if and only if b is not preferred to a (where the w_i are the tradeoffs assessed as for the Geoffrion-Dyer-Feinberg approach).

The above property of pseudoconcave value functions can be used to enhance the Geoffrion-Dyer-Feinberg algorithm, by a process of generating *"tradeoff cuts"* (Musselman and Talavage, 1980) to reduce the decision space following each iteration. The Geoffrion-Dyer-Feinberg algorithm as described in Figure 6.3 is modified by restricting the search for $\mathbf{y}^k$ at each iteration to ever smaller sets based on these cuts.

Specifically, we define $\mathbf{X}^0 = \mathbf{X}$, and then after the tradeoffs are obtained at each iteration, we define $\mathbf{X}^k = \mathbf{X}^{k-1} \cap \{\mathbf{x} \in \mathbb{R}^n | \sum_{i=1}^{m} w_i^k(f_i(\mathbf{x}) - f_i(\mathbf{x}^k)) \ge 0\}$. The search for $\mathbf{y}^k$ is then restricted to $\mathbf{y} \in \mathbf{X}^k$.

6.4.2.   Methods using direct comparisons

Instead of requiring tradeoffs between criteria, preference information can also be obtained from the decision maker in the form of direct comparisons between two or more decision alternatives (represented in terms of their attribute vectors $\mathbf{z}$). Such information can be used either directly to draw inferences concerning the form of the underlying value function, or indirectly to eliminate parts of the decision space (as with the tradeoff cuts above). The methods which we shall describe here were developed within the context of multiple objective linear programming.

Although the methods can be generalized to some extent, we shall restrict our presentation here to the MOLP case, i.e. to the case in which the $f_i(\mathbf{x})$ are linear functions of $\mathbf{x}$, and in which $\mathbb{X}$ can be expressed as $\mathbb{X} = \{\mathbf{x} \in \mathbb{R}^n | \mathbf{Ax} \le \mathbf{b} \; ; \; \mathbf{x} \ge \mathbf{0}\}$ (as illustrated by the game reserve planning problem). Let us define $\mathbb{Z}$ as the set of attribute vectors which are attainable, i.e. such that $\mathbf{z} \in \mathbb{Z}$ if and only if $\mathbf{z} = \mathbf{f}(\mathbf{x})$ for some $\mathbf{x} \in \mathbb{X}$. An important property of MOLP is that the set $\mathbb{Z}$ is convex.

Let $\mathbf{x}^*$ be the alternative maximizing $V(\mathbf{f}(\mathbf{x}))$ over $\mathbb{X}$, and let $z^* = \mathbf{f}(\mathbf{x}^*)$ be the corresponding attribute vector. If we again make the assumption that the value function is pseudoconcave, it then follows from the convexity of $\mathbb{Z}$ that there exist non-negative weights $w_1, w_2, \ldots, w_m$ such that $\mathbf{x}^*$ maximizes $\sum_{i=1}^m w_i f_i(\mathbf{x})$ over $\mathbb{X}$. This simple observation is the basis for a number of interactive procedures, in which preference information obtained from the decision maker is used to increasingly restrict the set of allowable weight vectors, which implicitly then also restricts the range of solutions to be considered.

One such scheme is that introduced by Zionts and Wallenius (1976, 1983). The basic concepts underlying their approach are summarized in Figure 6.4. An illustration of the basic Zionts-Wallenius procedure, as described by the numbered steps in Figure 6.4, is presented in Example Panel 6.9, where it is applied to the game reserve problem, which was formulated in Section 3.5.2 as a multiple objective linear programming problem.

The original Zionts-Wallenius algorithm, described by the numbered steps in Figure 6.4, are intuitively appealing, and will in fact lead to the maximum value solution if the value function is linear and the decision maker expresses preferences consistent with this linear value function at all times. In practice, however, tradeoffs will vary across the decision space, with the result that the inequalities in the w_i generated from local preferences at one point in the decision space might not apply at the true optimum point, leading to termination at a sub-optimal solution. With this and other practicalities in mind, Zionts and Wallenius (1983) introduced some refinements to the basic procedures, and these are also briefly described in Figure 6.4. Since the refined procedure either moves to an improved solution at each step, or terminates with a basic solution which is locally optimal amongst adjacent basic solutions, the result must be the extreme point of the simplex which has the largest value of $V(\mathbf{f}(\mathbf{x}))$.

A more direct, but also rather more heuristic approach to the same problem of finding the "correct" weight vector $\mathbf{w}$ has been suggested by Steuer and co-workers (see for example Steuer, 1986, Chapters 13 and 14). Steuer proposes a number of variations to the basic idea, perhaps

198 MULTIPLE CRITERIA DECISION ANALYSIS

1. Let $\mathcal{W}$ represent the initial set of feasible weight vectors, i.e. $\mathcal{W} = \{\mathbf{w} \in \mathbb{R}^m \,|\, w_i \geq 0; \sum_{i=1}^m w_i = 1\}$, where $\mathbf{w}$ represents the m-dimensional vector of weights w_i. Select a $\mathbf{w} \in \mathcal{W}$ (e.g., $w_i = 1/m$ for all i) and maximize $\sum_{i=1}^m w_i f_i(\mathbf{x})$ over $\mathbf{x} \in \mathbb{X}$. Let $\mathbf{x}^0$ be the solution, and define $\mathbf{z}^0 = \mathbf{f}(\mathbf{x}^0)$.

2. Search for one or more basic solutions adjacent to $\mathbf{x}^0$, say $\mathbf{x}^r$ $(r = 1, 2, \dots)$, which satisfy the following conditions (where $\mathbf{z}^r = \mathbf{f}(\mathbf{x}^r)$):

   (a) For each r there exists a $\mathbf{w} \in \mathcal{W}$ such that $\sum_{i=1}^m w_i z_i^r > \sum_{i=1}^m w_i z_i^0$, which is easily confirmed by checking whether the LP maximizing $\sum_{i=1}^m [z_i^r - z_i^0] w_i$ subject to $\mathbf{w} \in \mathcal{W}$ gives a positive solution; and

   (b) There exists at least one r for which the decision maker prefers $\mathbf{z}^r$ to $\mathbf{z}^0$.

   If no solution can be found satisfying the above conditions, then STOP. Otherwise, let the decision maker provide as far as possible a preference ordering amongst the attribute vectors examined $(\mathbf{z}^0, \mathbf{z}^1, \mathbf{z}^2, \dots)$. At very least, the most preferred outcome should be identified, which we shall denote by $\mathbf{z}^+$.

3. Restrict the set $\mathcal{W}$ to those weight vectors satisfying $\sum_{i=1}^m [z_i^a - z_i^b] w_i > 0$ for each pair of attribute vectors for which the decision maker stated in the previous step that $\mathbf{z}^a \succ \mathbf{z}^b$. If $\mathcal{W}$ is now empty, then start deleting the oldest constraints of this form until $\mathcal{W}$ is no longer empty.

4. Select a $\mathbf{w} \in \mathcal{W}$. [Zionts and Wallenius suggest using the "middlemost" solution, viz. the weight vector which maximizes the minimum slack over all constraints.] For this $\mathbf{w}$, maximize $\sum_{i=1}^m w_i f_i(\mathbf{x})$ over $\mathbf{x} \in \mathbb{X}$; let the solution be $\mathbf{x}^*$, and define $\mathbf{z}^* = \mathbf{f}(\mathbf{x}^*)$.

5. If the decision maker prefers $\mathbf{z}^*$ to $\mathbf{z}^+$, then set $\mathbf{x}^0 = \mathbf{x}^*$ and $\mathbf{z}^0 = \mathbf{z}^*$; otherwise set $\mathbf{x}^0 = \mathbf{x}^+$ and $\mathbf{z}^0 = \mathbf{z}^+$. Include an additional constraint, as in step 3, consistent with the preference stated between $\mathbf{z}^*$ and $\mathbf{z}^+$, and return to step 2.

Some refinements. The above describes the essence of the algorithm as presented in Zionts and Wallenius (1976). The later (1983) paper included some further refinements as follows:

- If no distinctly different adjacent basic solutions are found in step 2 to be preferred by the decision maker to $\mathbf{z}^0$, then ask whether the decision maker finds any of the tradeoffs in the directions of these adjacent solutions to be desirable. If such a desirable tradeoff is found, then a similar constraint can be added to that of step 3, and the procedure continues.

- If step 2 yields neither a preferred adjacent solution nor a preferred tradeoff, then the search is extended to edges and adjacent solutions which are efficient but not optimal for any $\mathbf{w} \in \mathcal{W}$.

- If only preferred tradeoffs are identified (but no preferred adjacent solutions), then there will be no $\mathbf{z}^+$ solution. In this case, the comparison in the last step is made between $\mathbf{z}^0$ and $\mathbf{z}^*$; if $\mathbf{z}^0 \succ \mathbf{z}^*$, then the procedure terminates. The solution is, however, only locally optimal (the optimal vertex), but more preferred non-basic solutions exist. Strictly speaking, adjacent facets of the LP should be investigated to find a truly globally optimal solution.

Figure 6.4: Outline of the Zionts-Wallenius algorithm

Recall that the game reserve planning problem was formulated in terms of four maximizing objectives, defined by (3.4) on page 70, and the set of linear constraints given by (3.5). Starting with $w_1 = w_2 = w_3 = w_4 = 0.25$, the solution maximizing $\sum_{i=1}^{4} w_i z_i$ subject to the constraints (3.5) is $z_1^* = 4.8$; $z_2^* = 0$; $z_3^* = 120$; $z_4^* = 33$.

The identification of *all* adjacent basic solutions can become a complicated process. However, a useful set of adjacent solutions can easily be identified by examining the standard sensitivity analysis from the LP package being used. This provides the ranges of values for each w_i for which the current solution is optimal. Re-running the LP with each w_i in turn set fractionally outside of these ranges generates a number of adjacent solutions. Three distinctly different basic solutions may be found in this way, so that the decision maker would be presented with the four solutions displayed in the following table (in which each row represents a distinct solution, and the columns the corresponding values for each criterion).

r	z_1^r	z_2^r	z_3^r	z_4^r
0	4.80	0.00	120.00	33.00
1	10.56	0.00	43.20	68.20
2	0.60	1.20	120.00	33.00
3	4.80	0.00	57.60	68.20

To proceed further, we need the decision maker's preferences. For illustration purposes, let us suppose that the decision maker's preference ordering is given by: $\mathbf{z}^1 \succ \mathbf{z}^3 \succ \mathbf{z}^0 \succ \mathbf{z}^2$ (so that $\mathbf{z}^+ = \mathbf{z}^1$). Note that two of the adjacent solutions are preferred to $\mathbf{z}^0$. The preference for $\mathbf{z}^1$ over $\mathbf{z}^3$ implies the constraint $5.76w_1 - 14.4w_3 > 0$. For purposes of determining the "middlemost solution", it useful to standardize the constraint to unit norm, as we did in the previous section. This is achieved by dividing through by 15.51 (the square root of $5.76^2 + 14.4^2$) to give $0.371w_1 - 0.928w_3 > 0$. In the same way, the other two preferences generate $-0.871w_3 + 0.491w_4 > 0$ and $0.962w_1 - 0.275w_2 > 0$. The "middlemost solution" is then found by the LP maximizing Δ subject to the constraints:

$$
\begin{aligned}
w_1 + w_2 + w_3 + w_4 &= 1 \\
w_i - \Delta &> 0 \quad \text{For } i = 1, \ldots, 4 \\
0.371w_1 - 0.928w_3 - \Delta &> 0 \\
-0.871w_3 + 0.491w_4 - \Delta &> 0 \\
0.962w_1 - 0.275w_2 - \Delta &> 0
\end{aligned}
$$

The middlemost weights obtained in this way are $w_1 = 0.472$; $w_2 = 0.091$; $w_3 = 0.091$; $w_4 = 0.346$. Maximizing the weighted sum of the z_i's using these weights yields the solution $z_1^* = 13.2$; $z_2^* = 0$; $z_3^* = 8$; $z_4^* = 77$. Suppose that the decision maker states that this is not as good a solution as $\mathbf{z}^+$. Then $\mathbf{z}^+ = \mathbf{z}^1$ becomes the $\mathbf{z}^0$ for the next iteration.

Example Panel 6.9: Application of the Zionts-Wallenius procedure to the game reserve planning problem

200 MULTIPLE CRITERIA DECISION ANALYSIS

the easiest of which to describe being that which he terms the "interactive weighted-sums/filtering approach" (Steuer, Section 13.5). This process characterizes the range of feasible weight vectors in terms of simple lower and upper bounds on each w_i. These bounds are set initially to 0 and 1 respectively. Thereafter, in the light of specific choices made by the decision maker, the lower bounds are gradually increased and/or the upper bounds gradually decreased, until such time as all remaining feasible weight vectors generate the same solution to the MOLP. An outline of the algorithm is given in Figure 6.5.

---

1 Define the set of feasible weight vectors at any step of the procedure by $\mathcal{W} = \{\mathbf{w} \in \mathbb{R}^m \,|\, \ell_i \leq w_i \leq \mu_i; \sum_{i=1}^m w_i = 1\}$. Initialise the bounds to $\ell_i = 0$ and $\mu_i = 1$ for all criteria i; let $W = 1$ be the width of the interval for each i.

2 Randomly generate a specified number of vectors from $\mathcal{W}$, and then filter these to obtain a smaller number of widely dispersed vectors. For each vector generated, find the corresponding attribute vector maximizing $\sum_{i=1}^m w_i f_i(\mathbf{x})$ over $\mathbf{x} \in \mathbf{X}$. Filter these solutions again to generate a specified number of attribute vectors which are as widely dispersed as possible. Let these be $\mathbf{z}^1, \mathbf{z}^2, \ldots, \mathbf{z}^P$ say.

3 Let the decision maker select the most preferred of $\mathbf{z}^1, \mathbf{z}^2, \ldots, \mathbf{z}^P$. Denote by $\mathbf{z}^0$ the vector which is chosen, and let $\mathbf{w}^0$ be the weight vector for which $\mathbf{z}^0$ is optimal.

4 Replace W by rW, where $r < 1$ is a chosen reduction factor. Select new values for ℓ_i and μ_i such that $\mu_i - \ell_i = W$, and such that w_i^0 is positioned as close as possible to the centre of the interval while ensuring that $\ell_i \geq 0$ and $\mu_i \leq 1$.

5 Return to step 2.

Figure 6.5: Outline of the Steuer's interactive weighted sums and filtering algorithm

The basic philosophy, of course, is that the correct weights for generating the optimal solution are more likely to be in the vicinity of the weights generating the most preferred of a sample set of solutions, than in the vicinity of weights generating the less preferred solutions. The approach is essentially heuristic, as no guarantee of convergence to the optimal solution can be given, but has a degree of plausibility, is easy to explain to users, encourages systematic exploration of the decision space, and is generally quick and easy to apply.

Steuer and Choo (1983) applied the same general idea to the context in which solutions are generated not by maximizing a linear approximation to the value function, but by minimizing the distance from the ideal solution according to an augmented weighted Tchebycheff norm. Thus, if for each criterion i we define $z_i^{**} = \max_{\mathbf{z} \in \mathbf{Z}}\{z_i\}$, then the solution corresponding to a weight vector $\mathbf{w}$ is obtained by minimizing $D - \epsilon \sum_{i=1}^m z_i$

for some small $\epsilon > 0$, subject to:

$$D \geq w_i(z_i^{**} - z_i) \quad \text{for } i = 1, 2, \ldots, m.$$

This is closer in spirit to the methods based on aspiration levels which are described in Chapter 7 (Section 7.3).

The procedures of Zionts and Wallenius and of Steuer and co-workers described above systematically reduce the decision space *indirectly* by means of restrictions placed on the parameters of a linearized approximation to the value function in the region of the optimal solution. Comparisons between alternatives can also be used to place direct constraints on the decision space itself. This is particularly useful in the case of discrete alternatives where linear value functions can exclude efficient solutions (which are "convex dominated", i.e. dominated by a hypothetical alternative formed by the convex combination of the attribute values of two other alternatives). The basic ideas are relatively simple, making use of what are termed "convex cones", but are quite lengthy to describe, and the reader is referred to Korhonen et al. (1984) and Ramesh et al. (1989) for details.

6.4.3. A generalized interactive value function approach

The methods of Zionts and Wallenius and of Steuer are technically quite complicated to apply, requiring non-standard use of linear programming packages. They also have two potential shortcomings, namely:

- The use of linear weighted sums restricts solutions to basic solutions in the case of linear programming, even though there may exist non-basic solutions which are more preferred. If the methods are applied with appropriate modifications to discrete choice problems, the situation may be even worse as solutions which are convex dominated are excluded even though non-dominated and thus potentially optimal. The problems can be overcome by additional search strategies, but this complicates the algorithms further. In Stewart (1987) we have demonstrated that the use of piecewise linear value functions, as described in the previous sections of this chapter, can largely eliminate the problem in an easily implementable manner with as few as 3 or 4 piecewise segments for each criterion.

- The inequalities implied by preference orderings may be substantially less informative than simply obtaining tradeoffs in the vicinity of each solution examined. (See Stewart, 1993, for discussion.)

As indicated in Stewart (1993), the above observations suggest a somewhat simpler interactive procedure which can be implemented in a stan-

202 MULTIPLE CRITERIA DECISION ANALYSIS

dard linear programming package (for example, the optimizer supplied with spreadsheet packages). This procedure is based broadly upon the methods described in Section 6.2. Prior information available before start of the interactive phase of the algorithm, for example restrictions on functional shape (such as concavity of the partial value functions), or bounds on relative importance weights, can be incorporated as constraints such as those given by (6.6).

The interactive phase is based on ascertaining desirable tradeoffs between criteria in the vicinity of a particular solution to the MOLP. Suppose that the decision maker is presented with a feasible solution represented by the attribute vector $\mathbf{z}^r$, say. One of the attributes (which, with no loss in generality, we can denote as attribute 1) is selected as a reference. The DM is then asked to state for each of the other attributes in turn, the maximum amount (say δ_i^r for attribute i) that could be sacrificed in order to gain a specified increase (say δ_1^r) on the reference attribute. Such tradeoff information generates additional constraints on the value function model, as specified in the following box.

Using the definition of the c_{ij} terms from (6.4), the approximate indifference between the pair of attribute values $\{z_1^r, z_i^r\}$ and $\{z_1^r + \delta_1^r, z_i^r - \delta_i^r\}$ which is implied by the stated tradeoff generates the following inequalities on the model parameters:

$$\sum_{j=1}^{\nu_i}[c_{ij}(z_i^r - \delta_i^r) - c_{ij}(z_i^r)]d_{ij} + \sum_{j=1}^{\nu_1}[c_{1j}(z_1^r + \delta_1^r) - c_{1j}(z_1^r)]d_{1j} \approx 0. \quad (6.15)$$

As with our earlier formulations, the constraints (6.15) would be standardized to a consistent norm, for example such that the sum of squares of the coefficients sum to one.

At any iteration of the procedure to be described below, we seek to find a set of d_{ij} values which are maximally consistent with any prior information represented by constraints of the form (6.6), while approaching equality as closely as possible in each of (6.15). Following similar procedures to those of Section 6.2, we can achieve this by means of linear programming. A minimum slack, say Δ_1, is defined over all inequalities by constraints:

$$\Delta_1 < d_{ij} \quad (6.16)$$

for all i and j, and

$$\Delta_1 \leq \sum_{i=1}^{m}\sum_{j=1}^{\nu_i} \phi_{ijt}d_{ij} \quad (6.17)$$

Value Function Methods: Indirect and Interactive 203

for all prior inequality constraints t. A second set of deviations, related to the approximate equality constraints in (6.15), is then defined as in the following box, together with an associated maximum deviation Δ_2.

The approximate equalities in (6.15) implied by the tradeoffs are replaced by the following:

$$\sum_{j=1}^{\nu_i}[c_{ij}(z_i^r - \delta_i^r) - c_{ij}(z_i^r)]d_{ij} + \sum_{j=1}^{\nu_1}[c_{1j}(z_1^r + \delta_1^r) - c_{1j}(z_1^r)]d_{1j} +$$

$$\sigma_{ir}^+ - \sigma_{ir}^- = 0 \quad (6.18)$$

for all $i \neq 1$, and for each set of points $\mathbf{z}^r$ at which tradeoffs are assessed. Then a maximum deviation is defined by the constraints:

$$\Delta_2 - \sigma_{ir}^+ - \sigma_{ir}^- \geq 0 \quad (6.19)$$

for all i and r.

Clearly we wish to maximize Δ_1 (to obtain maximum slack on the inequalities), but to minimize Δ_2, with the latter being of greater priority. The estimates are thus found by solving the LP maximizing $\Delta_1 - M\Delta_2$ for some $M > 1$ (e.g. $M = 10$ appears to work well), subject to the constraints (6.16), (6.17), (6.18) and (6.19), and to a normalization such as:

$$\sum_{i=1}^{m}\sum_{j=1}^{\nu_i} d_{ij} = 100.$$

The resultant procedure is summarized in Figure 6.6

---

1 Select an arbitrary initial guess for the d_{ij} values, consistent with any assumed properties for the value function. Set $r = 1$.

2 Maximize the estimated value function defined by the current d_{ij} subject to $\mathbf{x} \in \mathbb{X}$. Denote the solution by $\mathbf{x}^r$, and define $\mathbf{z}^r = \mathbf{f}(\mathbf{x}^r)$. If (in the judgement of the DM) the results have appeared to stabilize, then STOP.

3 If $\mathbf{x}^r$ has not previously been seen by the DM, then set $\mathbf{z}^* = \mathbf{z}^r$. Otherwise identify a solution $\mathbf{x}^*$ which has not previously beeen seen and set $\mathbf{z}^* = \mathbf{f}(\mathbf{x}^*)$. [For discrete sets, $\mathbf{x}^*$ may be the currently highest ranking alternative not yet seen by the DM; otherwise randomly generate feasible sets of d_{ij} parameters until the corresponding optimal solution is distinctly different to any of $\mathbf{x}^1, \ldots, \mathbf{x}^r$.]

4 Obtain tradeoffs at the point $\mathbf{z}^*$ for each criterion relative to the reference criterion, and adjoin the corresponding equations (6.18) to the constraint set for the d_{ij}.

5 Solve the LP described above to obtain the most consistent parameter estimates. Set $r = r + 1$ and return to step 2.

Figure 6.6: Outline of the generalized interactive procedure

204   *MULTIPLE CRITERIA DECISION ANALYSIS*

Numerical results reported by Stewart (1993), which were based on discrete sets with quite large numbers (50-100) of alternatives, indicated that the above procedure tended to converge after about 6 or 7 iterations. At this stage, results were effectively identical to that obtained by *a priori* fitting of a value function in the standard manner. It is worth noting that in the case of multiple objective linear programming, the piecewise linear approximations generate many more basic solutions, reducing the need for detailed exploration of an optimal facet (in the manner proposed by Zionts and Wallenius, 1983).

Application of this generalized interactive procedure to the game reserve problem is described in Example Panel 6.10.

6.5. CONCLUDING COMMENTS

This concludes a fairly extensive discussion on value function models in MCDA. In the previous chapter we saw how a value function model can be constructed in close collaboration between analyst and decision maker, often in some form of group decision workshop. It was stressed that the process is constructive rather than prescriptive, with the emphasis on using the model to assist decision makers in understanding their own preferences and the impacts these have on the available choices.

The present chapter has discussed a number of ways in which similar value functions can be built up from imprecise and indirect information such as holistic comparisons or partial tradeoffs. The process does involve some technically quite sophisticated procedures aimed at identifying models which are most consistent with the available incomplete information. In general, the ensuing analysis will result in the identification of a number of potentially "optimal" solutions and/or in a number of coherent preference orders, rather than in a precise recommendation. This would tend not to completely resolve the original decision problem, so that a return to the direct analysis of Chapter 5 may well be necessary to reach a final decision. Nevertheless, the methods discussed in this chapter may be of use particularly in the following contexts:

- Where the analyst needs to conduct some form of preliminary analysis, possibly with a view to generating a shortlist of options for more detailed evaluation (perhaps using the approaches described in Chapter 5): Decision makers may be able to provide the analyst with partial preference information and/or some definite preferences between a subset of the alternatives;

- Where decision makers, or perhaps more particularly specific groups of stakeholders such as special interest groups, wish to conduct a relatively "quick-and-dirty" evaluation of options in order to clarify

Value Function Methods: Indirect and Interactive   205

For illustrative purposes, let us suppose that the decision maker's true preferences can be represented by the following additive value function:

$$-e^{-0.4167(z_1-4.8)} - e^{-0.1603z_2} - 0.3e^{0.04167z_3} - 0.3e^{-0.1119(z_4-33)}. \qquad (6.20)$$

The maximum of this function subject to the given constraints turns out to give the following values of the objectives: $z_1 = 13.30$; $z_2 = 16.14$; $z_3 = 26.06$; $z_4 = 46.73$.

In order to apply the generalized interactive methods, we use four piecewise segments for each criterion, and suppose that the partial value functions are all concave, so that $d_{ij} - d_{i,j+1} > 0$ for $j = 1, 2$ and 3. For convenience, we shall select the breakpoints so that the range from minimum (m_i) to maximum (M_i) values for each criterion i as shown in Table 3.2 is divided into four equally spaced intervals, of length $s_i = (M_i - m_i)/4$. For example, for the first criterion, the breakpoints will be 4.8, 7.8, 10.8, 13.8 and 16.8, with $s_1 = 3$. In order to obtain the optimal solution corresponding to a fixed set of d_{ij} parameters, we express each objective in the form $z_i = m_i + z_{i1} + z_{i2} + z_{i3} + z_{i4}$, where we define $z_{ij} \le s_i$ to be the contribution from the j-th interval to the total objective function value. Thus, for example, for the first criterion we would have $z_1 = 4.8 + z_{11} + z_{12} + z_{13} + z_{14}$, where $z_{1j} \le 3$ for $k = 1, \ldots, 4$.

The maximum of the piecewise linear approximation to the value function is obtained by maximizing $\sum_{i=1}^{4} \sum_{j=1}^{4} d_{ij} z_{ij}$ subject to the above constraints, the constraints defining the z_i as given by (3.4), and the physical constraints defined by (3.5). *Note* that we do need to ensure for $j = 2$, 3 and 4, that $z_{ij} > 0$ only if $z_{i,j-1}$ is already at is maximum value of s_i. Fortunately for concave value functions, as we are assuming here, this last condition is automatically satisfied as a result of the fact that $d_{ij} < d_{i,j-1}$. For other value function shapes, however, binary variables will need to be included into the LP formulation in order to force the correct order of entry of the z_{ij} into the solution.

Before any tradeoffs have been assessed, middlemost parameter values are given by $d_{i1} = 10$, $d_{i2} = 7.5$, $d_{i3} = 5$ and $d_{i4} = 2.5$ for each i. Maximizing the corresponding piecewise linear approximation to the value function, we obtain the solution: $z_1 = 13.8$; $z_2 = 15.6$; $z_3 = 11.43$; $z_4 = 55.33$. Approximate tradeoffs based on (6.20) at this point, corresponding to a gain of 1 unit in z_1, would be 0.6 for z_2, 1 for z_3, and 2.5 for z_4. The first of these tradeoffs states that a 1 unit gain in the 4th segment of attribute 1 (corresponding to 1/3 of the length of the interval) approximately compensates for an 0.6 unit loss in the 2nd segment of attribute 2 (corresponding to 0.6/7.8, or 0.077, of the length of the interval). The desired equality is therefore: $0.333d_{14} - 0.077d_{22} \approx 0$. Standardizing the coefficients to unit norm, this gives the first constraint in (6.18) as $0.975d_{14} - 0.226d_{22} + \sigma_{21}^+ - \sigma_{21}^- = 0$. In similar manner for the other two tradeoffs we obtain: $0.995d_{14} - 0.0995d_{31} + \sigma_{31}^+ - \sigma_{31}^- = 0$, and $0.829d_{14} - 0.558d_{42} + \sigma_{41}^+ - \sigma_{41}^- = 0$.

Example Panel 6.10: Application of the generalized interactive procedure to the game reserve planning problem

> The LP solution to find the most consistent parameter values then yields the following estimates:
>
> $$
> \begin{array}{llll}
> d_{11} = 7.466 & d_{12} = 5.917 & d_{13} = 4.369 & d_{14} = 2.821 \\
> d_{21} = 13.718 & d_{22} = 12.170 & d_{23} = 2.643 & d_{24} = 1.095 \\
> d_{31} = 28.208 & d_{32} = 4.191 & d_{33} = 2.643 & d_{34} = 1.095 \\
> d_{41} = 5.739 & d_{42} = 4.191 & d_{43} = 2.643 & d_{44} = 1.095
> \end{array}
> $$
>
> Maximizing the corresponding piecewise linear value function yields $z_1 = 13.11$; $z_2 = 15.6$; $z_3 = 30$; $z_4 = 45.65$, which is extremely close to the solution stated earlier to correspond to the maximum of (6.20) subject to the constraints.

Example Panel 6.10: Application of the generalized interactive procedure to the game reserve planning problem (Continued)

in their own minds where the primary focus of future evaluation needs to be;

- Where implications of the decisions are relatively minor, so that decision makers seek a satisfactory solution with less effort than needed for a full workshop: in this case, decision makers might well be willing to make a final holistic choice between the potentially optimal solutions with little further analysis.

The aims implied by the above contexts may often also be realised by the goal programming and outranking methods discussed in the next two chapters. We shall return to consideration of the relative roles of the different methodologies of MCDA in Chapter 11. Nevertheless, it is worth noting here that one of the advantages of value measurement approaches is the clear audit trail of value judgements which have led to the final conclusion. In this sense, preliminary evaluation using the models of this chapter may be helpful as a consistency check on more detailed value function models constructed at a later stage.

Summary of main notational conventions for chapter

Symbol	Meaning	Page
$c_{ij}(z_i)$	Function relating κ_{ij}^a to attribute values	170
d_{ij}	Increment from $u_{i,j-1}$ to u_{ij}	169
h_{ij}	Reciprocal of segment length used in computing function slope	172
T	Total number of constraints on model parameters	177
T_k	Number of constraints up to and including those specific to the parameters of the partial value model for criterion k	180
$u_i(a)$	Scaled partial value of alternative a for criterion i	163
$u_i(z_i)$	Scaled partial values for criterion i expressed as a function of z_i	166
u_{ij}	Scaled partial value for criterion i at breakpoint or category j	166
$v_i(a)$	Standardized partial value of alternative a for criterion i	163
$V(a)$	Total (aggregate) value of alternative a	163
w_i	Importance weight for criterion i	163
$\mathbf{w}$	m-dimensional vector of weights w_i	198
$\mathcal{W}$	Set of feasible weight vectors	198
$\mathbf{x}$	Vector of decision variables in a multiobjective mathematical programming models	195
z_i	Measurable attribute value relevant to criterion i	166
z_{ij}	Attribute value defining breakpoint j for criterion i	166
$\mathbf{z}$	m-dimensional vector of attributes z_i	195
Δ, Δ_a, Δ_1, Δ_2	Measures of maximum deviations from constraints	182
κ_{ij}^a	Coefficients relating $u_i(a)$ to the increments d_{ij}	170
λ_{ij}	Values of d_{ij} corresponding to the lower bound on feasible partial value functions	183
μ_{ij}	Values of d_{ij} corresponding to the upper bound on feasible partial value functions	183
ν_i	Number of piecewise segments or index of most preferred category for the partial value function for criterion i	166
σ_t, $\sigma_{(r,s)}$, σ_{ir}	Measures of deviations from specific constraints	180
τ_{ik}	Number of constraints specific to the parameters of the partial value model for criterion k	180
ϕ_{ijt}	Coefficient of d_{ij} in the t-th constraint on the model parameters	177

Chapter 7

GOAL AND REFERENCE POINT METHODS

7.1. INTRODUCTION

Goal programming and its variants represent in many ways the earliest attempts at providing formal quantitative decision aid for complex multiple criteria decision problems. The approach was described by Charnes and Cooper (1961), while the reviews of Romero (1986) and of Schniederjans (1995) cite references dating back to the 1950s. A number of textbooks have appeared over the years, for example Lee (1972) and Ignizio (1976; 1985). A comprehensive recent review may be found in Lee and Olson (1999).

The underlying principles may be linked to Simon's (1976) concept of satisficing, as described in Section 4.4, in the sense that emphasis is placed on achieving satisfactory levels of achievement on each criterion, with attention shifting to other criteria once this is achieved. Initially, goal programming was formulated within the context of linear programming problems, but the principles carry through to nonlinear and nonconvex (including discrete choice) problems. Ignizio (1983) describes the view of goal programming as simply an extension of linear programming (a view in effect implied by many texts in management science) as a "common misconception". More recently, other variations have emerged, including a variety of "interactive" methods, and the reference point methods as formulated by Wierzbicki (1980, 1999), which Korhonen and Laakso (1986) have termed "generalized goal programming". In this chapter, we review these methods in a unified framework, noting that all share the following features:

1 Each criterion needs to be associated with an attribute defined on a measurable (cardinal) scale; it is not usually meaningful to apply goal

210   *MULTIPLE CRITERIA DECISION ANALYSIS*

programming to situations in which one or more criteria are represented on constructed or categorical scales. Thus, as before, we shall denote by $z_1(a), z_2(a), \ldots, z_m(a)$ the attribute values corresponding to the m criteria for alternative a.

2 The DM is required to express value judgements in terms of "goals" or "aspiration levels" for each criterion, defined in terms of desirable levels of performance for the corresponding attribute values. These goals or aspiration levels will be denoted by g_i for $i = 1, \ldots, m$.

The first feature limits the application of the goal programming techniques and related methodologies to problems in which there exists a natural and meaningful association between criteria and measurable attribute values. We need, nevertheless, to note that even when the problem structure includes criteria which are only expressible qualitatively (for example, issues such as security of tenure in the land use planning case study of Section 3.5.3), the goal programming approach may often remain a valuable tool for purposes of background screening of alternatives, to generate a shortlist for a more detailed evaluation which includes the less quantifiable issues. In other words, goal programming may assist in the development of "policy scenarios", i.e. in the design phase of the analysis of complex problems, as described in Sections 2.2.3 and 3.5.3.

It is the second feature that requires particularly careful consideration. We need to recognize that the interpretations of the goals will differ from context to context, from criterion to criterion, and from decision maker to decision maker! Differences may arise in at least two ways:

Direction of preference: Three cases are often distinguished:

(*a*) The attribute is defined in a maximizing sense, with the goal representing a minimum level of performance which is deemed to be satisfactory;

(*b*) The attribute is defined in a minimizing sense, with the goal representing a maximum level of performance which is deemed to be satisfactory;

(*c*) The goal represents a desirable level of performance for the attribute which the DM wishes to achieve as closely as possible.

The first two cases are easily interpreted in satisficing terms, as the goals clearly represent a point of "satisfaction" beyond which attention will be turned to other criteria. The third case is rather more problematical from the point of view of the MCDA approach espoused in this book. The stating of a desirable "best" point along an axis indicates that the attribute is related to two underlying criteria which are in conflict with each other, similar to the situation discussed in Section 5.2.2 in which we noted that an attribute such as distance from the railway station when buying a house may represent concerns about both noise avoidance (the further the better) and reduction of travel time (the nearer the better). The statement of a specific goal in this context suggests that the DM has already determined a most desirable compromise between the two opposing criteria, contrary to our view that the purpose of MCDA is to learn and to explore without prior prejudice. We would recommend further problem structuring in an attempt to elucidate these underlying conflicting criteria related to the same attribute. Many examples illustrating case (c) suggest highly repetitive tasks, and we conjecture that it is only in such situations that intermediate goals should be accepted without question (in contrast to what appears to be implied in many standard presentations of goal programming, namely that case (c) is almost the standard case).

For the purposes of this chapter we shall thus restrict attention to cases (a) and (b) only. Insofar as case (c) may be justified, we can model this in a formal algebraic sense by using the attribute and its negative as two separate attributes. In fact, in order to describe the goal and reference point methodologies in this chapter, we can restrict attention to case (a), by supposing without loss of generality that all attributes are defined in a *maximizing* sense, so that goals represent a minimum acceptable level for the attribute in question.

Strength or urgency of aspiration: Even when the direction of preference is known, different DMs may express goals which may range from the highly unrealistic and optimistic (thus representing an ideal towards which to strive, but with little expectation of reaching it) to a non-negotiable bottom line that allows no further concession. The point along this spectrum of possibilities at which the stated goals in any specific context lie can be difficult to ascertain, and is strongly dependent upon the decision-making culture and/or the psychology of the DM at that time. Clearly, methods need to be robust to these uncertainties, but most variants of goal programming appear to perform best when the set of goal levels selected are moderately, but not excessively, optimistic (i.e. when there is no alternative satisfying all goals simultaneously, but when individual goals, and even subsets of goals are feasible).

212   MULTIPLE CRITERIA DECISION ANALYSIS

Generally, the application of these procedures will be implicitly or explicitly iterative, in the sense that initial goals are modified in the light of earlier results with the procedure. In this way, goal programming can be a valuable learning tool. Unfortunately, however, there seems to have been very little research on the means by which people adapt and modify goals, although it might be suspected that well-known psychological biases observed in related cognitive tasks such as probability assessment (for example, the anchoring and availability biases, Kahneman et al., 1982) would also play a role here. The anchoring bias, in particular, might result in users making smaller adjustments than would be desirable for adequate learning (and some evidence for this is presented in Stewart, 1988), and the analyst may need to intervene in order to force a greater degree of exploration.

A number of the procedures to be described in this chapter start by presenting to the decision maker a "payoff table" of the form illustrated for the game reserve planning problem by Table 3.2 on page 71. In general terms, the payoff table is constructed by maximizing each attribute in turn. Thus, for each criterion i, let a^i be the alternative (or feasible solution in a mathematical programming framework) which maximizes $z_i(a)$ over the decision space, and let $z_i^* = z_i(a^i)$ be the corresponding optimal performance measure for this criterion. The values $z_1^*, z_2^*, \ldots, z_m^*$ are then termed the *ideal values*. Now let $z_{ij} = z_i(a^j)$, i.e. the performance level for criterion i when criterion j is being optimized, and define:

$$z_{*i} = \min_{j=1}^{m} z_{ij}.$$

The values z_{*i} are often termed the *pessimistic values*, and are typically used to give some indication of likely lower bounds on goals for each criterion. This has to be interpreted with care, however. Weistroffer (1985) demonstrated that there may exist efficient (and thus potentially optimal) solutions in which some attributes take on values lower than the pessimistic values; ironically, however, the pessimistic values obtained as defined above can also turn out to be worse than the minimum values in the efficient set. Both problems are easily illustrated by means of a simple example. Suppose that we have seven alternatives $(a^1, \ldots, a^7)$, evaluated in terms of three attributes as given in the following table:

Attrib:	a^1	a^2	a^3	a^4	a^5	a^6	a^7
z_1	10	3	3	9	9	x	10
z_2	3	10	3	9	x	9	y
z_3	3	3	10	x	9	9	y

Two observations may then be made:

- If $y = 3$ (so that a^1 and a^7 are identical), the unique pessimistic values are $z_{*i} = 3$ for each attribute. On the other hand, the actions a^4, a^5 and a^6 are efficient for any value of x. Consequently, if $x < 3$, the minimum values amongst efficient solutions will be x, no matter how small this value is.

- If $y < 3$, then a^7 is dominated by a^1. Nevertheless, the maximization of z_1 may generate either a^1 or a^7; in the latter case, the pessimistic values for the second and third attributes would be defined by y, even though the smallest values amongst efficient solutions remains 3. This problem, however, may be avoided by replacing maximization of z_i by maximization of:

$$z_i + \epsilon \sum_{k \neq i} z_k$$

for a suitably small value of ϵ. This was in fact done when generating the payoffs in Table 3.2.

In spite of the above problems, however, the ideal and pessimistic values together provide a simple and often useful summary of the extent of the decision space, expressed in terms of levels of achievement of each criterion.

7.2. LINEAR GOAL PROGRAMMING

As indicated at the start of the chapter, goal programming developed initially in the 1950s and 1960s within the context of linear programming. Some of the earliest applications related to personnel planning, in which very large numbers of goals could be identified. Many standard textbooks on management science still tend to equate goal programming (and in some cases even MCDA generally) with the multiple objective linear programming implementation. Although, as we have discussed, the principles are more generally applicable, it is useful to start by examining this particular formulation, in which solution procedures are often quite elegant.

The standard multiple objective linear programming structure can be expressed formally as "maximizing":

$$z_i = \sum_{j=1}^{n} c_{ij} x_j \tag{7.1}$$

214   *MULTIPLE CRITERIA DECISION ANALYSIS*

for $i = 1, \ldots, m$, subject to:

$$\sum_{j=1}^{n} a_{rj} x_j \leq b_r \quad \text{for } r = 1, \ldots, p \tag{7.2}$$

and to the non-negativity constraints $x_j \geq 0$ for $j = 1, \ldots, n$. The x_j denote in the usual way the decision variables or activities, which need to be chosen subject to the stated constraints.

The performance levels for each criterion are thus assumed to be expressible as linear functions of the activity variables, defined by (7.1). According to our conventions, the functions are, for purposes of presenting and describing the methodology, assumed without loss of generality to be defined in a maximizing sense. The problem is thus to choose the activity variables so as to maximize each of the z_i as far as is possible subject to the constraints. Note that the game reserve planning problem structure as described in Section 3.5.2 is precisely of the above form. We shall thus be using this example to illustrate a number of the concepts in this chapter.

The constraint set defined by (7.2) is assumed to contain only true hard constraints (i.e. externally imposed freedom of action). Constraints imposed by management policy (such as required quality or service levels) should preferably be reformulated as "criteria" or goals. It may thus be arguable in the case of the game reserve problem whether the final constraint of (3.5) ($X_{31} + X_{32} \geq 3$, representing a lower bound on stock sizes for one species) should not have been re-cast as a criterion of maximizing stocks of this species (by introducing a $z_5 = X_{31} + X_{32}$, to be maximized with a goal, see next paragraph, of achieving at least $z_5 = 3$). This would be a valid question if the constraint simply represented management's view of what was good practice. On the other hand, if this constraint were set as a condition for funding by some external, e.g. governmental, agency then retention as a hard constraint would be justifiable.

Value judgements of the decision maker are incorporated into the model in the form of "goals" or "aspiration level", say g_i for criterion i. These quantitative goals are interpreted as a desire to find a feasible solution such that:

$$z_i \geq g_i \; \forall i.$$

For realistically defined goals, there is probably no feasible solution satisfying this requirement, and for this reason we introduce non-negative *deviational variables* δ_i which measure the degree to which the achieved performance measures fall short of the goals. These are defined implicitly by the introduction of constraints of the form:

$$z_i + \delta_i \geq g_i. \qquad (7.3)$$

Recall that, according to our convention, we define the z_i in a maximizing sense, so that if $z_i \geq g_i$ the decision maker's goal for this criterion is fully satisfied and $\delta_i = 0$ satisfies (7.3). On the other hand, $z_i < g_i$ implies that the decision maker's goal for this criterion has not been met, and correspondingly the minimum value for δ_i satisfying (7.3) is the underachievement $g_i - z_i$.

It should be noted that the more traditional formulation is to define deviation variables on both sides of the aspiration level, through the following equality constraint:

$$z_i - \delta_i^+ + \delta_i^- = g_i.$$

This formulation provides a direct link to the three cases for direction of preference as defined in the introduction to the chapter. Case (a) corresponds to minimization of δ_i^-; case (b) corresponds to minimization of δ_i^+; and case (c) corresponds minimizing both deviations. Using our conventions, however, there is no need to differentiate these three cases, while the simpler form given by (7.3) emphasizes links with our more general MCDA formulations, in the spirit of an integrated view of MCDA.

In a formal sense, we introduce the above concepts by extending the variables in the MOLP to include m "deviational" variables δ_i, indicating the degree to which is goal is under-achieved. The constraint set of the MOLP is then extended to include the following additional constraints:

$$\sum_{j=1}^{n} c_{ij}x_j + \delta_i \geq g_i \qquad (7.4)$$

The formulation of the game reserve planning problem structure in terms of deviational variables is shown in Example Panel 7.1.

The aim then is to find a feasible solution which minimizes the magnitude of the vector of deviational variables according to some appropriate norm. If a solution can be found for which $\delta_i = 0$ for all criteria, all goals are achievable, and this solution will be found using any reasonable norm. More generally, however, this will not be the case, and the choice of norm will have a strong influence on the solution obtained. The simplest norm is the weighted sum of the deviations, *viz.*:

$$\sum_{i=1}^{m} w_i \delta_i.$$

216 *MULTIPLE CRITERIA DECISION ANALYSIS*

In the case of the game reserve planning problem, suppose that the decision makers have stated the following goals for the four objectives defined in (3.4): $g_1 = 13$, $g_2 = 25$, $g_3 = 60$ and $g_4 = 55$. Comparing these with the values shown in the payoff table, we note that the first two goals are set towards the upper end of the ranges (between ideal and pessimistic values), while the second two are at the mid-points of their respective ranges.

The four objectives given in (3.4) would then be replace by additional constraints of the form:

$$
\begin{aligned}
2Y_{12} + 2Y_{13} + 3Y_{22} + 6Y_{23} + Y_{32} + 8Y_{33} + \delta_1 &\geq 13 \\
5Y_{12} + 3Y_{22} + 2Y_{32} + \delta_2 &\geq 25 \\
20Y_{11} + 15Y_{21} + \delta_3 &\geq 60 \\
3X_{11} + 15X_{21} + 6X_{31} + X_{12} + 5X_{22} + 2X_{32} + \delta_4 &\geq 55
\end{aligned}
\tag{7.5}
$$

Example Panel 7.1: Formulation of deviational variables in the game reserve planning problem

Use of this norm gives rise to what we shall term *Archimedean goal programming*. This is mathematically equivalent to the use of an additive value function, where the marginal value function is linear and strictly increasing for $z_i \leq g_i$, and constant thereafter. Many of the potential problems associated with the incautious use of value functions thus apply here just as well. The weights need again to be interpreted in a trade-off sense, while the "partial functions" are linear for all $z_i < g_i$. The adverse effects of over-linearization may be ameliorated here, since with well chosen goals, the ranges of values taken on by the δ_i will be relatively small. Nevertheless, the piecewise linear structure can still be prone to generating solutions in which $\delta_i = 0$ for most criteria, and relatively large for the remaining one or two (which will probably not be construed as a satisfactory compromise in many cases). This property demands caution in the interpretation of results.

The Archimedean goal programming formulation for the game reserve planning problem is set out in Example Panel 7.2, where it is also contrasted with more simplistic solutions to the multiple objective linear programming problem.

Much of the earlier work in goal programming, especially when concerned with large numbers of goals, emphasized the classification of goals into priority classes (where each class may consist of one or more goals). Following this prioritization, minimization of the weighted sum of deviations is restricted initially to goals in the first priority class only. Once this solution is obtained, deviations for goals in the second priority class are minimized subject to the additional constraint that the weighted sum

Suppose that the weights to be used for the Archimedean goal programming solution for the game reserve problem are assessed by a form of "swing-weighting" (Chapter 5), in the sense that the weights are expressed by $w_k = I_k/R_k$, where R_k denotes the range of values found for this objective in the payoff table 3.2, and I_k is a measure of the relative importance of the swing from best to worst values in the table. For illustrative purposes, let us suppose that $I_1 = 10$, $I_2 = 8$ and $I_3 = I_4 = 6$. This gives: $w_1 = 0.833$, $w_2 = 0.256$, $w_3 = 0.05$ and $w_4 = 0.134$.

Before proceeding to the goal programming solution, let us first examine what happens if we follow the naive route of simply maximizing $0.833z_1 + 0.256z_2 + 0.05z_3 + 0.134z_4$ subject to the stated constraints. The solution to this LP gives $z_1 = 16.8$, $z_2 = 30$, $z_3 = 0$ and $z_4 = 33$, i.e. the same solution obtained when simply maximizing z_1 on its own! Variations in the weights do not help much either. The LP range analysis shows, for example, that the solution does not change with increasing values for w_3 until this weight is increased beyond 0.1473. At this value for w_3, the solution jumps to $z_1 = 4.8$, $z_2 = 0$, $z_3 = 120$ and $z_4 = 33$, i.e. the same solution obtained when simply maximizing z_3 on its own! It is clear, therefore, that simply maximizing a weighted sum of objectives does not easily generate solutions which are balanced between the criteria.

Returning to the Archimedean goal programming approach in an attempt to find better compromise solutions (using the goal levels specified in Example Panel 7.1), we solve the LP minimizing $0.833\delta_1 + 0.256\delta_2 + 0.05\delta_3 + 0.134\delta_4$ subject to the constraint sets (3.5) and (7.5), and to non-negativity of the variables. This yields the solution giving $z_1 = 13$, $z_2 = 20$, $z_3 = 0$ and $z_4 = 55$. The solution still appears somewhat unbalanced, although not quite as badly as that from the naive LP solution with the same weights. More importantly, however, it now turns out that the solutions change much more gradually as (say) w_3 changes, as indicated in the following table:

w_3	Solution obtained:			
0.05	13.0	20.0	0.0	55.0
0.066	13.0	19.0	4.0	55.0
0.074	13.0	20.5	7.1	50.4
0.076	13.0	20.5	38.0	33.0

It is also useful to study sensitivity to changes in the goal levels specified. Increases in g_3 turns out to have no effect. In order to achieve improvements in the third goal by adjusting goal levels, it is necessary to relax the other goals. For example, reducing the other goals to $g_1 = 11$, $g_2 = 18$ and $g_4 = 50$ respectively, results in the solution in which these three goals are precisely achieved, with $z_3 = 18.9$.

Example Panel 7.2: Weighted sum and Archimedean goal programming solutions to the game reserve planning problem

218   *MULTIPLE CRITERIA DECISION ANALYSIS*

of deviations from goals in the first priority class should not exceed that obtained in the first step. The process is continued through each priority class in turn, as illustrated in Example Panel 7.3. We shall refer to this approach, especially for cases in which each priority class consists of a single criterion, as *preemptive goal programming*. In early days of linear programming, an important advantage of the preemptive approach was that the solution could be generated in a single run of a cleverly modified version of the standard simplex method for linear programming. With modern algorithms and computing hardware, however, there is no longer any real advantage in attempting to create special purpose software; the solution is easily obtained by repeated calls to a standard LP solver.

Within preemptive goal programming, deviations from goals within the same priority class are *compensatory* (i.e. a reduction in one deviation compensates for an increase in another), while deviations between classes are *non-compensatory*. The classification into priority classes is an important step if the associated goals are truly non-compensatory, but prioritization as a means of avoiding the difficult task of assessing weights (trade-offs) would be poor preference modelling practice and should be avoided. Given the highly subjective nature of the goal levels, it is difficult to visualize many situations in which deviations are not to some extent compensatory. Naturally, available tradeoffs can (and should) be explored by means of sensitivity analysis on the specified goals, but this may be a somewhat *ad hoc* procedure, in which the influence of simultaneous changes on a number of goals would be difficult to assess. Our view is thus that use of preemptive goal programming with its strictly prioritized criteria may not be the best modelling practice in many situations (contrary, perhaps, to what may be implied by the goal programming chapters in many standard texts on Management Science, which often still put emphasis on the specially modified simplex method).

Preemptive and Archimedean goal programming tend to be presented in many texts on management science as the only forms of goal programming. Evidently, however, other norms of the vector of deviations can be minimized. The most obvious of these is the Tchebycheff norm, which does emerge in related work such as "compromise programming" (e.g. Zeleny, 1982), reference point methods (see Section 7.3.2), and the interactive methods discussed by Steuer (1986, Section 10.9 and Chapters 14 and 15)). Using the Tchebycheff norm, the aim is to minimize the maximum weighted deviation, $\max_i\{w_i\delta_i\}$. In many ways, this is closer in spirit to the "satisficing" concept of Simon, which is often invoked as a motivation for goal programming: in seeking to minimize the maximum deviation, attention is at all times focused on the relatively worst per-

> In specifying the weights used earlier for the game reserve planning example, we assumed that the priority ordering of the objectives was: z_1 first; z_2 second; z_3 and z_4 jointly third. Use of preemptive goal programming would accordingly seek first (a) to minimize δ_1; then (b) to minimize δ_2, subject to δ_1 being maintained at its optimum value; and finally (c) to minimize $0.05\delta_3 + 0.134\delta_4$ (assuming the same relative weights as before within the same priority class), subject to δ_1 and δ_2 being maintained at their optimum values.
>
> It is easily found that $\delta_1 = 0$ and $\delta_2 = 0$ are simultaneously achievable. Minimizing $0.05\delta_3 + 0.134\delta_4$, subject to $\delta_1 = \delta_2 = 0$ yields $\delta_3 = 60$ and $\delta_4 = 9.89$, i.e. the solution given by: $z_1 = 13$, $z_2 = 25$, $z_3 = 0$ and $z_4 = 45.11$. Although the first two aspirations are fully met, the other two objectives (especially z_3) are well below their aspiration levels, and it is difficult to conceive of this as being a particularly satisfactory solution. Progressively relaxing the first two goals in an attempt to find more satisfactory solutions does yield further solutions as follows:
>
> $$z_1 = 12.0 \qquad z_2 = 22.0 \qquad z_3 = 0.0 \qquad z_4 = 51.7$$
> $$z_1 = 11.0 \qquad z_2 = 18.0 \qquad z_3 = 10.0 \qquad z_4 = 55.0$$

Example Panel 7.3: Preemptive goal programming solution to the game reserve planning problem

formance areas, and is shifted from one criterion to another as soon as the performance of the former becomes adequate relative to other goals.

Tchebycheff goal programming remains amenable to solution by standard linear programming. All that is required is to introduce a new variable, say Δ, to represent the maximum weighted deviation, and then to minimize Δ subject to the previous constraints and to $\Delta \geq w_i \delta_i$ for all criteria. In order to avoid the possibility of non-efficient solutions in certain degenerate cases, it is, however, advisable to amend the objective function to be minimization of $\Delta + \epsilon \sum_{i=1}^{m} w_i \delta_i$, where ϵ is a small positive constant (such as $0.01 \leq \epsilon \leq 0.05$). This approach is illustrated in Example Panel 7.4, in which is demonstrated the more balanced nature of the solution relative to the other goal programming methods. The solution is also robust to moderate changes in the specific values chosen for the goals and weights, which is useful given the somewhat imprecise nature of these inputs. Nevertheless, the solution is still clearly guided by such value judgements, and management may well want to re-assess some of these in the light of the answers. We shall comment further on this in the next section.

220 *MULTIPLE CRITERIA DECISION ANALYSIS*

The Tchebycheff goal programming solution (with $\epsilon = 0.02$) is obtained by maximizing $\Delta + 0.01667\delta_1 + 0.00512\delta_2 + 0.001\delta_3 + 0.00268\delta_4$ subject to (3.5) and (7.5), to non-negativity of the variables, and to:

$$\Delta - 0.833\delta_1 \geq 0$$
$$\Delta - 0.256\delta_2 \geq 0$$
$$\Delta - 0.05\delta_3 \geq 0$$
$$\Delta - 0.134\delta_4 \geq 0$$

The solution turns out to be $z_1 = 11.1$ (against a goal of 13), $z_2 = 18.8$ (goal of 25), $z_3 = 28.2$ (goal of 60) and $z_4 = 43.1$ (goal of 55). None of the goals are fully met, but the more important goals are evidently more closely met than the less important, and the solution is clearly more balanced in terms of goal achievement than any of the previous we have seen.

The solution also turns out to be highly robust to changes in the inputs. For example:

- A 50% increase in the weight on z_3 up to 0.075 leads to the solution: $z_1 = 10.8$, $z_2 = 17.8$, $z_3 = 35.4$ and $z_4 = 41.2$;

- Relaxation of the first two goals to 12 and 22 respectively leads (using the original weights) to the solution: $z_1 = 10.4$, $z_2 = 16.7$, $z_3 = 33.0$ and $z_4 = 44.9$.

Example Panel 7.4: Tchebycheff goal programming solution to the game reserve planning problem

7.3. GENERALIZED GOAL PROGRAMMING

7.3.1. Interactive Goal Programming

In the previous section we have examined the concept of goal programming within the context of multiple objective linear programming. In many cases, goal programming is used as a first step in the MCDM process, the purpose being to identify a relatively small shortlist of alternatives (from a large or even an infinite set) for more detailed evaluation. In such cases, linearization and continuity assumptions may not be unreasonable, giving rise to the wide use of linear goal programming. Evidently, however, the same general principles are applicable to other problem types (e.g. discrete problems, non-linear problems), as the constraints (7.3) defining the deviational variables δ_i for each criterion are not dependent upon any structure of the measured level of achievement z_i. Non-linear or linear integer programming problems may create technical difficulties in performing the optimization steps, but do not introduce any fundamentally new principles. For discrete problems, the algorithms are in fact mathematically quite elementary, as all that is required is to evaluate the relevant norm of the vector of deviations for each discrete alternative, and to sort these, which can often simply be done on a spreadsheet. The general principles of goal programming still remain the same.

In this section, we shall be examining the use of goal programming for more extensive and systematic analyses of decision maker preferences over the decision space. In view of the comments in the previous paragraph, we shall no longer be restricting discussion to the linear programming structure. We shall, in particular, make some references to discrete choice problems.

In applying goal programming concepts, an important practical problem (apart from that of specifying weights, which goal programming shares with other MCDM approaches) is in the selection of appropriate goals. As has previously been mentioned, the results of goal programming are strongly dependent upon choice of the goals, but the selection by decision makers of realistic goals requires a considerable level of understanding of the options and tradeoffs which are available. In highly repetitive problems, decision makers may be in a position to state realistic goals, but more usually the very purpose of undertaking formal multicriteria analysis is to allow decision makers to explore their preferences and these options and tradeoffs. Even the process whereby decision makers select goals has not been subject to the same level of psychological research as has been applied to the axioms of utility theory for example.

With this background, goal programming is almost inevitably applied in an interactive manner, in the sense that the results of goal programming for one set of goals will be presented to the decision maker(s) for evaluation and possible re-assessment of goals, before the algorithm is re-applied. The process is continued until decision makers are satisfied that they have explored the solution space sufficiently, and that the current solution represents an acceptable compromise between the conflicting goals. This can be done in the relatively informal or *ad hoc* manner, as illustrated in the examples of the previous section, or a formal interactive algorithm can be employed, giving rise to what may thus be termed *interactive goal programming* (see also Section 6.4 for further discussion on the concept of interactive methods in MCDA). An important practical problem which emerges, however, especially with the more informal approach, is the lack of a well-established behavioural theory as to how goals are formed and modified by decision makers, and this lack has an important bearing on the validation of goal programming methods. For this reason, it is useful to discuss the underlying problems in general terms before turning to the description of specific methodologies. To this end, let us suppose that after the setting of an initial set of goals a solution to the goal programming problem is presented to decision

222   MULTIPLE CRITERIA DECISION ANALYSIS

makers. The decision makers' value trade-offs at this point may be such that substantial sacrifices on one of the criteria may be deemed to be acceptable in order to obtain even quite modest gains in another. Under such circumstances, we conjecture (although it is difficult to secure empirical evidence for this) that the decision makers would adjust the goal for the latter criterion upwards, and that for the former downwards. If, however, the decision space is such that gains on the latter criterion are only possible at the expense of even much more extreme sacrifices on the other (a fact which may not easily be evident to DMs), then the result of the next application of the goal programming algorithm may either be no change to the solution at all, or a solution which the DM perceives to be decidedly worse than before. The procedure may well then be terminated at this point, even though better solutions may remain undiscovered. Some empirical evidence of such early termination at sub-optimal solutions is given in Stewart (1988), while simulations of this effect are reported in Stewart (1999b).

In spite of the above problems of validation, a number of interactive goal programming ideas have been developed, while anecdotal and some practical experience by one of the authors suggests that users often find the approach easy to understand and to use.

The first two approaches which we discuss are perhaps not generally considered to be goal programming, but do nevertheless share many of the features described above. These are Zeleny's (1982) concept of "compromise programming" linked to the theory of the "displaced ideal", and STEM (Benayoun et al., 1971). In both cases, the "goals" are not subjectively specified by the decision maker, but are simply the ideals relative to the currently specified decision space, i.e. $g_i = z_i^*$ as defined previously. The distinctive feature of these two approaches is that the interaction with the decision maker is not directly to modify the goals, but in effect to place constraints on the decision space, which in turn modifies the set of alternatives under consideration (and so also the ideals z_i^*).

The concept of compromise programming is simply to minimize norms of the form:

$$\left[\sum_{i=1}^{m} [w_i \delta_i]^p \right]^{\frac{1}{p}} \tag{7.6}$$

for various $p \geq 1$. With $p = 1$ this gives the Archimedean norm, while as $p \to \infty$, the above expression tends to the Tchebycheff norm:

$$\max_i [w_i \delta_i]$$

so that compromise programming can be seen to be a generalization of goal programming. The weights w_i in this case serve primarily to ensure a comparable scaling for all criteria (e.g. to normalize all deviations to the [0,1] interval), so that preference or value judgements are expressed entirely in terms of changes in aspiration or goal levels. By minimizing the above norm for a variety of values for the exponent p, a number of efficient solutions can be identified and presented to decision makers. In the light of the solutions generated, decision makers are encouraged to eliminate clearly undesirable options (either by eliminating specific alternatives in the discrete case, or by inserting lower bounds on achievement for certain criteria in continuous problems). This leads to a shift, or "displacement" of the ideal, after which the process is repeated with adjustment of the goals to the new ideals. The process ultimately terminates when the difference between the ideal and the compromise solutions are found to be acceptably small. Example Panel 7.5 illustrates this process in the context of the game reserve planning example used in the previous section.

The STEM (or Step Method) approach of Benayoun et al. (1971) was formulated in the multiple objective linear programming context, but can be generalized to other problems, and shares many features with the compromise programming approach. Once again the ideal is used as the goal for each criterion, and deviations are minimized in this case using the Tchebycheff norm only. Weights are again not subjectively selected, but are automatically generated for each criterion as the product of $(z_i^* - z_{*i})/z_i^*$ (representing the relative ranges of values available on each criterion) and a term which standardizes the objective functions (in the case of linear programming by standardizing the objective function coefficients for each criterion to unit Euclidean norm). Suppose that the resulting values for each criterion are given by $\widehat{z_i}$ for $i = 1, \ldots, m$. Decision makers are required to classify these values into those which are "satisfactory" and "unsatisfactory" respectively, in the sense that "unsatisfactory" values need to be improved, while some sacrifices on the "satisfactory" criteria can be tolerated. The decision makers are thus also required to specify, for each "satisfactory" criterion, an amount (say Δz_i) which they would be prepared to sacrifice on this criterion, in order to achieve gains in the "unsatisfactory" criteria. Decision alternatives are then constrained to satisfy:

$$z_i \geq \widehat{z_i} - \Delta z_i \quad \text{for "satisfactory" criteria}$$

and

$$z_i \geq \widehat{z_i} \quad \text{for "unsatisfactory" criteria.}$$

224 *MULTIPLE CRITERIA DECISION ANALYSIS*

We continue with the game reserve planning problem from Section 3.5.2, which was used to illustrate linear goal programming in the previous section. Recall that the payoff table was given by Table 3.2 as follows:

Objective being maximized	Values obtained for:			
	z_1	z_2	z_3	z_4
z_1	16.80	30.00	0	33.00
z_2	12.60	31.20	0	33.00
z_3	4.80	0	120.00	33.00
z_4	4.80	0	22.50	77.67

The ideals z_i^* are given by the diagonal entries in the payoff table. The weights required to standardize the deviations for each criterion in compromise programming are given by the reciprocals of the value ranges in each column of the payoff table, i.e. $w_1 = 1/12$, $w_2 = 1/31.2$, etc. Minimization of (7.6) for $p = 1$ or ∞ corresponds to Archimedean and Tchebycheff goal programming respectively, using the ideals as goals. The case of $p = 2$ follows essentially the same procedure as for Archimedean goal programming, but with a quadratic objective (which is an option provided in most LP solvers, including spreadsheet optimizers). The results turn out to be as follows:

p	z_1	z_2	z_3	z_4
1	14.9	10.8	0	68.2
2	13.6	13.0	20.4	54.9
∞	9.3	11.6	44.5	49.6

Suppose now that in the light of these results, together with the original payoff table, the user specifies minimally acceptable levels of 10, 12, 15 and 45 respectively for the four criteria, i.e. the decision space is restricted to solutions for which $z_1 \geq 10$, $z_2 \geq 12$, $z_3 \geq 15$ and $z_4 \geq 45$. The elimination of solutions violating these restrictions leads to a shift in the ideals to new values of $z_1^* = 14.65$, $z_2^* = 21.84$, $z_3^* = 50.06$ and $z_4^* = 64.78$. Repeating the compromise programming solution with these new ideals as goals, and with weights defined by the new ranges, yields the following solutions:

p	z_1	z_2	z_3	z_4
1	14.6	19.7	15.0	45.0
2	13.3	15.6	22.5	49.7
∞	11.5	15.1	26.1	51.2

Clearly there is already quite a strong convergence, and the decision maker may well be prepared to stop at this point, possibly with the second of the three solutions.

Example Panel 7.5: Compromise programming applied to the game reserve planning problem

This eliminates some alternatives, and leads to a shift in the ideal, after which the process is repeated, but with zero weight placed on the "satisfactory" criteria (in the sense that as long as $z_i \geq \hat{z}_i - \Delta z_i$, no further improvements are sought at the expense of the "unsatisfactory" criteria). At each iteration, either no further criteria are deemed to be satisfactory (in which case the process immediately stops), or at least one further criterion achieves satisfactory status. It thus follows that the process must terminate in at most m iterations, with either a solution in which all criteria are "satisfactory", or a definite conclusion that no such solution exists (in which case attempts are presumably needed to create or discover new alternatives). The application of STEM to the game reserve planning problem is illustrated in Example Panel 7.6.

Applying the Tchebycheff goal programming procedure, using the ideals as goals and the weights as defined in the description of the STEM method, we solve the resulting linear programming problem to obtain the initial STEM solution which is as follows: $\hat{z}_1 = 6.6$; $\hat{z}_2 = 16.3$; $\hat{z}_3 = 59.6$; and $\hat{z}_4 = 55.2$. Suppose now that the decision maker declares the second and third objectives to be "satisfactory", and specifies lower bounds of 15 and 25, say, on z_2 and z_3 respectively (i.e. $\Delta z_2 = 1.3$ and $\Delta z_3 = 34.6$). With the addition of these restrictions, the ideal and pessimistic values for the remaining two criteria can be calculated to be as follows:

	z_1	z_4
Ideal	14.30	55.15
Pessimistic	6.64	33.00

The STEM solution based on these ideals, and with zero weight on the second and third objectives, turns out to be: $\hat{z}_1 = 12.3$; $\hat{z}_2 = 15.0$; $\hat{z}_3 = 25.0$; and $\hat{z}_4 = 50.9$. Judging from our previous solutions with goal programming, this might well be considered satisfactory. It is worth noting, however, that our illustrative value of $\Delta z_3 = 34.6$ at the first step represents a quite substantial reduction from the initial solution. Anchoring biases might perhaps make such a concession unlikely, with the result that the value obtained in the second step may be found much less satisfactory. For example, if the satisfactory bound for z_3 had been set at 40 ($\Delta z_3 = 19.6$) then the resulting values at the next step for z_1 and z_4 would have become 10.3 and 44.6 respectively. In using STEM, therefore, it is probably advisable to encourage the decision maker to backtrack from time to time, to re-assess some of the earlier bounds placed on "satisfactory" values.

Example Panel 7.6: Application of STEM to the game reserve planning problem

The key to the success of both STEM and compromise programming is the ability of decision makers either to specify what constitutes a satisfactory level of performance or to identify alternatives which can be

226   *MULTIPLE CRITERIA DECISION ANALYSIS*

eliminated. As previously discussed, the process by which this is done is not well-understood, but requires at least a substantial degree of global understanding of the available tradeoffs, and there must always be some question of the extent to which this is true. In problem settings which are relatively familiar to the decision maker (for example, in the selection of investment portfolios, which is likely to be a repetitive task) the choices may well be justifiable, and it is in such contexts that methods such as STEM and compromise programming may well offer an efficient means of decision support.

An approach which has been termed interactive multiple goal programming (IMGP), although it does not fit precisely into our general definition of goal programming, was introduced by Spronk (1981). As with compromise programming and STEM, IMGP is also based on a pruning of the decision space, and it is interesting to note that the primary applications of IMGP appear to have been in capital budgeting and financial planning, i.e. the type of familiar decision context to which we have referred. In essence, the IMGP approach is based on presenting the decision maker at each stage of the iterative process with what is termed by Spronk a "potency matrix", which consists simply of the vectors of ideal and pessimistic values, based on the current decision space. In this case the pessimistic values represent effectively a set of guaranteed lower bounds on performance for each criterion. At each iteration, the decision maker is asked which criterion should first be improved, and a lower bound greater than its current pessimistic value is tentatively imposed, and a new set of ideals calculated. If the decision maker is satisfied that any consequent losses in the ideal values are worth the gain in the guaranteed performance bound for the chosen criterion, then the tentative lower bound is made permanent and the process repeats. If the decision maker is not satisfied, then provision is made for backtracking (i.e. relaxing the tentative lower bound until the decision maker is satisfied with the tradeoffs).

IMGP thus, to a greater extent than the previous two methods, does seek to assist the decision maker to explore the decision space, by providing direct feedback on the effects of increasing aspirations on one criterion on performance levels for the other criteria. This feedback is very approximate however, as the loss in ideal may be substantially different to the actual (local) tradeoffs which are available. Nevertheless, in some practical experience reported in Stewart (1988), IMGP appeared to generate more satisfying results than other interactive goal programming methods, even though convergence was somewhat slower.

A more systematic form of interactive goal programming, and one which is closer to the spirit of the original goal programming, is the

Interactive Sequential Goal Programming (ISGP) proposed by Masud and Hwang (1981). Their approach was formulated in the context of a multiobjective mathematical programming framework, but the principles seem equally appropriate to discrete choice problems. The core of the approach is essentially based on an Archimedean goal programming formulation (although there appears to be no reason why this should not be replaced by the Tchebycheff formulation), but contains certain interesting variations:

- As in many of the other methods discussed above, weights are not subjectively assessed to reflect relative importances, but merely ensure comparable scaling. In this case, the weights have the effect of re-scaling the criterion values so that the difference between g_i and z_i^* is the same for all criteria.

- Apart from the standard goal programming solution, the method also generates a further m solutions by solving a sequence of additional goal programming problems, in each of which the goal for one of the criteria is replaced by a hard constraint (i.e. $z_i \geq g_i$, with no deviations allowed). This set of solutions is meant to inform the decision maker when re-assessing the goals (see below).

- Non-dominated solutions are avoided by including maximization of over-achievement of goals as a second-order objective in a preemptive goal programming sense. (We shall see later, however, that with the use of the Tchebycheff norm it is possible to achieve the same end without recourse to the preemptive approach.)

At each iteration of the ISGP process, the decision maker is presented with the $m + 1$ solutions and the payoff table. If one of the $m + 1$ solutions is satisfactory, the process terminates; otherwise the decision maker is asked to revise the goals in the light of the solutions presented. The idea is that the multiple solutions which are presented inform the decision maker concerning available trade-offs, thus contributing to more realistic goal specifications. The value of the procedure nevertheless still depends fundamentally on the ability of the decision maker to specify meaningful goals. Application of ISGP to the game reserve planning problem is illustrated in Example Panel 7.7.

7.3.2. The Reference Point Approach

The fundamental goal programming paradigm discussed up to now has been based in effect on minimization of underachievement of (maximizing) goals. A problem can arise if the goals are unduly modest, to the extent that feasible solutions exist satisfying all goals simultaneously. If

228 MULTIPLE CRITERIA DECISION ANALYSIS

Let us start with the same goals as in the previous section, namely $g_1 = 13$, $g_2 = 25$, $g_3 = 60$ and $g_4 = 55$. The standard Archimedean goal programming solution using weights inversely proportional to $z_i^* - g_i$ (to ensure the constant scaling of the range) yields a solution in which the first two goals are satisfied (see table below). Thus only two further solutions need to be generated, namely those including the constraints $z_3 \geq 60$ and $z_4 \geq 55$ respectively. The three solutions presented to the decision maker are thus:

z_1	z_2	z_3	z_4
13.0	25.0	0.0	45.1
10.8	15.0	60.0	33.0
13.0	20.0	0.0	55.0

The serious conflict between achievement of the third goal and achievement of the other goals is evident. Suppose that the decision maker accordingly reduces g_3 to 40, with smaller adjustments to the second and fourth goals as follows: $g_2 = 20$ and $g_4 = 50$. Two distinct solutions are now generated by ISGP as follows:

z_1	z_2	z_3	z_4
13.0	20.0	9.5	50.0
12.8	20.0	40.0	33.0

Values for the first two objectives have stabilized, and the main remaining conflict is between the last two. Further reductions in g_3 would allow us to explore the available tradeoffs between z_3 and z_4 in more detail. For example, $g_3 = 30$ generates the further solution given by: $z_1 = 13.0$, $z_2 = 20.0$, $z_3 = 30.0$ and $z_4 = 38.5$.

Example Panel 7.7: Application of ISGP to the game reserve planning problem

the goals are truly levels of universal and objective satisfaction, then this may be a pleasing result, but more typically the decision maker would not be satisfied with this outcome, especially since the solution generated by the algorithm may then be dominated. We have noted that the ISGP procedure does recognize this problem, by including maximization of over-achievements as a low priority objective (to be sought only after minimization of under-achievement has been completed). The *reference point* approach introduced by Wierzbicki (1980, 1999) addresses the problem more directly. In this approach, the term "goal" is replaced by the neutral term "reference" level, indicating levels of achievement currently viewed as a good starting point for further exploration of the decision space. Wierzbicki introduces the concept of a "scalarizing function", which may be viewed as a surrogate value function to be applied in the vicinity of the reference point. Maximization of the scalarizing function then produces an efficient solution which is in a sense closest to the reference point. In the spirit of interactive goal programming, the decision maker is required to judge whether the solution found is satisfactory (in which case the process terminates), and if not to revise the reference values.

Although Wierzbicki discussed a number of different types of scalarizing function, the most commonly applied form is closely allied to the Tchebycheff norm for goal programming, in the sense that its maximization is equivalent to *minimization* of:

$$\max_{i=1}^{m} w_i \delta_i \ + \ \epsilon \sum_{i=1}^{m} w_i \delta_i. \tag{7.7}$$

The main difference to the goal programming formulation is that in the reference point method, the deviational variables δ_i are permitted to be negative, in which case $-\delta_i$ becomes a measure of over-achievement. Clearly the min-max term dominates in (7.7), so that while any δ_i remains strictly positive, we have effectively the Tchebycheff goal programming formulation. Only once all reference levels have been exceeded does the process continue to maximization of overachievement. The summation term is weighted by the small constant ϵ, and serves primarily to ensure that solutions are non-dominated in cases when the min-max solution is not unique.

Some of the concepts of the reference point approach were adapted by Korhonen and Laakso (1986) in a multiple objective linear programming context, and by Korhonen (1988) for the discrete choice problem, to provide a visual interactive graphical procedure for applying goal or aspiration level procedures. Starting from a particular efficient solution, say $\hat{z}_1, \hat{z}_2, \ldots, \hat{z}_m$ (which we shall denote in vector notation by $\hat{\mathbf{z}}$), a "reference direction" vector $\mathbf{d}$, rather than a single reference point, is chosen by the decision maker. A sequence of reference points of the form $\hat{\mathbf{z}} + \theta\mathbf{d}$ is then generated (by choosing a sequence of positive values for θ), and projected on to the efficient frontier by maximizing Wierzbicki's scalarizing function (i.e. minimizing an expression of the form (7.7) for each point). This generates a path along the efficient frontier which the decision maker can examine to find a best point. A new reference direction can be chosen at this point from which the process can restart. Korhonen and Laakso suggest a visualization of the process in terms of a "Pareto race", i.e. driving along the Pareto frontier, which has been incorporated into a software package called VIG (or VIMDA for discrete choice problems). The current solution at any point is represented by bar

230   *MULTIPLE CRITERIA DECISION ANALYSIS*

graphs representing levels of achievement for each attribute. By pressing on a "accelerator" or "brake" keys the user can see how these levels change as one moves along the current direction, and can reverse direction by changing "gears". At any stage, the user can change direction by requiring greater emphasis on a specified attribute.

Another practical implementation of the reference point approach in an interactive framework, with associated software, is AIM (Aspiration-level Interactive Model) as described in Lotfi et al. (1992). This implementation was originally designed for discrete choice problems, but a generalization to multiple objective linear programming problems has also been proposed by Lotfi et al. (1997). The initial reference level is generated as a median value for each criterion, and information is provided to the decision maker at each stage as to the feasibility of the current reference level (expressed in terms of proportions of the decision space satisfying the references levels singly and jointly), and the relative sizes of the increments which are available for each criterion. Overall, however, the problem remains in any implementation, that if the decision maker indicates shifts in reference level which are strongly at odds with available tradeoffs, then the solution generated at the next iteration may be perceived to be worse than before and the process may be terminated prematurely. The idea from ISGP, of generating a $m + 1$ feasible solution for comparison, could be incorporated here with some benefit.

7.4.   CONCLUDING COMMENTS

Goal programming and reference point methods are particularly well-suited to application in problems with very large or infinite numbers of decision alternatives (e.g. problems with a mathematical programming structure), either to produce a solution to the MCDM problem directly, or as a first stage in a decision making process to produce a shortlist of options for more detailed evaluation by other methods. They can be viewed as algorithmic implementations of a natural human decision-making heuristic, namely that of "satisficing".

It is worth commenting at this stage that these methods do extend quite easily to treatment of relatively large numbers of criteria as well, as it is only necessary for the decision maker to state desirable aspiration levels for each criterion, without the need for evaluations of trade-offs, strengths of preference, etc. This facilitates, for example, the formal incorporation of the treatment of uncertainty through use of scenarios (cf. Section 3.4.4). The set of criteria can be extended to include levels of achievement for each of the natural criteria under each scenario. For example, instead of setting one goal for a criterion such as percentage growth in a porfolio investment problem, one could set goals for such growth under each of three (or more) economic scenarios. Such goals may possibly (but need not necessarily) differ for each scenario. It appears to be relatively easy to set such differentiated goals, while the allocation of weights to the extended set of criteria could be based on the product of the weights associated with the original criteria and the relative weights to be placed on the scenarios (in the sense of the importance to be attached to seeking satisfactory performance under each scenario). Use of goal programming assists in genenerating robust solutions which perform satisfactorily on all criteria across all scenarios.

Goal programming and reference point methods may thus be expected to be an intuitive and comfortable approach to use for multi-objective decision problems of a repetitive nature, or for those which in some other sense are sufficiently familiar to the decision maker so that there is a clear perception of what is "satisfying" performance. In problems of strategic decision making, where very unfamiliar options and consequences are being explored, it may be much more difficult to specify goals in such a way that the resulting solutions are truly the most satisfying option according to the decision maker's preference structure. In such cases, the use of goal or reference point methods may be effective primarily at the level of preliminary screening, to generate a shortlist of alternatives (perhaps by using a number of different sets of goal levels). This preliminary screening mode may be particularly relevant when an initial set of options has to be generated by a staff group, or a technical committee, to be presented for consideration by a higher level decision making group (as, for example, in the case of the landuse planning problem described in Sections 2.2.3 and 3.5.3). One of the advantages of using aspiration-level methods in this context is the evidence presented by Stewart (1999b), to the effect that results are highly robust to violation of preferential independence conditions, so that analysis can be undertaken at an early stage of the structuring process.

The various approaches to interactive goal programming and related procedures offer in principle a user-friendly means for decision makers to explore the decision space systematically, expressing preferences primarily in terms of goals or reference (aspiration) levels. This provides a valuable tool for individual decision makers, especially in repetitive problem contexts, but may be less useful for more strategic problems involving many interest groups, as the sequence of changing reference levels may be difficult to document and to justify, while the final result will depend very much on these choices. A further potential problem with interactive procedures is that choice of goals which are substantially inconsistent with available tradeoffs between criteria may lead to

232 MULTIPLE CRITERIA DECISION ANALYSIS

early termination, leaving possibly more desirable options unexplored. This is an area deserving of further research, but some work (Stewart, 1999b) has suggested that automatic re-adjustment of goals towards directions of known improvement may be a useful decision support aid.

Goal and reference point methods require that the performance measures, z_i, be available in quantitative form. In practice, this means that the criteria need to be representable in terms of quantitative attributes, which limits these methods to problems which can naturally be formulated in this way, and largely excludes problems which include subjectively assessed criteria. (In principle, one could develop value functions to relate performance in such criteria to measurable attributes, but it may be difficult to set meaningful goals in terms of these values, and in any case, having gone this far, there seems no reason not to complete the analysis in value theory terms.) Once again, this creates problems in applying goal programming or reference point methods to strategic choice problems, but may well not be an issue when using them as a tool for preliminary screening. In fact, the requirements placed on staff groups or technical committees may well be that attention be restricted to the "objective" features of the problem.

A substantial area of uncertainty in applying goal- or aspiration-level methods relates to how people form and adjust goals. This remains an important area for future research, as the solutions generated by goal programming or reference point methods can be sensitive to unrealistic choices of goals. It is perhaps not so much of a serious concern when these approaches are used in the preliminary screening mode (in which a variety of aspiration levels will be selected as a means of generating a short list of alternatives), but must be seen as a source of practical concern, in which considerable caution needs to be exercised, when used as a tool for developing a final decision choice. In particular, for interactive goal programming, the uncertainty regarding the process of goal adjustment creates a corresponding uncertainty regarding how and where the process will terminate, leading to a potential for termination at sub-optimal solutions without the decision space having been fully explored.

Chapter 8

OUTRANKING METHODS

8.1. INTRODUCTION

As discussed in Chapter 4, on Preference Modelling, the outranking approaches differ from the value function approaches in that there is no underlying aggregative value function. The output of an analysis is not a value for each alternative, but an outranking relation on the set of alternatives. An alternative a is said to outrank another alternative b if, taking account of all available information regarding the problem and the decision maker's preferences, there is a strong enough argument to support a conclusion that a is at least as good as b and no strong argument to the contrary. The way in which the outranking relation is exploited by a method depends on the particular problematique (choice, sorting, or ranking; see Chapter 2).

Much of the literature on outranking methods has been in French, but some useful English language references are those of Roy (1996) and Vincke (1992, 1999). The two most prominent outranking approaches, the ELECTRE family of methods, developed by Roy and associates at LAMSADE (Laboratoire d'Analyse et Modélisation de Systèmes pour l'Aide à la Décision), University of Paris Dauphine, and PROMETHEE, proposed by Brans from the Free University of Brussels, are presented in this chapter. Roy, who must be credited for the initial and much subsequent work on outranking methods, was critical of the utility function and value function methods on the grounds that they require all options to be comparable. He developed the ELECTRE methods which he describes as providing weaker, poorer models than a value function, built with less effort, and fewer hypotheses, but not always allowing a conclusion to be drawn. Brans and Vincke give an appealing description

234   *MULTIPLE CRITERIA DECISION ANALYSIS*

of outranking methods as providing an enrichment of the dominance relation which is not excessive, as with a utility function, but realistic (Brans and Vincke, 1985; Vincke 1992).

The family of ELECTRE methods differ according to the degree of complexity, or richness of the information required or according to the nature of the underlying problem or problematique. We begin with a description of ELECTRE I, the earliest and simplest of the outranking approaches (first published in 1968), which provides a good basis for understanding the underlying concepts. Thereafter, we discuss in some detail the extensions termed ELECTRE II and ELECTRE III, before briefly mentioning more recent developments. Finally, we complete this chapter with an outline of the PROMETHEE approach.

Outranking methods focus on pairwise comparisons of alternatives, and are thus generally applied to discrete choice problems, such as the Business Location case study described in Sections 2.2.1 and 3.5.1, which we shall use to illustrate the outranking methods. The starting point for most outranking methods is a decision matrix describing the performance of the alternatives to be evaluated with respect to identified criteria. Such a decision matrix was given for the Business Location case study by Table 5.4 in Chapter 5. The performance measures shown in Table 5.4 were, however, developed in the context of applying value function concepts. Outranking methods often make use of less precise inputs, so that for purposes of illustrating these methods we shall introduce some variations to the assessments shown in Table 5.4.

8.2.   ELECTRE I

In order to describe the ELECTRE I method, we shall make use of the decision matrix for the Business Location case illustrated in Table 8.1 (using the seven cities and six criteria previously identified). The hierarchical structuring of Table 5.4 is not utilized by the ELECTRE methods, which tend to be based on a concise set of criteria, typically around 6-10. Note that all evaluations have been translated to a subjectively defined 5-point ordinal scale: Very Low (VL), Low (L), Average (Av), High (H), Very High (VH). A higher rating indicates a higher preference. The criteria have been allocated weights such that a higher value indicates a greater "importance". Unlike the weights used in value functions, however, these do not represent trade-offs. Their psychological interpretation is, in fact, not well-defined, although they have been interpreted as a form of "voting power" allocated to each criterion.

	Avail. of staff	Access. from US	Qual. of life	Business Potential		Ease of set up and operations
				Public sector	Private sector	
Weights	6	4	3	10	8	6
Paris	Av	H	H	H	VH	Av
Brussels	L	Av	Av	VH	Av	Av
Amsterdam	L	VH	Av	H	VH	H
Berlin	H	L	L	Av	H	H
Warsaw	H	L	L	Av	Av	L
Milan	H	L	VH	Av	H	H
London	Av	VH	H	Av	H	VH

Table 8.1. Decision matrix for the business location problem

Indices of concordance and discordance

The ELECTRE methods are based on the evaluation of two indices, namely the *concordance* index and the *discordance* index, defined for each pair of options a and b according to the principles discussed in Chapter 4. The concordance index, $C(a, b)$, measures the strength of support in the information given, for the hypothesis that a is at least as good as b. The discordance index, $D(a, b)$, measures the strength of evidence against this hypothesis.

There are no unique measures of concordance and discordance, and a number have been used. The concordance index used in ELECTRE I is defined by:

$$C(a, b) = \frac{\sum_{i \in Q(a,b)} w_j}{\sum_{i=1}^{m} w_j} \qquad (8.1)$$

where $Q(a, b)$ is the set of criteria for which a is equal or preferred to (i.e. at least as good as) b.

That is, the concordance index is the proportion of criteria weights allocated to those criteria for which a is equal or preferred to b. The index takes on values between 0 and 1, such that higher values indicate stronger evidence in support of the claim that a is preferred to b. A value of 1 indicates that a performs at least as well as b on all criteria (so that a dominates or is equivalent to b). The concordance indices for all pairs of options in the business location case study, based on the weights and assessments in Table 8.1, are illustrated in Example Panel 8.1.

236   MULTIPLE CRITERIA DECISION ANALYSIS

The discordance index initially suggested for ELECTRE I was given by:

$$D(a, b) = \frac{\max_{i \in R(a,b)}[w_i(z_i(b) - z_i(a))]}{\max_{i=1}^{m} \max_{c,d \in A}[w_i|z_i(c) - z_i(d)|]} \qquad (8.2)$$

where $R(a, b)$ is the set of criteria for which b is strictly preferred to a and A is the set of all alternatives.

The above discordance index for a compared to b is the maximum weighted value by which b is better than a, expressed as a proportion of the maximum weighted difference between any two alternatives on any criterion. This also takes on values between 0 and 1, with a high value indicating that on at least one criterion b performs substantially better than a, thus providing counter-evidence to the claim that a is preferred to b. However, the form of this index means that it is only appropriate if all evaluations are made on a cardinal scale and the weights render scales comparable across criteria, which are quite restrictive assumptions. An alternative approach is to define a veto threshold for each criterion i, say t_i, such that a cannot outrank b if the score for b on any criterion exceeds the score for a on that criterion by an amount equal to or greater than its veto threshold. That is:

$$D(a, b) = \begin{cases} 1 & \text{if } z_i(b) - z_i(a) > t_i \text{ for any } i \\ 0 & \text{otherwise} \end{cases} \qquad (8.3)$$

The application of this definition is illustrated in Example Panel 8.1.

Building an outranking relation

The concordance and discordance indices for each pair of options can be used to build an outranking relation, according to the process illustrated in Figure 8.1. We need to start by specifying concordance and discordance *thresholds*, C^* and D^* respectively. Alternative a is defined as outranking alternative b if the concordance coefficient $C(a, b)$ is greater than or equal to the threshold C^* and the discordance coefficient $D(a, b)$ is less than or equal to D^*. The values of C^* and D^* are specified for a particular outranking relation and they may be varied to give more or less severe outranking relations - the higher the value of C^* and the lower the value of D^*, the more severe the outranking relation, that is, the more difficult it is for one alternative to outrank another. If the outranking relation is made too severe, then almost all pairs of alternatives will be deemed to be "incomparable"; while if the outranking relation is not severe enough then too many alternatives will outrank too many others (i.e. most are deemed to be essentially equally good in the light of the current information). Since neither of these outcomes is particularly

Application of (8.1) to the evaluations displayed in Table 8.1 generates the concordance indices shown in the matrix below (i.e. values of $C(a,b)$, where a is the city given in the row of the matrix, and b the city given by the column):

	Par.	Bru.	Ams.	Ber.	War.	Mil.	Lon.
Paris	1.00	0.73	0.73	0.68	0.84	0.59	0.73
Brussels	0.43	1.00	0.51	0.46	0.84	0.38	0.27
Amsterdam	0.76	0.73	1.00	0.84	0.84	0.76	0.59
Berlin	0.32	0.54	0.32	1.00	0.84	0.92	0.65
Warsaw	0.16	0.38	0.16	0.62	1.00	0.54	0.43
Milan	0.41	0.62	0.41	1.00	0.84	1.00	0.73
London	0.51	0.73	0.51	0.84	0.84	0.76	1.00

For example, in order to determine C(Paris, Brussels), we note that Paris is equal or preferred to Brussels on all criteria except "business potential - public sector", thus:

$$C(\text{Paris, Brussels}) = \frac{6 + 4 + 3 + 8 + 6}{6 + 4 + 3 + 10 + 8 + 6} = \frac{27}{37} = 0.73$$

As all evaluations in Table 8.1 are made according to a qualitative 5 point scale, discordance must be defined by a veto threshold for each criterion as in (8.3). Suppose that we set this at 3 scale points for all criteria. Thus, alternative a cannot outrank b if b is 3 or more points higher on the 5 point scale on any criterion (for example, if b scores VH and a scores L or VL on any criterion). The discordance matrix is shown below, in which an entry of 1 in cell (a,b) indicates that alternative a cannot outrank b.

	Par.	Bru.	Ams.	Ber.	War.	Mil.	Lon.
Paris	–	0	0	0	0	0	0
Brussels	0	–	0	0	1	0	0
Amsterdam	0	0	–	0	1	0	0
Berlin	0	0	1	–	0	1	1
Warsaw	0	0	1	1	–	1	1
Milan	0	0	1	0	0	–	1
London	0	0	0	0	0	0	–

Example Panel 8.1: Values of concordance and discordance indices (ELECTRE I) for the business location problem

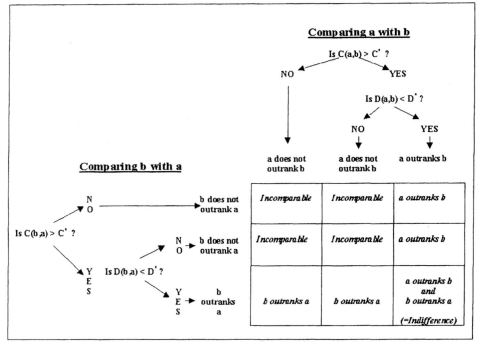

Figure 8.1: Building an Outranking Relation

useful, it is a matter of experimentation to find a C^* large enough (but not too large) and a D^* small enough (but not too small) in order to define an informative and useful outranking relation.

An outranking relation can be represented visually by an outranking graph, examples of which are shown in Example Panel 8.2. All alternatives are displayed in the graph; an arrow from alternative a to alternative b indicates that a outranks b. A double-headed arrow between alternatives a and b indicates that a outranks b and b outranks a.

Having built the outranking relation, the final step in the decision process is the exploitation of that relation, i.e. to make use of the relation for decision aid or support. The procedure in a particular case will depend on the nature of the problem, or the problematique, as discussed in Chapter 2. Is it to determine the "best" option, to rank the options, or to segregate them into, say, acceptable and unacceptable sets (e.g. Roy, 1996, Chapter 6)? The aim of ELECTRE I is to assist in the identification of a preferred alternative. This is achieved by determining the set of alternatives, referred to as the *kernel*, such that:

- Any alternative not in the set is outranked by at least one alternative which is in the set;

- All alternatives in the set are incomparable.

It should be noted that when there is a cycle in the outranking relation (for example, when a outranks b and b outranks a), there may be more than one kernel set, or there may not be such a set. Such a situation occurs with our example, as may be seen in Example Panel 8.2. A number of ways of dealing with this situation have been suggested (Roy and Bouyssou, 1993, Section 6.2), one of which is to treat all the alternatives which are part of the cycle as a single unit, assuming indifference between them. Later versions of ELECTRE modified the definition of the concordance index to reduce the likelihood of this problem occuring.

A mathematical aside

The outranking relation can be represented by a directed graph and the set just described is the kernel of that outranking graph. If the graph does not contain any circuits the kernel exists and is unique. However, if there are circuits in the graph the kernel is not necessarily unique and may not exist.

Sensitivity and robustness analysis

It is important to examine the impact of changes in the values C^* and D^* used to define the outranking relation. Is the kernel robust in the light of variations, or is membership sensitive to changes? Similarly, how are the results affected by changes in the weights assigned to criteria? If the result of these investigations is that several alternatives are identified as "robust" members of the kernel, then a more detailed analysis should be carried out to try to identified the preferred one.

Clearly sensitivity and robustness analysis is an important part of the decision process. Unfortunately, it is not possible to do this in any automated or interactive way, so that the analysis becomes an *ad hoc* investigation into the effect of changing values. One practical problem here may be the absence of clear operational and psychological meanings for the threshold levels C^* and D^*. They do not appear to have a clear interpretation in terms of decision maker values and preferences, being largely an *ad hoc* device to achieve an adequately rich set of outranking relations. In consequence, the sensitivity analysis might have limited value as a learning tool for the decision maker.

An illustration of the building and exploitation of the outranking relation in ELECTRE I is given by Example Panel 8.2.

The outranking relations for the business location case study can now be built from the concordance and discordance matrices shown in Example Panel 8.1. We start by setting $C^* = 0.7$ and $D^* = 0.1$ (since the discordance index takes only values of 0 or 1, all non-zero values of D^* are equivalent in this example), which gives the following outranking relation:

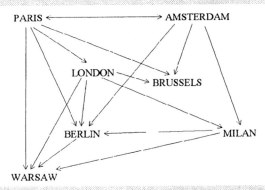

There is a circuit in the graph comprising Paris and Amsterdam (Paris outranks Amsterdam and Amsterdam outranks Paris) from which we might infer indifference between these two cities. Taken together these two alternatives outrank all others, although they are not incomparable and thus do not form a kernel. The subsets (Amsterdam and London) and (Paris and Milan) both satisfy the conditions for a kernel. Thus it would seem that at least Paris and Amsterdam, possibly also Milan and London merit more detailed consideration. Before focusing our attention on this smaller set of options, let us explore the effect of specifying different values for C^*. A stronger outranking relation, obtained by increasing C^* to 0.8, gives the following outranking relation, in which the kernel of the graph now comprises: Paris, Amsterdam, Brussels, London and Milan.

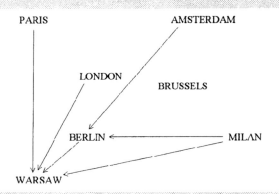

Reducing the value of C^* to 0.6 has only minor effect on the outranking relation obtained for $C^* = 0.7$ and no impact on the definition of the kernel.

Example Panel 8.2: Building and Exploiting the Outranking Relation for the Business Location Problem

8.3. ELECTRE II

ELECTRE II, developed shortly after ELECTRE I, differs from ELECTRE I in that it aims to produce a ranking of alternatives rather than simply to indicate the most preferred. The building of the outranking relation proceeds in essentially the same way as for ELECTRE I. However, two outranking relations are built using different pairs of concordance and discordance thresholds. These are referred to as the strong and weak outranking relations, the former having a higher concordance threshold and a lower discordance threshold.

Another small change from ELECTRE I was the introduction of an additional constraint in the test for outranking; in addition to $C(a,b) \geq C^*$ it is also required that $C(a,b) \geq C(b,a)$. The effect of this is to reduce the possibility of two alternatives each outranking the other. For example, in the concordance matrix illustrated in Example Panel 8.1, C(Amsterdam,Paris)=0.76, while C(Paris,Amsterdam)=0.73. These relatively large values for both concordance indices result from the fact that the two cities were deemed to be essentially equivalent as regards the two important business potential criteria. The result was that each city outranked the other when we set $C^* = 0.7$, as demonstrated by the first figure in Example Panel 8.2. However, the additional restriction introduced for ELECTRE II prevents Paris from outranking Amsterdam.

The exploitation procedure uses the two outranking relations to determine two rankings of the alternatives, as described below. The first starts with the "best" alternatives and works downwards, giving the descending order. The second starts with the "worst" and works up, giving the ascending order.

1 Specify concordance and discordance thresholds C^*, D^* for the strong outranking relation, and C^-, D^- for the weak outranking relation, defined such that $C^- < C^*$ and $D^- > D^*$.

2 Define A to be the full set of alternatives.

3 Determine the set of alternatives, say $F \subset A$, which are not strongly outranked by any other alternative in A.

4 Within F determine the set of alternatives, say F', which are not weakly outranked by any other member of F – this defines the first class of the descending ranking.

5 Delete the alternatives in F' from A, and repeat the procedure from step 3, continuing until all alternatives have been classified. This generates the *descending* order.

242   *MULTIPLE CRITERIA DECISION ANALYSIS*

6 Start again with A defined to be the full set of alternatives.

7 Determine the set of alternatives, say $G \subset A$, which do not strongly outrank any other alternative.

8 Within G determine the set of alternatives, say G', which do not weakly outrank any other member of G – this defines the first class of the ascending ranking.

9 Delete the alternatives in G' from A, and repeat the procedure from step 7, continuing until all alternatives have been classified. This generates the *ascending* order.

The ascending and descending orders should strictly be termed *complete preorders*, or *weak orders*, since only the classes are strictly ordered (with no preference ordering implied or inferred between alternatives in the same class). These two preorders should now be compared. Significant differences highlight unusual alternatives. For example, an alternative which neither outranks, nor is outranked by any other alternative will appear in the first class of both preorders (i.e. at the top of one and at the bottom of the other). The two orders can be combined to give a single one and a number of ways of doing so have been suggested. The most commonly used method is to determine the partial order (allowing for incomparabilities) defined by the "intersection" of the ascending and descending preorders. The intersection, R, of two outranking relations is defined such that aRb (i.e. a outranks b according to relation R) if and only if a outranks or is in the same class as b according to the preorders corresponding to both outranking relationships. This process is illustrated in Example Panel 8.3.

The intersection of the two preorders can further be refined to give a complete order (removing incomparabilities); in this context, the literature and the ELECTRE software make mention of the "median order" and the "final order". Where the descending and ascending preorders are relatively close, the intersection would seem to be a useful aggregation. However, information contained in the partial order may be lost in the refinement, so that understanding is not obviously enhanced, particularly if the two individual preorders differ substantially. We feel that if there remain worrisome incomparabilities, then the decision maker would learn more by detailed consideration of the alternatives in question than by relying on an algorithm to resolve the problem.

8.4.   ELECTRE III

ELECTRE I and II as described above work under the assumption that all criteria are "true" criteria, in the sense that any difference in

We can use the two outranking relations illustrated in Example Panel 8.2 as the weak and strong outranking relations required by ELECTRE II, corresponding to selection of $C^- = 0.7$ and $C^* = 0.8$. The additional constraint in the definition of the outranking relation means that Paris no longer outranks Amsterdam (since $C(\text{Paris, Amsterdam}) < C(\text{Amsterdam, Paris})$).

Following the procedure outlined in the text, we obtain the following preorders:

Descending Order	Ascending Order
[Amsterdam]	[Amsterdam]
[Paris]	[Paris]
[London]	[London, Milan]
[Brussels, Milan]	[Berlin]
[Berlin]	[Warsaw, Brussels]
[Warsaw]	

The intersection of these gives the following partial order:

[Amsterdam]
[Paris]
[London]
[Milan]
[Brussels][Berlin]
[Warsaw]

Example Panel 8.3: Determining a Ranking of Alternatives Using ELECTRE II

performance as measured by the $z_i(a)$ implies a corresponding difference in preference. It is assumed that indifference occurs only when two alternatives perform identically on a given criterion. Later developments sought to model preferences in increasing detail; in particular, ELECTRE III introduces the notion of indifference and preference thresholds as discussed in Section 4.5 of Chapter 4. A criterion modelled in this way is referred to as a "quasi-criterion".

The preference and indifference thresholds ($p_i(.)$ and $q_i(.)$ respectively) are utilised to construct a concordance index $C_i(a, b)$ for each criterion, defined by:

$$C_i(a, b) = \begin{cases} 1 & \text{if } z_i(a) + q_i(z_i(a)) \geq z_i(b) \\ 0 & \text{if } z_i(a) + p_i(z_i(a)) \leq z_i(b) \end{cases} \tag{8.4}$$

or by linear interpolation between 0 and 1 when $z_i(a) + q_i(z_i(a)) < z_i(b) < z_i(a) + p_i(z_i(a))$. By this definition, a is concluded to be at least as good as b according to criterion i (i.e. having a concordance index of 1 for this criterion) if there is not even weak preference for b

over a. If, on the other hand, b is strictly preferred to a on criterion i, then a definitely does not outrank b according to this criterion (i.e. the corresponding concordance index is 0). When alternative b is weakly but not strictly preferred to a, then the evidence remains somewhat ambiguous, and the value of the concordance index is set between 0 and 1. This is illustrated graphically in Figure 8.2.

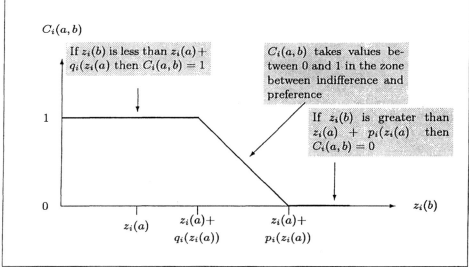

Figure 8.2: Definition of the Concordance Index for Criterion i

The overall concordance index is then defined as:

$$C(a,b) = \frac{\sum_{i=1}^{m} w_i C_i(a,b))}{\sum_{i=1}^{m} w_i} \qquad (8.5)$$

Discordance is defined similarly by the introduction of a veto threshold for each criterion, say $t_i(z_i)$, such the outranking of b by a is vetoed if the performance of b exceeds that of a by an amount greater than the veto threshold, i.e. if $z_i(b) \geq z_i(a) + t_i(z_i(a))$ for any i. A corresponding discordance index for each criterion is defined by:

$$D_i(a,b) = \begin{cases} 0 & \text{if } z_i(b) \leq z_i(a) + p_i(z_i(a)) \\ 1 & \text{if } z_i(b) \geq z_i(a) + t_i(z_i(a)) \end{cases} \qquad (8.6)$$

with linear interpolation between 0 and 1 when $z_i(a) + p_i(z_i(a)) < z_i(b) < z_i(a) + t_i(z_i(a))$. If no veto threshold is specified, then we define $D_i(a,b) = 0$ for all pairs of alternatives.

The overall concordance index and the discordance indices are combined to give a valued outranking relation. Alternative a is said to outrank b with *credibility* $S(a,b)$ defined as follows:

$$S(a,b) = \begin{cases} C(a,b) & \text{if } D_i(a,b) \leq C(a,b) \ \forall i \\ C(a,b) \displaystyle\prod_{i \in J(a,b)} \frac{(1 - D_i(a,b))}{(1 - C(a,b))} & \text{otherwise} \end{cases} \tag{8.7}$$

where $J(a,b)$ is the set of criteria for which $D_i(a,b) > C(a,b)$.

Note that $S(a,b)$ takes values between 0 and 1, and is thus closely allied to the concept of the membership function of a fuzzy set (see Section 4.6). If there are no discordant criteria (i.e. criteria on which the discordance index is greater than the overall concordance index) then the credibility index is equal to the overall concordance index. When there are discordant criteria the concordance index is modified to arrive at a credibility index which is lower in value. If $D_i(a,b) = 1$ for any criterion then the index of credibility, $S(a,b) = 0$.

Care should be taken in interpreting the credibility index. It is perhaps best read as an indicator of the "order of magnitude" of support for the claim that a outranks b. Roy and Bouyssou (1993, Section 6.4) warn that $S(a,b) > S(c,d)$ should not necessarily be interpreted as being a stronger argument in favour of a outranking b than that in favour of c outranking d. In view of the generally imprecise nature of the credibility index, such a conclusion requires that $S(a,b)$ be substantially larger than $S(c,d)$. Thus, Roy and Bouyssou suggest that if $S(a,b) = \lambda$, then the conclusion that a outranks b can be viewed as more firmly grounded than that of c outranking d if $S(c,d) \leq \lambda - s$, where $s = 0.3 - 0.15\lambda$. For example, a credibility index, $S(a,b) = 0.8$ suggests that there is a stronger argument in favour of a outranking b, than in favour of c outranking d, if $S(c,d)$ is less than or equal to $(0.8 - (0.3 - 0.15 \times 0.8)) = 0.62$.

The above rule is used by Roy and Bouyssou to define what they term λ-preference which is incorporated into the procedure for exploiting the valued outranking relation as follows: a is said to be λ-*preferred* to b if:

$$(1 - s)S(a,b) > S(b,a) \quad \text{and} \quad S(a,b) > \lambda$$

where $s = 0.3 - 0.15\lambda$. This concept of λ-preference is used to translate the valued outranking relation defined by $S(a,b)$ into a crisp outranking relation.

In common with ELECTRE II, ELECTRE III leads to a descending and an ascending order of the alternatives which are then combined. The two parts of the process are referred to as the descending and ascending distillation procedures, defined in the following.

246 *MULTIPLE CRITERIA DECISION ANALYSIS*

Descending distillation:

1 Determine the maximum value of the credibility index, $\lambda_{max} = \max S(a, b)$, where the maximization is taken over the current set of alternatives under consideration

2 Set $\lambda^* = \lambda_{max} - (0.3 - 0.15\lambda_{max})$

3 For each alternative determine its λ-strength, namely the number of alternatives in the current set to which it is λ-preferred using $\lambda = \lambda^*$

4 For each alternative determine its λ-weakness, namely the number of alternatives in the current set which are λ-preferred to it using $\lambda = \lambda^*$

5 For each alternative determine its *qualification*, which is its λ-strength minus its λ-weakness

6 The set of alternatives having the largest qualification is called the first distillate, D_1

7 If D_1 has more than one member, repeat the process on the set D_1 until all alternatives have been classified; then continue with the original set minus D_1, repeating until all alternatives have been classified.

Ascending distillation:

This is obtained in the same way as the descending distillation except that at step 6 above, the set of alternatives having the *lowest* qualification forms the first distillate.

Comments

ELECTRE III permits more sophisticated modelling of preferences on individual criteria than does ELECTRE II, which does however call for more work in modelling preferences with respect to each individual criterion before progressing to the building and exploitation of the outranking relation. The decision makers and analyst / facilitator now face the additional task of determining how each criterion should be modelled – is it a true criterion, or a quasi-criterion as defined above? If it is considered appropriate to model a particular aspect as a quasi-criterion then consideration must be given to the indifference, preference and veto thresholds. If these are non-zero, then what value should they take? Is the value of the threshold constant, or is it proportional to the value of the criterion? If it is proportional, is it defined relative to the more or less preferred alternative in the pair under consideration?

The use of quasi-criteria is intuitively appealing, perhaps more so than the seemingly stark and precise numerical judgements sought by the multi-attribute value function approach. However, the indifference, preference and veto thresholds do not have a clearly defined physical or psychological interpretation. The indifference threshold is possibly still an accessible intuitive concept, but we find it difficult to understand how to guide decision makers to interpret their preferences in terms of preference and veto thresholds. The literature is rather silent on this issue, implying that the thresholds are intuitively meaningful to many. Perhaps the exercise of specifying values for the thresholds does help decision makers to gain insight into their own preferences and values, even if numerical values for the thresholds are difficult to assess.

The aggregation procedure is somewhat opaque, based on a technically complicated algorithm which would be difficult for most decision makers to understand clearly. As such, the aggregation phase of the analysis may be less likely to help the decision makers gain further insight and understanding into the problem. A related problem is that the distillation procedure for ELECTRE III can yield some very strange and non-intuitive results (e.g. non-monotonicity, when an improvement in an element of an alternative's performance can lead to a poorer ranking position; see Perny, 1992). Furthermore, the results of the distillations are dependent on whole option set, so that the addition or removal of an alternative can alter preferences between remaining alternatives. Thus using the aggregation process face-to-face with a decision maker is somewhat hazardous for the facilitator or analyst, as these inconsistencies do not seem to be easily explainable. Our comments on ELECTRE III echo those of Vincke (1992, p67) and of Roy and Bouyssou themselves (1993, p426).

An illustration of the application of ELECTRE III to the business location case study is provided in Example Panel 8.4. Although it would have been possible to base this example on the five point scales used to measure performance against each criterion in illustrating ELECTRE I and II, these are not really adequate to support the more sophisticated definition of criteria incorporated in ELECTRE III. For this reason, the example is based on the more detailed measurements for the same case study as provided by Table 5.4 in Chapter 5. The same importance weights as in the earlier ELECTRE examples (Table 8.1) are again used here.

We conclude the analysis presented in Example Panel 8.4 at the point of the descending and ascending distillations. The two preorders show a fairly consistent pattern, which is also similar to that obtained using ELECTRE II, except that Milan is more favourably represented. Rea-

We consider the business location case study, using the data summarized in Table 5.4, and corresponding thresholds defined as follows (where a veto threshold of "none" implies that the criterion does not contribute to the discordance):

Criterion	q_i	p_i	t_i
Availability of staff	1	5	none
Accessibility from US	2	7	15
Quality of Life	1	2	none
Business potential - public sector	3	15	50
Business potential - private sector	3	15	50
Ease of set-up & operations	1	2	none

As illustration, consider the comparison between Berlin and the other cities in terms of the fourth criterion (*business potential - public sector*), for which Berlin's score was assessed as 50. The following diagram displays the concordance (solid lines) and discordance indices (broken line), expressed as functions of the score of the city to which Berlin is compared:

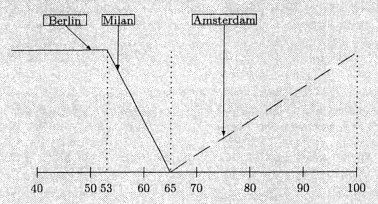

Looking at the scores for Milan (55) and Amsterdam (75), we observe that:

- In comparing Berlin with Milan, the concordance index is given by (55-53)/(65-53)=0.833, while the discordance index is still 0;

- In comparing Berlin with Amsterdam, the concordance index is 0, while the discordance index is given by (75-65)/(100-65)=0.286.

The concordance indices for Berlin compared to Milan, C_i(Berlin,Milan), for the six criteria turn out to be 0.5, 1, 0, 0.833, 1 and 1 respectively. Based on the weights from Table 8.1, the weighted average of these gives the overall concordance index as 0.793. The full set of **concordance indices** calculated in this way is given in the following table.

(Continued on the next page)

Example Panel 8.4: Application of ELECTRE III to the business location case study

Outranking Methods   249

		Concordance Indices					
	Par.	Bru.	Ams.	Ber.	War.	Mil.	Lon.
Paris	1.000	0.730	0.489	0.676	0.838	0.676	0.732
Brussels	0.516	1.000	0.514	0.550	0.838	0.468	0.532
Amsterdam	0.757	0.802	1.000	0.838	0.838	0.757	0.676
Berlin	0.414	0.649	0.324	1.000	0.838	0.793	0.811
Warsaw	0.162	0.591	0.162	0.577	1.000	0.383	0.477
Milan	0.608	0.708	0.405	1.000	0.959	1.000	0.892
London	0.514	0.730	0.514	0.883	0.838	0.676	1.000

Discordance indices need only be calculated for the 2nd, 4th and 5th criteria. We have noted that D_4(Berlin, Amsterdam)=0.286. Similarly, we may obtain D_2(Berlin,Amsterdam)=0.625 and D_5(Berlin,Amsterdam)=0.143. Since only D_2(Berlin,Amsterdam) exceeds C(Berlin,Amsterdam)=0.324, the corresponding credibility index is $0.324 \times (0.375/0.676) = 0.180$. The **credibility indices** for all pairs of cities are given in the following table.

	Par.	Bru.	Ams.	Ber.	War.	Mil.	Lon.
Paris	1.000	0.730	0.489	0.676	0.838	0.676	0.732
Brussels	0.516	1.000	0.514	0.550	0.838	0.468	0.532
Amsterdam	0.757	0.802	1.000	0.838	0.838	0.757	0.676
Berlin	0.414	0.649	0.180	1.000	0.838	0.793	0.811
Warsaw	0.138	0.413	0.017	0.577	1.000	0.383	0.228
Milan	0.608	0.708	0.170	1.000	0.959	1.000	0.892
London	0.514	0.730	0.514	0.883	0.838	0.676	1.000

We now proceed to the distillation steps based on these credibility indices. Starting with the full set of alternatives, we find that $\lambda_{max} = 1$, so that $\lambda^* = 0.85$ and $s = 0.15$. Thus a is λ-preferred to b provided $S(a,b) > 0.85$ and $0.85S(a,b) > S(b,a)$. The only cases in which this occurs give Milan λ-preferred to Berlin, Warsaw and London, so that Milan has the only positive qualification, and is ranked first in the descending distillation. After elimination of Milan, we obtain $\lambda_{max} = 0.883$, $\lambda^* = 0.715$ and $s = 0.168$. It now turns out that Amsterdam has the largest qualification, and is ranked second.

The ascending distillation starts as for the descending distillation, but now Berlin, Warsaw and London form the first distillate, since (as we have seen) Milan is λ-preferred to all three. When we restrict attention to these three, we find that Berlin and London are λ-preferred to Warsaw, so that Warsaw becomes the first distillate (i.e. lowest ranking alternative).

(Continued on the next page)

Example Panel 8.4: Application of ELECTRE III to the business location case study (Continued)

250   MULTIPLE CRITERIA DECISION ANALYSIS

Continuing as above, the rank orders generated by the ascending and descending distillations respectively emerge as follows:

Descending Distillation	Ascending Distillation
[Milan]	[Amsterdam]
[Amsterdam]	[Paris, Milan]
[Paris]	[Brussels]
[London]	[Berlin, London]
[Brussels, Berlin]	[Warsaw]
[Warsaw]	

There is a substantial similarity between the two rankings, with Amsterdam, Paris and Milan clearly indicated as the more preferred options, and Berlin and Warsaw the least preferred.

Example Panel 8.4: Application of ELECTRE III to the business location case study (Continued)

sons for the better position of Milan are not immediately evident, but may be due to the replacing of the qualitative assessments by the more detailed numerical measures.

As with ELECTRE II, software for ELECTRE III also includes the option of integrating the descending and ascending preorders into what are called the final and median preorders. The precise mechanism by which the final and median preorders are obtained is somewhat obscure, however, and as we noted in discussing ELECTRE III, it is perhaps arguable whether they add much to further understanding of the problem.

Clearly, effective software is essential if the method is to be used in practice. Sensitivity analysis should be carried out to determine the robustness of the preferred options, and the ability to do this is reliant on good software. As with ELECTRE I and II, this must take the form of an *ad hoc* investigation of the many parameters of the problem.

8.5.   OTHER ELECTRE METHODS

Our discussion of ELECTRE I, II and III presents a reasonably comprehensive picture of the concepts underlying outranking methods and details of the preference modelling approaches adopted by Roy and associates. There are, however, variations on each of the methods described, as well as extensions to the family as a whole. ELECTRE IV and ELECTRE TRI, which, although based on ELECTRE III, have significant distinguishing features, are described briefly below.

ELECTRE IV (Roy and Hugonnard, 1982) was developed for use in circumstances in which it is not possible to specify criteria weights. Initial preference modelling is as for ELECTRE III; that is, indifference, preference and veto thresholds are specified for each criterion. Outranking relations, of differing strength, are then defined by direct reference to the performance levels of alternatives. For example, a strong and weak outranking relation may be defined as follows:

- a strongly outranks b if there is no criterion on which b is strictly preferred to a and the number of criteria on which b is weakly preferred to a is less than the number on which a is strictly preferred to b.

- a weakly outranks b if:

  - there is no criterion on which b is preferred (strictly or weakly) to a, OR

  - b is preferred to a (strictly or weakly) on only 1 criterion, a is preferred (strictly or weakly) to b on at least half of the criteria, and b does not veto a on any criteria.

Exploitation of the outranking relation can then proceed using the descending and ascending distillation processes as described for ELECTRE III, the initial qualification of alternatives being determined by the strong outranking relation and the weaker relation being used to discriminate, if possible, between those alternatives having the same initial qualification value. The reader is referred to the original paper by Roy and Hugonnard or to Roy and Bouyssou (1993) for more detailed discussion of the method.

ELECTRE TRI is for use in classification problems. The original procedure was designed to allocate alternatives to one of three categories (hence the name) - acceptable, unacceptable or indeterminate. This has been extended for use in classification problems in which there are more than three different categories. A number of different approaches have been developed; we outline here the one described by Roy and Bouyssou (1993). At the root of the method is a procedure based on the ELECTRE III valued outranking relation defined by the credibility index $S(a,b)$. The categories are ordered and defined by a set of reference actions, or limiting profiles, $c^0, c^1, \ldots, c^k$, where c^x corresponds to a hypothetical alternative with performance levels $z_i(c^x)$ for $i = 1, \ldots, m$. Category C^x, which is bounded below and above by the limiting profiles c^{x-1} and c^x respectively, defines a higher level of performance than category C^{x-1}. The limiting profiles must be determined in consultation with the decision makers.

252 MULTIPLE CRITERIA DECISION ANALYSIS

An alternative for classification, a, is compared to each reference profile in succession to determine:

- the highest limiting profile such that a outranks c^x (i.e. $S(a, c^x) \geq \lambda$, where the threshold λ is a specified value such that $0 \leq \lambda \leq 1$), which determines the pessimistic allocation of a to C^{x+1}.

- the lowest limiting profile such that c^y is preferred to a (i.e. c^y outranks a and a does not outrank c^y), which determines the optimistic allocation of a to category C^y. Note that $y \geq x + 1$.

In the case where $y = x + 1$, a is unambiguously allocated to that category. Where $y > x + 1$, then either:

- a is indifferent to all intermediate limiting profiles: If $y = x + 2$, then a outranks c^x, is indifferent to c^{x+1}, and is outranked by c^{x+2}; If an alternative is indifferent to more than one limiting profile, then that is probably an indication that the categories are too narrowly defined.

- a is incomparable to all intermediate profiles.

Further extensions of ELECTRE TRI have been developed for the assignment of alternatives to nominal (i.e. not ordered) categories. The principles are similar to those outlined above, but the reference profiles define prototypical members of each category rather than upper and lower limits and comparisons are based on the indifference relation rather than the outranking relation. Perny (1998) provides a review and illustrative example of these approaches, which he refers to as filtering methods.

8.6. THE PROMETHEE METHOD

The PROMETHEE method, developed by Brans and co-workers (e.g. Brans et al., 1984; Brans and Vincke, 1985; Brans et al., 1986), is another outranking approach. As in our description of the ELECTRE methods we will take as the starting point the decision matrix of evaluations of alternatives against an appropriate set of criteria. The next step in the PROMETHEE method is to define what Brans calls a preference function for each criterion. Rather than the specification of indifference and preference thresholds as described in the context of ELECTRE III, the *intensity of preference* for option a over option b, $P_i(a, b)$, is described by a function of the difference in performance levels on that criterion for the two alternatives, i.e. on $z_i(a) - z_i(b)$ by our earlier notation. This function takes on values between 0 and 1, and a number of suggested standard shapes are illustrated in Figure 8.3, which is taken from the

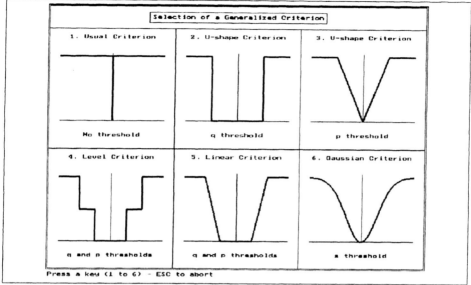

Figure 8.3: PROMETHEE preference functions

PROMCALC software. The decision maker selects the desired shape of function, and specifies any parameters that are then needed. As shown in the figure, the functions are symmetric around a difference of zero. Clearly, for positive differences ($z_i(a) > z_i(b)$), the function value gives $P_i(a,b)$ while $P_i(b,a) = 0$. Conversely, when $z_i(a) < z_i(b)$, $P_i(a,b) = 0$ and the chosen function from Figure 8.3 generates the required value for $P_i(b,a)$.

The next step is to determine a preference index for a over b. This is analogous to the credibility index defined in ELECTRE III and is a measure of support for the hypothesis that a is preferred to b. The preference index $P(a,b)$ is defined as a weighted average of preferences on the individual criteria:

$$P(a,b) = \frac{\sum_{i=1}^{m} w_i P_i(a,b)}{\sum_{i=1}^{m} w_i}$$

which is thus also defined to be between 0 and 1. As with the ELECTRE methods, the weights do not represent scaling factors, but some notion of global importance.

Note that:

- If all individual preference functions are crisply defined, taking only values 0 and 1 (type 1 or type 2 in Figure 8.3), then $P(a,b)$ is simply the proportion of criteria on which a is preferred to b, which is equivalent to the concordance index in ELECTRE I.

254 MULTIPLE CRITERIA DECISION ANALYSIS

- Individual preference functions of type 5 are equivalent to the preference models incorporated in ELECTRE III, with constant threshold values q_i and p_i. However, note that this does not lead to equivalence of the PROMETHEE preference index and the concordance index in ELECTRE III. The concordance index $C_i(a, b)$ only starts to fall below its maximum value of 1 when $z_i(b) - z_i(a) > q_i$, whereas for the preference index $P_i(a, b)$ drops below 1 as soon as $z_i(a) - z_i(b) < p_i$ (i.e. when $z_i(b) - z_i(a) > -p_i$. A little thought confirms that the two indices are related through $P_i(a, b) = 1 - C_i(b, a)$, which also implies that $P(a, b) = 1 - C(b, a)$.

The preference index thus defines a valued outranking relation, which, as in ELECTRE III, is exploited to determine an ordering of the alternatives. Two further indices, the positive outranking flow, and the negative outranking flow are defined as follows:

$$\text{The positive outranking flow for } a: \quad Q^+(a) = \sum_{b \neq a} P(a, b)$$

$$\text{The negative outranking flow for } a: \quad Q^-(a) = \sum_{b \neq a} P(b, a)$$

where the sums are taken over all alternatives under consideration.

The positive outranking flow expresses the extent to which a outranks all other options. The negative outranking flow expresses the extent to which a is outranked by all other options. Each of these indices defines a complete preorder of the alternatives, the intersection of which gives a partial order as follows:

- a outranks b if:

  i $Q^+(a) > Q^+(b)$ and $Q^-(a) < Q^-(b)$; or

  ii $Q^+(a) > Q^+(b)$ and $Q^-(a) = Q^-(b)$; or

  iii $Q^+(a) = Q^+(b)$ and $Q^-(a) < Q^-(b)$.

- a is indifferent to b if $Q^+(a) = Q^+(b)$ and $Q^-(a) = Q^-(b)$.

- a and b are incomparable if:

  i $Q^+(a) > Q^+(b)$ and $Q^-(b) < Q^-(a)$; or

  ii $Q^+(b) > Q^+(a)$ and $Q^-(a) < Q^-(b)$.

It should be noted here that the values of the positive and negative outranking flows depend on the complete set of alternatives under consideration. It is thus possible that the classification of the relationship between two alternatives a and b may be affected by the inclusion or exclusion of one or more other alternatives.

The above procedure is sometimes termed PROMETHEE I. In the extended PROMETHEE II approach, a complete preorder of alternatives is derived from the "net flow" for each alternative, defined as:

$$Q(a) = Q^+(a) - Q^-(a).$$

Then a outranks b if $Q(a) > Q(b)$, with indifference if $Q(a) = Q(b)$. Brans comments that this complete order is more disputable than the partial order derived by PROMETHEE I, as information is lost. Indeed, if this complete preorder were to be the goal of the analysis, then all the advantages claimed over value function methods are sacrificed.

The PROMETHEE method is attractive in that it combines the simplicity and transparency of the early ELECTRE methods with some of the increased sophistication of preference modelling incorporated in ELECTRE III. However, in common with ELECTRE III, the distillation process can yield results which are counter intuitive (see, for example, Roy and Bouyssou, 1993, Section 6.4; de Keyer and Peeters, 1996). The method is supported by the PROMCALC and Decision Lab 2000 software, which offer extensive facilities for investigating the sensitivity of the complete ranking of alternatives to changes in criteria weights. The application of PROMETHEE to the business location case study is illustrated in Example Panel 8.5.

GAIA

The PROMETHEE analysis may be used in conjunction with a procedure termed GAIA (Geometric Analysis for Interactive Aid) which provides a visual representation of the problem (Brans and Marechal, 1990). Principal components analysis is applied to the matrix of "normed flows", defined for alternative a and criterion i by:

$$\phi_i(a) = \frac{1}{n-1} \sum_{b \neq a} [P_i(a, b) - P_i(b, a)]$$

where n is the number of alternatives, and this is used to generate a two-dimensional plot in which the alternatives and criteria are represented in the same plane. Figure 8.4 displays such a plot, based on the normed flows calculated from the same preference indices used in generating the results in Example Panel 8.5. Here the plotting positions for the alternative cities are represented by the first three letters of each name, while the positions of each criterion is linked by lines from the origin indicating the different axes of preference.

256 MULTIPLE CRITERIA DECISION ANALYSIS

We suppose that preferences in the business location case study are judged to be well represented by the type 5 form of function in Figure 8.3, with indifference and preference thresholds identical to those used in the ELECTRE III example (Example Panel 8.4). The preference indices are easily obtained from the ELECTRE III concordance indices as follows:

	Par.	Bru.	Ams.	Ber.	War.	Mil.	Lon.
Paris	0.000	0.484	0.243	0.586	0.838	0.392	0.486
Brussels	0.270	0.000	0.198	0.351	0.409	0.292	0.270
Amsterdam	0.511	0.486	0.000	0.676	0.838	0.595	0.486
Berlin	0.324	0.450	0.162	0.000	0.423	0.000	0.117
Warsaw	0.162	0.162	0.162	0.162	0.000	0.041	0.162
Milan	0.324	0.532	0.243	0.207	0.617	0.000	0.324
London	0.268	0.468	0.324	0.189	0.523	0.108	0.000

The leaving and entering flows are then calculated as follows:

	Leaving Flow	Entering Flow	Net Flow
Paris	3.029	1.859	1.170
Brussels	1.791	2.583	-0.792
Amsterdam	3.592	1.333	2.259
Berlin	1.477	2.171	-0.694
Warsaw	0.851	3.648	-2.797
Milan	2.248	1.427	0.821
London	1.880	1.847	0.033

The resulting outranking relationships imply the following partial ordering, where the cities bracketed together are viewed as incomparable and not indifferent:

[Amsterdam]
[Paris, Milan]
[London]
[Berlin, Brussels]
[Warsaw]

This gives a very similar ordering to that of the ELECTRE methods. Use of the net flows to resolve the incomparabilities results in Paris being ranked ahead of Milan, and Berlin ahead of Brussels, both by relatively small margins. This seems rather arbitrary, and does not add much insight into the decision problem.

Example Panel 8.5: Application of PROMETHEE to the business location case study

Outranking Methods 257

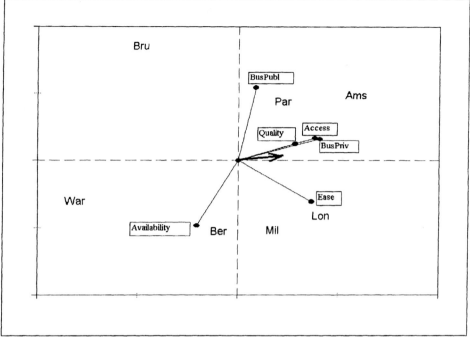

Figure 8.4: GAIA-type plot for Example Panel 8.5

The use of multivariate statistical techniques for visualization of MCDM problems was proposed in Stewart (1981), and is described in Section 10.2 of Chapter 10, where the interpretation of plots such as Figure 8.4 is further discussed. It is important to note that the key feature of such plots is the direction of departure from the origin, so that (for example) the direction from the origin towards the plotted position of Milan is linked with the directional axes of *ease of set-up and operations* and of *availability of staff*.

The original proposals in Stewart (1981), to be described in Section 10.2, concerned the application of multivariate techniques directly to the performance levels for each criterion (the decision matrix) without any prior preference modelling. As indicated above, the approach in GAIA is to apply such techniques to the matrix of normed flows. However, as was noted in Stewart (1992), and as will be demonstrated in Section 10.2, the results from principal components analysis applied to either the decision matrix or to the matrix of normed flows are to a large extent indistinguishable. Thus the incorporation of the PROMETHEE preference functions into GAIA does not appear to add much to the geometrical representation. The real value of GAIA lies in the idea of

258 MULTIPLE CRITERIA DECISION ANALYSIS

projecting a weighted sum of the normed flows as a direction vector in the plane in which the criteria and alternatives are plotted. The large arrow head shown in Figure 8.4 indicates a direction of preference corresponding to equal weights on the criteria. This suggests no clearly identifiable best alternative, although Amsterdam, Paris and London lie closer to this direction than do the other cities. In the GAIA software, the user may interactively adjust the relative weights, leading to a rotation of the preferred direction, and the identification of which alternatives tend to be associated with different sets of weights. It takes a short while to become familiar with the presentation and how to interpret it, but having done so it seems to be a useful means to feed back information to decision makers.

8.7.   FINAL COMMENTS

The primary appeal of all of the outranking methods is in the avoidance of what are perceived to be the overly restrictive assumptions of value- or utility-based approaches. The opportunity is provided to make use of a richer array of preference models that appear to more faithfully capture the manner in which decision makers think. There is no doubt that concepts such as incomparability of alternatives and the existence of grades of preferences (e.g. "weak" and "strong") are part and parcel of the decision making process. By focussing on the manner in which decision makers naturally form preferences, outranking methods give considerable insight into what conclusions can be supported concerning whether or not one alternative should be judged as more preferred than another.

The various forms of outranking relationships, expressed either graphically as in ELECTRE I or by the credibility matrix of ELECTRE III, go a long way towards facilitating that insight, and are generally relatively simple to interpret at a qualitative level. The power of outranking methods lies in the pairwise comparison of alternatives in this manner. The major drawbacks of outranking methods arise from the many rather non-intuitive inputs that are required, such as: concordance and discordance threshold levels; indifference, preference and veto thresholds; and the preference functions of PROMETHEE. Such drawbacks assume particular importance when efforts are made to extend the outranking methods to produce explicit preference orderings over the full set of alternatives. The impacts of the various inputs are difficult to appreciate intuitively, and the algorithms themselves tend to be complicated for unsophisticated decision makers fully to understand. As a result of this, results can be counter-intuitive, with unexpected changes in rank orderings arising in response to changes in the threshold levels or to addition or deletion of alternatives.

In view of the complexities and potential for counter-intuitive results mentioned in the previous paragraph, it seems that the outranking methods may not generally be suitable for use in the decision workshop mode to be discussed in Chapter 9, in which the decision analyst works directly with the decision makers. Outranking methods may thus be more appropriate to "backroom" analyses by analysts and/or by support staff to the final decision makers. In this role, time can be taken to experiment with various inputs, in order to generate a richer understanding of the arguments that can be made for and against the various decision alternatives. Large numbers of incomparabilities may indicate the need for more detailed preference information and/or for the creation of or search for new decision alternatives. Such understanding will provide decision aid in the form of clearly documented reports with recommendations for action that can be considered by political decision makers for example. The concept of distinctly different roles for different methodologies in the decision making process will be taken up again in Chapter 11.

Chapter 9

IMPLEMENTATION OF MCDA: PRACTICAL ISSUES AND INSIGHTS

9.1. INTRODUCTION

In previous chapters we looked in detail at how to build multicriteria models and at types of problems which can be usefully modelled in this way. In chapters 4 to 8 we have presented a variety of procedures, describing the technical details, underlying assumptions and practicalities of implementing each of these. However, as we emphasised in Chapter 1, the real challenge to MCDA is to help an individual, or perhaps more commonly a group of people with responsibility for planning or decision making, to make effective and defensible decisions. Just as the gift of a car to someone who cannot drive does not help them to explore the country, the mere existence of methods for MCDA does not help people make better considered decisions. In both cases, support is called for. The new car owner can either be chauffeured around the country, or taught how to drive themselves. People wishing to explore their problem using MCDA can be supported by an analyst or facilitator, or they can be taught how to use the methods as a part of their own investigation. The successful use of MCDA methods in practice thus relies on effective facilitation by a decision analyst, or on the ability of individual users to learn how to make effective use of an approach for themselves without becoming a technical expert in the field of MCDA. In this chapter we focus on the implementation of MCDA, by which we mean the management of a process which involves the use of an MCDA method (not to be confused with "implementation" in the sense of putting into practice a solution or action plan resulting from such a process). Thus we shall be examining the broader practicalities of providing multicriteria decision support. In this we go beyond the specific details relating to individual

262 *MULTIPLE CRITERIA DECISION ANALYSIS*

procedures, focusing on broader issues such as the interactions between the analyst or facilitator and the decision makers. Such issues are clearly not specific to MCDA, but are equally relevant in the wider context of OR/MS and much can be learned from others' experience, particularly in the use of the so-called "soft" OR methods.

The implementation of MCDA spans a range of different ways of working, having differing implications for the nature of support required in terms of facilitation and analysis, as well as attendant facilities and technology. Before going on to discuss aspects of this in detail we should highlight three factors which are of particular relevance.

The parties involved and their role in the decision: Important considerations here are: the size of group involved; the extent to which members of a group share a common goal; and their role in the process – are they participating as advisors, as experts, or with a responsibility for the decision? At one extreme the "decision maker" may be a single individual with sole responsibility for a personal decision, for example a person buying a house, or assessing a career move. It may be a small, relatively homogeneous group of individuals sharing more-or-less common goals, for example a sales team evaluating marketing strategies. It may be a larger group representing different elements of the same organisation, thus sharing the same broad corporate objectives but also bringing their own agendas, for example a hospital management team comprising representatives of clinical staff, nursing staff, non-clinical support staff, administrators, etc., tasked with evaluating alternative plans for expansion. Or, at the other extreme, decision making may call for consensus across highly diverse interest groups with very different agendas, as described in the land use planning case study. The group may share corporate responsibility for a decision, it may be tasked with investigating an issue with a view to making a recommendation to a decision making authority (e.g. a higher level of management or a government body), or it may have been assembled for the explicit purpose of exploring alternative perspectives without any executive power. Members of the group may possess knowledge relating to, and assume responsibility regarding, all aspects of a problem or they be involved as experts with knowledge pertaining only to a part of the problem.

The nature of interaction between individuals: The way in which a group works is likely to be influenced by the nature of the group, by practical considerations and by the objective of the process. If the aim is to achieve a shared understanding of an issue and to arrive at a consensus on the way forward, then it is important that the group works together to explore the issue. Although technology now permits virtual groups to work together, the need to meet face-to-face is considered still to be highly important. In such situations, the majority of activity may take place in a decision workshop, or workshops, which bring together the members of the group on one or more occasions. If, however, the aim is to bring together the views of experts in different subject areas and to present these in a unified framework to a higher authority then there is less need for the group to meet. In this case, individuals or specific interest groups may undertake their own evaluations initially, prior to a group decision workshop or as input to a broader political process.

Facilitated or unfacilitated process: The multicriteria analysis may be guided by one or more expert facilitators (or decision analysts), or may be carried out by an individual or group of decision makers in a "do-it-yourself" (DIY) mode without the support of a facilitator. The DIY mode of working is probably only appropriate for individuals, or small homogeneous groups in which the group dynamics are not problematic and individual agendas do not undermine the group process. Clearly this mode of operation calls for knowledge and understanding of the method adopted. We believe, however, that an analysis conducted under the guidance of a facilitator is the most appropriate and most likely way of working for the majority of non-trivial problems. Much of the discussion in this book has assumed this mode of working and this will continue to be the emphasis in the present chapter.

In Sections 9.2 to 9.5 we focus on the practical issues facing the facilitator or decision analyst in implementing MCDA, either in the context of a group decision workshop or via one-to-one consultations with involved parties. Most of the issues to be addressed are not specific to MCDA, but are relevant to any facilitated process of group decision support. Our own practice is very much informed by the work of Phillips and Phillips (1993), Eden and Ackermann (1998) and other contributors to the edited book on "Tackling Strategic Problems" (Eden and Radford, 1990). The reader is referred to these excellent writings and those of Schein (1987, 1988 and 1999) on Process Consultation, and Schwartz (1994) on Facilitation, for a more in depth discussion of many of the issues which will be addressed here. We focus in particular on:

- Initial contacts – establishing the contract.

264 MULTIPLE CRITERIA DECISION ANALYSIS

- The nature of modelling and interactions with participants – sharing, aggregating or comparing? Which approach is most appropriate in a given context and environment?

- Organising and facilitating a decision workshop – managing group dynamics and trivialities of process

- Working "off-line" or in the backroom

Although much of the discussion relates to a decision workshop setting, we do believe that it is important to acknowledge this is not the only context for MCDA. An individual or small group may work with a facilitator or analyst in a more informal setting, or they may work without facilitator support in what we have termed "DIY" mode. In Section 9.6 we discuss these ways of working, emphasizing the need for appropriate supporting software for such DIY decision analysis. Such software should be able to act partly in place of an expert facilitator to ensure that the process is fully understood and correctly pursued, thereby avoiding the appearance of a "rational" solution which hides substantial inconsistency and bias.

This point leads naturally on to Section 9.7 which is concerned with software for MCDA: what are the requirements of a facilitator / analyst and what is available? It is our view that good software is essential for the effective conduct of MCDA in practice and that it should facilitate not drive the process, illuminate not intrude.

The availability of good supporting software, which can rapidly synthesise information and judgements and effectively reflect back the tentative outcomes to decision makers, is key in generating understanding, learning and creativity. Throughout the book we have emphasized that the process of MCDA is not simply one which seeks to find the "right answer", that analysis and insight does not end with the initial output of a model, but that this should be viewed as a starting point for challenging intuition, enhancing learning and increasing confidence in and commitment to a way forward. In Section 9.8 we seek to illustrate further what we mean by this by sharing some of our insights from MCDA practice.

As a final point in this chapter we return once again to the critical practical issue of the meaning and understanding of importance weights in MCDA. We do so because we wish to stress that it is an issue not only for theoretical debate but also of significant practical relevance. Furthermore, we are very much aware from our own experiences both in using MCDA to support decision making in organizations and in teaching the subject, that the concept of a criterion weight is one which is both difficult to grasp and regularly misunderstood. We introduced the discussion of the meaning of criteria weights in Section 4.7 in the context of preference modeling and have returned to it during the presentation of specific MCDA methodologies. As has been indicated, people (decision makers) often have intuitive concepts of the relative level of importance of different criteria which may differ substantially from the algorithmic meaning of the weight parameters in the MCDA model being used. Since these importance weights are central to the finding of a consensual outcome to the multicriteria problem, it is essential that both the psychological and algorithmic meanings of the weights are understood by both analyst and decision maker, and that the process used avoids the many traps of misunderstanding and misinterpretation into which both can fall. Critical to the implementation of MCDA, therefore, is the need for decision analysts to guide clients towards a clear understanding of importance weights *within the context of the model being used.*

9.2. INITIAL NEGOTIATIONS: ESTABLISHING A CONTRACT

We are not concerned here with the marketing of MCDA, or how to arrive at a position where a client recognises the potential benefit of engaging an expert decision analyst (although this itself introduces significant challenges to the MCDA community). Our concern is rather with the process from that point of recognition onwards. The process of contracting – deciding whether and how to work together – is discussed in detail by Schwarz (1994) in "The Skilled Facilitator"; he comments that ineffective contracting almost invariably leads to problems later in the facilitation. Eden and Ackermann (1998, chapter P5) discuss in detail the issues to be considered in designing an intervention together with a client, or, in their terminology, in getting started on a "journey". Schein (1999 Part IV) also addresses initial interventions, the building of a relationship through an exploratory meeting and the concept of the "psychological contract".

Clearly a first step is for the facilitator and client to explore whether or not working together is going to be mutually beneficial. It is important that the client understands what is involved in the process and the nature of the anticipated outcomes. What can or cannot be expected from the results? MCDA does not diminish the role of managerial judgement, and (the other side of the same coin) MCDA will not relieve the decision maker of responsibility for the decision. It must be made clear to potential clients that MCDA will aid and support, and is there to enhance their decision making skills. The process will generate recommendations for action, but these will always be subject to the assumptions and simplifications made as part of seeking and generating understanding, so

266 *MULTIPLE CRITERIA DECISION ANALYSIS*

that decision makers still need to apply their minds to the proposed solutions generated. Conveying this message will be straightforward if the client has previous experience of working with MCDA, but if that is not the case then it is useful to have "stories" of previous interventions to illustrate the process, the potential benefits and the possible pitfalls. An agreement should be reached reflecting expectations about how the process will unfold. This should cover issues such as: Who should be involved? How much time will they be required to give? What is the nature of their involvement? How often, for how long, and where, should meetings be held? What are the objectives of these meetings?

However, the facilitator should be aware that an intervention may evolve differently from the original intention. For example, one recent study evolved into a series of six short workshops – involving the development of new ways of working. We could not have predicted this at the outset. Under such circumstances, it is important that changes to the original plan be negotiated with the client as soon as their need becomes evident. Schein (1999, pp236-7) comments on the difficulty of being completely explicit about expectations up front, stressing the need for an openness which acknowledges uncertainty about how the process might evolve.

9.3.   THE NATURE OF MODELLING AND INTERACTIONS

9.3.1.   Modelling issues

It is not possible to separate considerations of modelling issues from the nature of interactions between participants; the two are inextricably intertwined. Different ways of working will be more appropriate in different contexts with decisions about, requirements for, and constraints on one aspect significantly influencing the other. In this section we focus on modelling issues, going on in Section 9.3.2 to consider the implications for interactions with participants. In Section 9.3.3 we comment briefly on the role of the facilitator.

When discussing MCDA methods in earlier chapters, the implicit assumption was perhaps that the decision making group by and large operated as an individual, i.e. a single model and set of judgements was derived and uncertainties about structure, or about particular values, were handled through iteration and sensitivity analysis. We pause now to consider other possible ways of working. Belton and Pictet (1997) propose a framework for the consideration of how individual viewpoints are synthesised in a process of multicriteria decision support. The structure of a model (criteria and alternatives) and evaluation method may be specific to the individuals involved or common to the group, and each judgement, or element, in a common model may be determined by a process of sharing, comparing or aggregating, as follows:

- *Sharing* aims to obtain a common element by consensus, through a discussion of the views of individuals and the negotiation of an agreement; it addresses any differences and tries to reduce them by explicitly discussing their cause.

- *Aggregating* aims to obtain a common element by compromise, through a vote or calculation of a representative value; it acknowledges differences and tries to reduce them analytically without explicitly discussing their cause.

- *Comparing* aims to reach an eventual consensus based on negotiation of independent individual results. It begins by obtaining individual views on each element, acknowledging differences but not necessarily trying to reduce them unless they mitigate against overall consensus.

This framework allows for a wide spectrum of ways of working. At one extreme, separate interested parties - individuals or stakeholder groups - may consider an issue from their own perspectives using their own procedures to evaluate courses of action which they believe to be feasible. There are no common elements; in essence there is no group decision making here, only negotiation. Moving slightly away from this extreme, independent expert groups may be asked to evaluate a common set of alternatives, possibly using a common approach to modelling, but incorporating criteria and judgements which reflect their individual perspectives. These independent viewpoints are then synthesised by an over-arching decision making group. At the other extreme a group seeks to work together throughout, defining a common model and shared judgements, as is the case in the Decision Conferencing approach described by Phillips (1989, 1990), and by Phillips and Phillips (1993). An intermediate approach is for a group to work together to develop a common model (an agreed criteria structure and set of alternatives, or an agreed specification of objectives and constraints) and then for individuals or sub-groups independently to use that model to evaluate alternatives or explore possible solutions, coming together again to compare results. Although we acknowledge that aggregation of individual evaluations using a voting mechanism or analytical procedure is a feasible way to proceed, we do not advocate it as a way of resolving decisions that matter. While such mechanisms or procedures may be useful for individuals or homogeneous groups to explore potential compromises between their own conflicting criteria, they tend to lead to compromise (often uncomfort-

268   *MULTIPLE CRITERIA DECISION ANALYSIS*

able and unstable) rather than the creative seeking of consensus between groups with divergent interests.

As this discussion illustrates, there are significant choices to be made about the way of working and the facilitator plays a key role in guiding that choice. The most appropriate choice will depend on the specific situation - the extent to which the aim is to arrive at a shared understanding and shared commitment to a way forward, the extent to which expertise is shared or distributed, and the extent to which the group has responsibility for making a decision or is advisory. It will also depend on the time the group can devote to working together, as well as on the availability of appropriate supporting technology. Furthermore, the way of working has significant implications for the management of process. For example, *sharing* is very demanding of the facilitator, who has to be constantly managing the interaction between group members. Such is the importance of process facilitation in this context that it is recommended that a facilitator should be dedicated to this role, assisted by a second facilitator focused on content or technical support (Eden, 1990). By comparison, *comparing* focuses demands on the facilitator at particular stages in the process, and may subtly change the nature of skills required. For example, if acting contingently on the basis of judgements emerging, then the facilitator needs to be skilled in judging how to direct discussion.

As illustration of the above considerations faced by the facilitator, suppose that in using a multi-attribute value tree for evaluation it turns out that there is agreement at the higher levels of the tree despite significant differences at the lower levels. Should the facilitator seek to explore these differences? Will this be helpful in achieving an ultimate sharing of vision, or will it detract from reaching consensus? Or, if there are substantial differences in the evaluation of alternatives on a criterion to which everyone has assigned a low weight should these differences be discussed?

The interplay between decision modelling and the group interactions extends in particular to choice of MCDA methodology itself. Where the terms of reference and culture of the group are such that a clear recommendation for action is required, especially when such a recommendation requires clear justification in an open forum, then it is our view that the value function methods described in Chapter 5 are particularly well-suited. Where such terms of reference and/or culture are, on the other hand, such that the primary aim is to provide succinct summaries to higher level decision makers concerning the pros and cons of alternative courses of action, then the outranking methods of Chapter 8 may be particularly useful. For a relatively homogeneous group seek-

9.3.2. Interactions with and between participants

ing a rapid solution to their own satisfaction, not requiring justification to outsiders, the interactive and aspiration-level methods of Chapters 6 and 7 are useful tools. We shall take up this theme in Section 11.2 when discussing the integration of methodologies of MCDA.

9.3.2. Interactions with and between participants

As already mentioned, the approach to modelling adopted has implications for the interactions between participants in the decision making process, and for the way in which the facilitator is involved in and manages those interactions. On the one hand, the facilitator may work largely with individuals, bringing the group together only occasionally; at the other extreme, all work may be carried out with the whole group. The group may meet only once, for an intensive period, or may meet on a number of occasions over a longer period of time. The best way of working depends on the nature of the issue and on the roles of the participants.

If the aim is to create a shared understanding of the issue, enabling participants to learn from each other, then working together as a group is essential. One such approach is the decision conference pioneered by Peterson, a decision analyst, in the late 1970's (see Phillips, 1989, 1990). A decision conference is an intensive, 2 or 3 day, problem solving session supported by a facilitator and an analyst who make use of on-the-spot computer modelling. The facilitator attends to process and structure of the decision conference; the analyst is concerned with computer modelling to support the facilitator. The issue under consideration must be a live one and the views of all important stakeholders must be represented in the group. Consultation of previously prepared material is not allowed during the problem formulation stage. Intensive workshops involving key actors, although possibly spread over a longer period of time, are also characteristic of other approaches to problem solving, for example, SODA and Strategic Choice, and represent the authors' preferred way of working with MCDA (e.g., Belton 1985, 1993, and Stewart and Joubert, 1998).

However, even in circumstances where the aim is to create shared understanding it may be appropriate to do some work with individual members of the group. This may be justified on the basis of limited time available for working together as a group. The time spent working together as a group might be early on at the stage of problem structuring and model building, leading to the development of a common model which can be used by individuals to evaluate options (for comparison). Alternatively, the group may come together to validate and make shared use of a common model developed by merging together a series of models

270   *MULTIPLE CRITERIA DECISION ANALYSIS*

built by working with individuals. This is an approach adopted within the SODA methodology (Eden, 1989) and one which has been used successfully by one of the authors to build a value tree to be used in a group workshop. Other reasons for working initially with individuals may be to ensure that everyone has full opportunity to contribute their ideas, or to ensure that they give the issue full attention without being able to hide in the group.

If, however, the aim is to allow an overarching decision making group to benefit from the differing expertises of individuals, as described above, then it may not be necessary for those individuals to meet, or at least not for all to meet at the same time. The facilitator can work on a one-to-one basis with the individual experts, or with small expert or special interest groups, bringing their analyses together for a decision workshop involving only the overarching group.

In summary, then, the facilitator or decision analyst may need to work with relatively large groups involving representatives of all important stakeholders, smaller groups with specialized expertise or specific interests, or groups involving only those responsible for the final decision. In some cases, these "groups" may be very small, involving perhaps only two or three people, or may even be a single person, so that we have an essentially one-to-one relationship between decision analyst and client. We return to consideration of this context later.

In the context of more substantial groups, we must pay more attention to the group dynamics in implementing MCDA. Cropper (1990) describes the process of group decision support as a "happy medium between facilitation of human interaction and intellectual analysis of a problem", the synthesis of a process of assisting social interaction and commitment making together with an analytical or intellectual process of problem-solving. Phillips, in a number of conference presentations, has spoken of the socio-technical process of decision analysis. Thus, facilitation calls for management of the social process alongside analytical expertise. Just as decisions about process and analysis are interrelated, the skills of process management and content analysis must be effectively interwoven to achieve successful intervention (Eden, 1990). The two skills become interdependent, with process management being informed by analysis of content, which in turn is influenced by the analysis of process issues.

9.3.3.   The role of the facilitators

Before continuing with a discussion of the tasks a facilitator faces in organizing and conducting a workshop we feel it is appropriate to spend a little time discussing the nature of their role. An important question is the extent to which the facilitator / analyst should have and should bring to the table some "content knowledge", i.e. of the technical or organizational issues surrounding the decision problem. It is widely accepted that facilitators, although needing to concern themselves with methods for content analysis, should not contribute directly to content. To do so would negate their neutral stance and may also threaten their position as a facilitator. On the other hand, content knowledge on the part of at least someone in the decision analysis team can assist considerably in reflecting back questions to the group, in understanding the technical jargon, in assisting with the formulation of scenarios, and in recognizing problems such as double counting or preferential dependencies. Analysts or facilitators should, however, consciously guard against becoming part of the decision making process itself, even when clients urge them to express opinions (as has happened to us on occasions).

The facilitator's role is very much that of a process consultant (Schein 1999), one which seeks to help the client(s) to perceive and understand a situation in order to take action to improve it for themselves. A facilitator or decision analyst cannot tell a client what is the "preferred" solution to their problem, and can only help them find it for themselves. It is the client who owns both the problem and the solution, as Schein (1999, p20) writes "It is not my job to take the clients problems onto my own shoulders, nor is it my job to offer advice and solutions for situations in which I do not live myself".

9.4. ORGANIZING AND FACILITATING A DECISION WORKSHOP

9.4.1. Preliminary issues

As discussed in the preceeding section, the implementation of MCDA tools and thinking processes often needs to be carried out within some form of decision workshop. Decision analysts thus need to concern themselves not only with the technicalities of the MCDA and associated software, but also with the practicalities of organizing and running the workshop itself. The facilitators' responsibilities include ensuring that an appropriate venue is available and that it is appropriately equipped. Eden and Radford, in their introduction to the volume on "Tackling Strategic Problems" (1990), comment that designing and creating an effective environment for the group to work in is "... probably the most important element in the success of group decision support." We discuss this, often overlooked, aspect of facilitation in more detail below.

It is important that clear expectations are set at the outset of the meeting. If necessary, introductions should be made, and the role of the

272 *MULTIPLE CRITERIA DECISION ANALYSIS*

facilitator(s) and any other non-participants (e.g. observers, or those operating computer systems) should be made clear. The facilitator should explain the proposed process, including any ground rules, review the agenda and timetable and before moving on, check for concerns of group members. Note that although it is useful to have an initial agenda, it is also important to be able to work flexibly according to the group's needs in order to maintain interest and creativity.

9.4.2. Running the workshop

As indicated earlier, many approaches to decision support that adopt a decision conferencing / workshop approach make use of two facilitators during the meeting. The main aim of this is to allow one facilitator to attend principally to process issues, ensuring that all group members are given an opportunity to contribute and drawing in those who may be less socially skilled or less dominant than others. In order to be able to intervene effectively the facilitator must endeavour to understand what is going on in the group, observing verbal and non-verbal cues. Meanwhile, the second facilitator is concerned with capturing and managing content, often using a computer model. This means that group members should not feel a need to take notes, leaving them free to contribute ideas and reflect on issues. During the meeting the role of the facilitators is to guide the analytical process whilst paying attention to social process. Phillips and Phillips (1993) cite six ways in which a facilitator can intervene to assist the group process; these, together with our own experiences, form the basis of the following list:

Pacing the task - making sure the group achieves its objectives in the time available: It is important to ensure that the initial schedule is a realistic one and does not seek to achieve too much in the time available. However, flexibility is important and the facilitator should be able to adapt to meet the group's needs. It is also important that the facilitator consults the group about significant departures from the proposed schedule and that they accept responsibility for changes. Some group support systems advocate the allocation of strict time allowances to particular activities and the use of an alarm to signal the time to move on. Whilst this might be beneficial in workshops, or parts of a workshop, focusing on creativity (e.g. while generating alternative courses of action), we do not feel it is appropriate when seeking to negotiate shared understanding and commitment which is key feature of primary interest in decision analysis.

Directing the group - e.g. by introducing a new activity: It is important that the group members are engaged by the workshop, so that they do not lose interest. This can be achieved by building in changes in the way in which the participants are engaged in either or both of the content and process. Different forms of process include: open discussion; individual idea generation; Post-It exercises (cf. Section 3.3); round robins; breakout groups; idea writing; and simple voting.

The nature of the content of the workshop activity will change as the process of modelling progresses from problem structuring, through model building and model use. It can be varied further by asking participants to adopt particular perspectives, as we have described in Section 3.3 for the structuring phase, for example:

- *A particular stakeholder viewpoint:* For example, considering the viewpoint of another stakeholder, think of how might they weight criteria? How might they might respond if a particular alternative were chosen?

- *Positive or negative review:* What might turn out particularly well, or particularly badly if a specific alternative were chosen?

- *Angel's or devil's advocate:* Have all the strong / weak points of a specific alternative been considered?

These activities and processes are not mutually exclusive; they can be combined to yield many different ways of working and designed to suit the nature of the issue, the group and the facilities available. Breakout groups (dividing the group into smaller groups which report back to the main group) are particularly useful when working with larger groups and provide a way of ensuring that everyone remains actively engaged in the process. We have used them with success at all stages of an analysis, for example:

- to identify stakeholders;

- to consider a decision from the perspective of a particular stakeholder;

- to brainstorm alternatives;

- to "trial" an MCDA model with a small set of alternatives.

Questioning - e.g. asking for explanation of jargon, or clarification of ambiguous concepts: Sometimes this can also help members of the group, who feel they ought to be familiar with certain concepts and do not want to lose face by asking for clarification. Any potential ambiguities are likely to mask a number of different interpretations which come to light when a more explicit description is sought. Examples of such ambiguities may be criteria definitions such as: quality

274   *MULTIPLE CRITERIA DECISION ANALYSIS*

of service, a good image, a friendly working atmosphere, an equitable solution. Even definitions of decision alternatives may contain ambiguities such as timing of implementation.

Reflecting back what the group is saying: This may be simply in the form *"what you seem to be saying is ......"*, for example, *"what you seem to be saying is that there are actually two client groups with differing needs, so we cannot simply think in terms of satisfying client needs, we have to consider the two groups separately – is that right?"*. This helps the facilitator make sense of an issue and helps to ensure that the issue is interpreted in the same way by all participants. It prompts reflection on what has been said and may challenge the participants' thinking. Encouraging such reflection is important at all stages of an analysis. Similar to this is handing back in changed form – that is, reflecting back what has been observed from a different perspective.

Summarising - possibly verbally, or by a diagram or drawing: This is another form of reflecting back, possibly in a different manner, but at a more holistic level. A facilitator should also be aware of differing preferred communication styles – whilst some group members may be able to quickly assimilate a verbal summary, others may be more comfortable with visual presentations (such as flowcharts or decision trees), and vice-versa. Effective software tools can facilitate visual representation of information, but the facilitator must take care not to confuse some participants and thereby cause them to disengage from the process.

9.4.3.   Closing the meeting

It is important that the facilitator brings the meeting to a proper close. This should include a review of the workshop activity and how it relates to the objectives and agenda specified at the start, plus a review of the agreed action plan. The nature of the action plan will depend on how the workshop fits within the larger decision making process. A final workshop may plan for implementation of a decision, or presentation of recommendations to another body. An interim workshop should make clear who is responsible for what actions in preparation for the next workshop, as well as scheduling the next meeting and agreeing an agenda if appropriate. In both cases, some work may fall to the facilitator, some to members of the decision making group.

9.4.4. Housekeeping Issues

To conclude this discussion of facilitating decision workshops, we turn briefly to what may be perceived as the trivialities of the process. These, as Huxham (1990) points out, are important elements of a successful intervention – particularly in the workshop setting – but are often overlooked. The issues are common to any facilitated process rather than specific to MCDA and the detail will depend on ones preferred mode of working. Factors to consider include:

- The space available and the organisation of seating: Can participants see and hear each other and the facilitator? Is there space for breakout groups to work?

- Is there plenty of wall space to which flipcharts can be attached? Is this easily seen by all participants?

- If computer software is to be used, are there good projection facilities available?

- Does the facilitator have a good set of pens for writing on flipcharts or the whiteboard? Is the facilitator's writing legible?

- Heating, lighting and ventilation: If the participants are uncomfortable they will not be focused on the workshop. Working in an artificially lit room for long periods can be very tiring.

9.5. WORKING "OFF-LINE" AND IN THE BACKROOM

We use the term backroom work to describe work by the facilitator or analyst, developing or using the multicriteria model, but not involving direct interaction with participants. This may take place during meetings or workshops as described in the previous section, either "off-line" by one of the facilitators, or during meeting breaks (overnight, or over lunch, dinner or coffee breaks). Alternatively, more comprehensive background work might be undertaken between meetings with decision makers or stakeholder groups. Backroom work may be appropriate at any stage of an intervention in order to progress, reflect on, or explore the process or model. For example:

- To progress the process: for example, to merge individual models into a group model or to move from the loosely structured output of an idea generation process to an initial model.

276 *MULTIPLE CRITERIA DECISION ANALYSIS*

- To reflect on the product of a meeting or workshop: for example, to consider alternative models of, or ways of modelling, a particular situation.

- To transfer a model from paper to an appropriate software.

- To explore a model, in order to identify key sensitivities, either in preparation for a meeting (in order to be able to guide participants in exploring these and ensure they are not overlooked), or during the meeting (to identify what additional information may need to be collated between meetings), or subsequent to a meeting (to ensure that no key sensitivities were overlooked).

- To identify implications of model inputs, to highlight implicit trade-offs such as monetary values implied by the weightings used.

- To prepare alternative displays of information, possibly to develop new modelling capabilities.

- To prepare documentation.

Clearly the amount of such backroom work carried out during meetings will be limited, but is nonetheless crucial to "reflecting back" to participants their judgements and the implications thereof, while issues are still fresh in their minds. In particular, the transfer of verbal and written models into available software, and the use of software to generate robustness and sensitivity analyses and implied tradeoffs, is crucial to the success of a decision workshop.

The time between meetings or workshops provides opportunity for examining the same decision problem from the point of view of different decision aiding methodologies. The present authors have found the use of the value measurement approaches described in Chapter 5 to be particularly useful in guiding analysis and evaluation during decision workshops. By definition, key stakeholders are present and can directly express value judgements (qualitatively where necessary) relating to the issues at hand, so that value function model of preferences can be constructed relatively quickly and easily. These judgements may, of course, still be "fuzzy" in the sense that the numerical judgements need not be precise, while any ordering of the alternatives based on the value function models subject to sensitivity analysis may not be complete, and this provides scope for further analysis between workshops.

Such background work between workshops will, to an even greater extent than that carried out during meetings, be done by the analyst without the opportunity for direct interaction with the decision maker. This work will, nevertheless, be informed by the partial orderings of alternatives and other value judgements emerging from the workshop. Besides the general background work described above, more detailed analyses using other methodologies might include the following:

Identifying justifiable conclusions: What conclusions concerning the adoption or rejection of specific alternatives can be justified at this stage in the light of the available preference information? The outranking methods described in Chapter 8, particularly the various ELECTRE methods, are directed towards precisely this type of question. The entire concept of outranking is based on the strength of evidence supporting a claim that one alternative is at least as good as another, and thus the various partial rank orders generated by the ELECTRE approaches highlight directly what conclusions may or may not be justifiable on the basis of the current information. Such evidence may be useful either in preparing motivations to be placed before final decision makers, or as summaries to be tabled at the next meeting or workshop where attention can be focused specifically on incomparabilities which are revealed. What further information or construction of preferences would be needed to resolve these in the search for a final decision?

Generating other alternatives: During a meeting or decision workshop, it is usually necessary to focus on a relatively small number of decision alternatives ("7±2", or even fewer), as the comparisons between these need to be discussed in considerable depth, often involving a substantial level of subjective judgement. In many instances, these may only be representative of a much larger range of options, as discussed for the land-use planning case study in Section 3.5.3. When the results emanating from one meeting or workshop do not clearly identify a single "winning" alternative, there may be scope for a search amongst a wider set of alternatives, to identify additional options which offer potentially better compromises. Various MCDA models can be utilized for this purpose. However, since in this "backroom" mode of work the analyst will not have direct access to value judgements from decision makers regarding any new alternatives generated, the multicriteria analysis will need to be applied only to the more objectively quantifiable attributes of alternatives. Nevertheless, preferences expressed during the meeting or workshop regarding the alternatives under consideration at that time can often still be utilized in conjunction with the more objective data. Preferences expressed by the group might well have included quite intuitive holistic judgements, but provided that there exists some degree of correlation between such subjectively determined preferences and the measurable

278   *MULTIPLE CRITERIA DECISION ANALYSIS*

attributes, the observed preferences will provide information to guide the search for other alternatives. Approaches which may be adopted in this context include the following:

- *Model fitting from inverse preferences:* The UTA method and related approaches discussed in Section 6.3 are aimed at extrapolating preferences from a reference set of alternatives (in this context those considered at the meeting) to a wider set, and are thus directly relevant to identifying potentially good alternatives that have not previously been evaluated in detail. In some cases, it may be useful to extrapolate the preferences to a full multiple objective linear programming structure, by using ranges of piecewise linear value functions consistent with the preferences in the reference set, much as illustrated in Section 6.4.3. In essence, instead of using the trade-off information described in that section, the holistic constraints (6.13) would be applied to limit the range of feasible value functions.

- *Outranking methods:* One approach may be to screen sets of possible alternatives that have not been directly evaluated by decision makers or stakeholder groups, in order to find those that may outrank one or more of the two or three best alternatives identified in the workshops or meetings. Alternatively, it may be useful to use the results from the meetings to generate profiles of good alternatives, and to evaluate the as yet unevaluated alternatives against these, to obtain a classification much as in the ELECTRE TRI method discussed in Section 8.5.

- *Goal programming or reference point methods:* The results of a decision workshop will include some indication of the characteristics of desirable outcomes, either by identifying one or two "satisfactory" alternatives, or by recording how the best alternatives need to be modified in order to become more "satisfactory". The resultant benchmarks can then serve as reference points or goals, when applying goal programming or reference point methods to the wider set of alternatives.

Other participants in meetings or decision workshops (decision makers, advisors, stakeholders) should also be encouraged to do work on the problems between meetings (and not just to rely on the outputs of the MCDA process). They might be urged, *inter alia* (a) to reflect on the product of a meeting (for example, do the value tree and criteria fully capture all important issues?); (b) to generate inputs for a model (for example, to seek information on the performance of alternatives or to evaluate performance against agreed scales); or (c) to explore the decision models for themselves (where appropriate software is available).

9.6. SMALL GROUP INTERACTION AND DIY ANALYSIS

As indicated in the Introduction to this chapter, the discussion up to now has largely been in the context of workshops or decision conferences in which the facilitator assists a group representing many stakeholders, or at least some divergence of interests, to find a satisfactory consensus or compromise solution to a decision problem. For decisions of significant consequence, involving multiple parties, we believe that this will always be the most appropriate way of working; the facilitator takes responsibility for directing the analytical process as well as managing the group. It is important in such contexts for facilitators to ensure confidence and trust in their impartiality, in the process adopted and in the decision models used. In particular, there is a need for transparency in the processes and models, and for a clear audit trail recording value and preference judgements made, and the means by which these are translated into the final recommendations. Much of our discussion has been directed towards this end.

There are situations, however, in which the decision analyst is called in to assist a small and relatively homogeneous group, or even a single individual, in reaching conclusions about which course of action best satisfies their goals or objectives. Such persons or groups (the "clients") may need simply to reach their own decisions about important but individual issues that do not affect others. Alternatively, it may be that the client is trying to formulate a position to take in discussions or negotiations with a broader group. In either case, clients wish to gain understanding of their problem, and to be satisfied that the best answer has been reached, but may not need to document the full procedure or to justify the decision to others.

In such contexts, the relationship between decision analyst and client can generally be a lot less formal than discussed earlier in this chapter. The success of the process will probably depend more on personal relationships between client and analyst, and on the latter's competence in the MCDA tools being used, than anything else. Quite frequently, the analyst will be a single individual, who will both guide and facilitate the exploration and thinking about the problems (structuring and analysis), and implement the MCDA methods using relevant software. Although in some cases we might use precisely the same approaches and methods as with a larger group, the informal atmosphere and reduced need for a detailed audit trail may lend itself to using some of the "interactive" methods described in Chapters 6 (value functions) and 7 (goal programming and reference point methods). These allow the client to play around with alternatives directly, making direct global comparisons between different feasible sets of outcomes, seeing how achievements on the different criteria shift in moving from one option to another.

For decision analysis at such small group or personal levels, it need not even be necessary for the process to be facilitated by an expert decision analyst. Knowledgeable individuals or small groups can use MCDA approaches to help themselves to explore an issue. We refer to such use as "DIY" (Do-It-Yourself) MCDA. While there is no doubt that decision makers at all levels do wrestle frequently with multicriteria problems, there is little evidence in the literature of formal analysis being conducted at this level, except by expert MCDA analysts exploring their own personal problems (e.g. Keeney, 1992). We conjecture that people very often use their own pet heuristics in dealing with multicriteria problems, such as the advice given by Benjamin Franklin to Joseph Priestley in a letter dated September 1772, seemingly first cited in the MCDA context by MacCrimmon (1973). These heuristics may well lead to the re-invention of wheels, but perhaps also to falling into many of the potential traps which we have attempted to identify. It is our desire that the present book will encourage better practice in the DIY framework, as we feel it is an area which deserves attention because of pressures in two directions. The first is the "dumbing down" of MCDA techniques which may make them more appealing to generalist audiences, as evidenced by the book on "Smart Choices" by Hammond, Keeney and Raiffa (1999), which addresses the difficult issue of trade-offs through the concept of the even-swap. Secondly, there is a pool of middle and senior managers which is becoming increasingly educated in formal methods of management through MBA programmes. These managers are more technologically aware and have had greater exposure to modelling than their predecessors, two factors which should make them more receptive to the use of multicriteria analysis. This predisposition may make them more inclined to use MCDA without the support of a trained analyst/facilitator.

However, the notion of DIY MCDA opens up new and interesting challenges, both in the education of potential users and in the development of appropriate softwares. There is a need to guide potential users through all stages of the process: in appropriately framing the decision; in determining an appropriate approach; in building good models – selecting a good set of criteria, or building an appropriate value tree; in defining valid scales for evaluating options; in explaining outcomes; and in ensuring that results are fully explored and that thinking is challenged. Without such guidance, there is a real danger that users may fall into the trap of accepting a solution because the "model says so", or because it is the "rational objective thing to do", when in fact it may be far from satisfying the decision maker's true goals. So called "intelligent" multicriteria decision support systems may come to play an important role here. Angehrn (1990) described an interesting early specification of such a system. Recent work with MBA students and other "naive" users of software to support multicriteria analysis indicates that intelligent features provided to support the interpretation of results to encourage sensitivity analysis, tend to be used in a systematic way as a checklist to ensure the thoroughness of an investigation (Belton and Hodgkin, 1999).

9.7. SOFTWARE SUPPORT

Good supporting software is essential for the effective conduct of multicriteria analysis in practice. Software to be used when working directly with decision makers should be visual and interactive to facilitate communication about the model and results. Interactivity permits information on evaluations, values and parameters of a model to be easily entered and changed - possibly using visual representations. Effective visual displays are also the means of reflecting back to the decision makers information provided and judgements made, and the synthesis of these through the multicriteria model. This synthesis may be, for example: an outranking graph, an aggregate evaluation, a visual representation of an inferred value function, or a move to a new region of decision space. Even if software is used only in backroom work the facilitator / analyst will benefit from an effective visual and interactive implementation.

The approach to modelling adopted and the nature of the interaction between the facilitator and individuals places further demands on the supporting software. If a sharing approach is adopted then the model required is essentially one for individual support - a single set of information is input. However, if an aggregating or comparing approach is adopted, then a separate set of inputs is generated for each individual and it must be possible to aggregate of compare these using the supporting software. In either approach, effective visualization tools may be helpful in creating understanding of the essence of the issues.

There are now many multicriteria software tools available, most of which focus on a single technique, and a list of those which we are aware are currently available is given in Appendix A. These include a number of commercially available packages for the most widely used techniques. Most software systems were originally designed for individual support, or a group working in the sharing mode – i.e. allowing the input of a single set of information. However, with rapidly developing technology a number of "group" systems have been developed, allowing more than one user to independently input their own evaluations and providing facilities for synthesising and displaying this information.

There are, in addition, a number of other tools which incorporate some form of multicriteria analysis or thinking (e.g. *Group Systems, STRAD*), even though they are not MCDA tools in the full sense as defined here. Unfortunately, some such systems incorporate little more than simple scoring and weighting models, paying little attention to issues of scaling and measurement, or to the meaning of the weights employed (as discussed, for example, in Section 9.9). Care, and critical evaluation should thus precede the use of such systems.

If customised software is not available most multicriteria analyses can be carried out with the help of a spreadsheet. Furthermore, most commercially available tools allow data to be imported from or exported to a spreadsheet. This allows easy access to existing data and enables the design of customised data displays based on the output of a model.

As computer technology continues to develop at a rapid rate we will certainly see developments in supporting software. Multimedia input – photographs, videos, sound – can now be incorporated to enhance models. Networked systems allow multiple users to access a shared model; they may be working at the same time in the same place using a local area network (in a decision conference or workshop), or they may be geographically dispersed and accessing the model at different times over the Internet.

There are, of course, a number of practical issues relating to the use of software for multicriteria analysis. It should be used to support the process, not as the driving or dominating force. The use of technology can be very low key with the majority of the modelling and elicitation of values being paper-based (or using post-its and flipcharts or whiteboard), relying only on the software for the synthesis of information and presentation of "results". On the other hand, the use of technology can be central, with each participant entering their own evaluations directly into the computer and subsequently exploring the results. The facilitator needs to take account of the skills and attitudes of the participants, as well as the available technology, in designing an appropriate process.

It should go without saying that the facilitator should be highly skilled in the use of supporting software as well as the underlying multicriteria method.

9.8. INSIGHTS FROM MCDA IN PRACTICE

In Chapter 1 we expressed our view that MCDA is not about finding the "right answer", but is a process which seeks to help decision makers learn about and better understand they problem they face, their own values and priorities, and the different perspectives of other stakeholders. Throughout the book we have sought to present the approaches we have described as relevant at each stage of the overall process in this context, emphasizing the ever-present potential for new insights, from the qualitative stages of problem structuring through to sensitivity and robustness analyses founded on sophisticated mathematical analysis. Such insights may simply derive from conversations about the issue with others participating in the process, or from trying to resolve a misunderstanding or ambiguity; or they may arise as a result of a challenge to intuition posed by an outcome of the model. However, in our writing so far we have focused on illustrating different approaches through reference to the case studies and have only hinted in passing at the possibility of such insights. In this section we seek to give a fuller picture, based on our own experience of and knowledge about the use of MCDA in practice, of the ways in which decision makers are supported in their thinking and the nature of insights which can arise − in essence, a picture of some of the learning which occurs.

Analysis versus intuition

We have mentioned at several points that we see one of the roles of the MCDA model as a sounding board against which the decision makers can test their intuition, a view also strongly expressed by Goodwin and Wright (1997) and other decision analysts. Some of our most memorable interventions in organizations have been those in which the multicriteria analysis has brought about a strong challenge to the decision making group's intuition. That is not to say that studies in which this has not happened have not been equally valuable to the participants. Such a challenge usually emerges when the option which the group intuitively preferred does not emerge as "best" in the initial analysis, or even more seriously, not even anywhere near best. This may cause the group to question their intuition and possibly to revise their opinion, or to revisit the model, questioning whether important criteria were omitted. The following are examples of such challenges and how they were resolved.

284 MULTIPLE CRITERIA DECISION ANALYSIS

Choosing an aerial policing facility

A regional police force had obtained financial approval to invest in a means of policing from the air (this could be a helicopter or fixed-wing aircraft, or other less conventional means), but being a publicly funded body were required to carry out a full evaluation of the costs and benefits of different options. The senior officers responsible for the decision all felt, intuitively, that the most expensive option – a twin-engined helicopter – must represent the best value option. A multicriteria evaluation considered costs and performance, defined by the operational needs, ease of operation and public perception (each of which was broken down into greater detail). The analysis revealed that not only was the twin-engined helicopter not the most highly rated option, but it did not even appear on the efficient frontier of benefits versus costs, as illustrated in Figure 9.1. Furthermore, this position was insensitive to changes in criteria weights. Each of the senior officers had carried out the analysis individually, using an agreed value tree to rate the same set of options, and each had obtained the same result. The analysis caused the decision makers to realize that even though the twin-engined helicopter was the only option able to support some very sophisticated, but relatively rare aspects of their work, a regular fixed-wing aircraft was better for the majority of routine needs. This led them to the decision to purchase a fixed-wing aircraft, but only after putting in place a facility to loan a helicopter on the occasions it was needed. Thus the challenge to intuition was resolved through a partial revision of preferences as a consequence of a clearer understanding of the problem, but accompanied by the creative re-definition of one of the options.

Strategic futures for Management Services group

A large parent company had taken a decision to divest itself of part of its central management services provision. However, it wished to support the group in establishing a future for itself. Initially three options were considered, an independent spin-off company, a wholly owned subsidiary of the parent company and continuation in-house but at a much reduced level of operation. A steering group comprising a representative of the parent company and senior personnel in the management services group (who would become the senior management in the new organization) was formed to give full consideration to the decision. After some months spent researching the options the group wished to hold a decision workshop to formalize their evaluation, this was facilitated by one of the authors. The workshop very quickly went through the process of evaluating the three options and confirming the

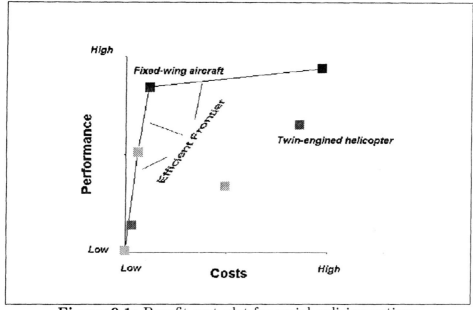

Figure 9.1: Benefit-cost plot for aerial policing options

group's preference for the spin-off company. However, this process did not appear to have really challenged their thinking and the facilitators suggested that the group consider two other possible futures: a merger with another group in a similar position, or selling out to a large management consultancy. The group agreed to this and revisited the evaluation to include the two new options, which, to their surprise and discomfort, outperformed the preferred spin-off company. A lively discussion ensued, which accepted the superior performance of the new options on the criteria initially considered, in particular with respect to reducing risk and maximizing opportunities for future business, but led to the identification of new factors relating to anticipated challenges and interest in setting up and running a new business, and the ensuing power which would be enjoyed by many of the members of the decision making group. They went back to the parent organization, not with their intuitive view confirmed, but with new options to consider and hard thinking to be done.

The ideal office location?
The director of a small start-up company spun-off from university research activity, faced with the decision about the nature of office accommodation to rent, was considering a number of typical options. These

286   *MULTIPLE CRITERIA DECISION ANALYSIS*

included accommodation on a newly developed and very prestigious "science park", in a specialized university facility for start-up companies, and several city-centre premises including recently refurbished accommodation in a spacious and elegant Victorian building of architectural interest. It was to this latter type of accommodation that the director was particularly attracted. However, an analysis taking into account costs, the nature of the accommodation, accessibility and image failed to rate the Victorian building, or any of the city-centre premises highly. The director was forced to admit to himself that his romantic / idealist notion of an office location was not compatible with practical issues and opted for one of the other options. Note that this was not the inevitable outcome; it is possible that he may have decided that it was worth paying the extra rent and sacrificing some of the advantages of the other offices (in particular accessibility and flexibility with respect to space) in order to satisfy his ideals.

Confirming intuition

The examples reported above should not be taken as a denial of the value of analyses which confirm intuitions. We have been involved in many such cases, which have furthered learning, leading to greater shared understanding, greater confidence in the decision and a greater commitment from the decision making group as a whole. Comments such as, "I didn't know that was an option for us", "I didn't know they were thinking of doing that" or "I didn't know that you were concerned about that", are common. The following is a quote from the CEO of a small company involved in a decision workshop to explore alternative strategic futures: "The decision support process was pivotal in that this was the first time that everyone was together talking about issues. It was a surprise that not everyone knew about some of the issues. Communication wasn't as good as we thought, even though it's such an open company."

What if there is no intuition?

We have also worked on cases in which we believe that there was no prior intuition regarding a preferred option. This is particularly so if initial analysis is carried out by expert subgroups, meaning that no individual has a detailed overview of the issue. The decision about how to allocate the tender to develop a financial information system, described by Belton (1985), is one such case. The responses to an invitation to tender were narrowed down to a shortlist of three companies which were subjected to extensive investigation from a team appointed to make the decision, with individuals or smaller groups focusing on specific aspects of the project, such as systems issues or strategic issues. Although broadly aware of each others' work it was only when all the analyses were brought together in a workshop intended to arrive at a final recommendation that it became apparent that one of the three options was virtually dominated by the other two. This option had only one strength, on a factor which was not felt to be of high importance, and on all other criteria it was outperformed by the other two proposals. It was dropped from further consideration early in the workshop and attention was focused on the very difficult issue of choosing between the two remaining proposals.

What if a clear preference does not emerge?

A number of analyses have reached a stage when it is clear that it is not possible to identify a preferred option. This may be a situation such as the one just described, in which more than one option is judged to be almost equally good and it is difficult to choose between them. Or it may be the case that the best options available have very different strengths and weaknesses, but none is considered to be good enough on its own. Or it may be that an option that is generally good has a significant weakness in some respect. These are circumstances which may be highlighted by an analysis based on outranking methods, which can identify incomparable options, but would be masked by the overall scores of a value function analysis, illustrating the importance of a broader consideration of performance.

These are all circumstances we have encountered in analyses we have facilitated. In the first case, the difficulty of choosing between two similar, almost equally good options, we would suggest the use of other tools to aid thinking. In the case referred to above we used positive and negative reviews (angel's and devil's advocate) to focus the decision makers' thinking on what might be unexpected bonuses from each choice, or on things which could go disastrously wrong.

Faced with a situation in which the options being evaluated each demonstrated very different strengths, but none satisfied all requirements, we looked to explore other possible options. Convinced that there were no other options available, and faced with the need to take an immediate decision the decision makers decided to explore creating an option which was a combination of two of those available. The context was the need to employ a contractor to complete a complex, but urgent piece of work. Each of the contractors bidding for the work was highly skilled in some aspect of it, but less so in other areas. The outcome was to negotiate an agreement to work together between two contractors who between them had all the required skills.

288 *MULTIPLE CRITERIA DECISION ANALYSIS*

An illustration of a situation in which an option which was generally good had a significant weakness relative to other choices was encountered by an international company, trying to settle on an exclusive travel deal with one of the airlines which served the cities in which their main regional offices were located. The initial analysis revealed that the intuitively preferred option, which performed well in most respects, was significantly less attractive financially than some of the competitors. In this case the company used the analysis to confront the airline and managed to negotiate a better deal.

Highlighting stark trade-offs

In the cases described so far the ultimate outcome was generally satisfactory, in the sense that the analysis and intuition were reconciled, either by the creative generation of a new option, or reconsideration of preferences. However, it is not always possible to achieve such an outcome. Sometimes what the analysis achieves is to highlight a stark and difficult trade-off, which cannot be reconciled. Such a situation arose when an organization operating on two sites, located about 60 miles apart and each serving a distinct large centre of population, was looking to rationalize the provision of a key service offered at both sites. The decision facing the company was whether to concentrate the service on one or the other of the sites, or to continue with the current arrangement. Analysis of the options revealed that concentration on one or other of the sites would save of the order of half a million pounds per annum and was best in most other respects, except for the support from a key and very powerful group of personnel who wished to preserve the status quo. Lack of cooperation from this group could severely jeopardize the operation of a merged facility. No way was found of pacifying this group and the decision boiled down to whether or not to keep them happy or to save half a million pounds each year. It was a sufficiently powerful group that the decision was in their favour.

9.9. INTERPRETATION AND ASSESSMENT OF IMPORTANCE WEIGHTS

In the discussion of preference models in Chapter 4, we emphasized (Section 4.7) the important issue of the interpretation of the weight parameters in different preference models. We have attempted to provide the reader with some understanding of the role of these weight parameters in the implementation of the various methodologies of MCDA.

One of the critical challenges facing the analyst in guiding decision makers through the process of MCDA is to ensure that the clients do

Implementation of MCDA: Practical Issues and Insights 289

correctly understand and interpret the meaning of the weights when they are asked to provide numerical inputs. As was recognized in Section 4.7, weight parameters may have widely differing interpretations for different methodologies and different decision contexts, so that we must be wary of any approach to assessment of weights which does not take such differences into account.

Let us briefly recall some of the key points regarding weight assessment which have emerged from our discussion of the three broad schools of MCDA.

Value function methods: As indicated in Section 4.7, the algebraic meaning of the weight parameters is perhaps more clearly defined in value function methods than for any of the other MCDA methods. The weights are directly related to tradeoffs between the criteria, expressed in terms of the partial value scales generated, which has *inter alia* the implication that if the scales are changed, the weights need to be changed accordingly. On the other hand, as discussed in Section 4.7, people are prone not to adjust direct assessments of relative importance (sufficiently or at all) in the light of the changed ranges of outcomes which lead to a re-scaling of the partial value functions. The implication is that *directly assessed importance ratios are almost certainly inappropriate for use in a value function model!* It is for this reason that we introduced the concept of swing weights in Section 5.4. In the assessment of swing weights, questions are posed in such a way that the attention of respondents is focused directly on the available tradeoffs. As noted in Chapter 5, we explicitly avoid questions which involve the less well-defined notion of "importance weights" in the abstract, since these may generate highly misleading results if the intuitive notion of importance and the desired tradeoff ratios do not coincide. Unfortunately, not all users of value function methods appear to have taken heed of this advice. The rank reversal phenomenon in AHP (mentioned in Section 5.7) has its origin in the fact that the introduction of a new alternative results automatically in the rescaling of the value scores (because of the normalization of scores to sum to one), but the relative magnitudes of the weights are unchanged, having been assessed independently of the alternatives.

Generalized goal programming methods: In preemptive goal programming there are no weight parameters to assess, so that the problems of weight assessment are seemingly avoided. Even here, however, the prioritization of goals should take account of the ranges of achievable outcomes. It is not clear that this is always done, but the use of "swing-weighting" concepts would appear to be relevant here as well.

290 *MULTIPLE CRITERIA DECISION ANALYSIS*

Generally speaking, however, even in the other variations of goal programming (Archimedean and Tchebycheff) and in the related reference point methods, weight parameters do not play the same crucial role as in value function methodologies, since the importance of a criterion is expressed to a large extent by the goals or aspiration levels specified. Nevertheless, the weights do in effect represent a re-scaling of the attribute values, intended to ensure appropriate levels of trade-off between deviations, i.e. between attribute values in the vicinity of the goals or aspiration levels. Care is thus necessary to ensure that if deviations from aspiration levels on two attributes, say i and k, are such that $w_i \delta_i = w_k \delta_k$ (see Section 7.2 for definitions), then the two deviations δ_i and δ_k are indeed of approximately equal importance to the decision maker. The weights are thus ideally obtained from direct assessments of trade-offs between attributes in the vicinity of the aspiration levels. Fortunately, however, it appears that solutions from goal programming or reference point approaches are relatively insensitive to the choice of weights, provided that the goals or aspiration levels are realistically specified and the attributes are at least sensibly scaled to comparable magnitudes. This latter requirement can often be met by a simple re-scaling of attribute measures to a fixed range (e.g. 0-1).

Outranking methods: Importance weights are explicitly used in most definitions of concordance. As previously discussed, the weights represent some form of "voting power" associated with each criterion. In this sense, the concordance weights are less directly interpretable than for value function weights, so that it is also less easy to make definite recommendations as to how they should be assessed. Private discussions with practitioners of ELECTRE methods have suggested a prevailing view that outranking weights may be closer to the intuitive concept of relative importance than are the trade-off weights in value function and goal programming methods. It is difficult to see how such a view can be validated or falsified. Furthermore, it does seem that in contexts in which there is a very wide range of possible outcomes for one criterion, this criterion *should* have greater impact (i.e. greater weight) than when the range of outcomes is small, so that weights should still be adapted to the range of outcomes contrary to what is reported for example by Mousseau (1992) for intuitive weight assessment. Other private discussions with ELECTRE practitioners have indicated an approach to the assessment of concordance weights which is closely allied to the swing-weighting concept, which reinforces the view expressed in the previous sentence.

The overall responsibility of the decision analyst or facilitator, particularly with value measurement and outranking methods, is thus to ensure:

- that the decision makers are fully informed concerning the meaning of weights within the context of the model being used;

- that questions concerning importance weights are formulated in such a way that mismatch between responses and model requirements are avoided;

- that the potential effects of the cognitive biases arising from problem framing and hierarchical structuring of criteria (as discussed in Section 4.7) are fully explored as part of the sensitivity analysis.

There is scope for considerable further research on developing appropriate techniques for achieving these aims. Practical issues around weight assessment for value function methods have been described in some detail in Section 5.4. It is our belief that much of the advice presented there is appropriate to other methodologies, although literature is rather silent on this topic.

9.10. CONCLUDING COMMENTS

In this chapter we have focused on issues that are important when seeking to use in practice any of the approaches to MCDA described in this book. For the most part this will involve a facilitator or decision analyst interacting with an individual or group of decision makers, although on occasions a decision maker or small group may be self-supporting. The issues addressed span a broad range of considerations, from the conceptual underpinnings of modeling, through the important skills of facilitation, to the very practical issues of room layout. Many concerns are generic to any facilitated approach to problem solving, whilst others are specific to the use of MCDA. Although a number of these issues may appear to be "trivial" or "common-sense", for exactly that reason they may be overlooked when designing an intervention. Consideration of such matters is, however, essential in ensuring effective implementation and failure to do so can prevent realization of the potential of any MCDA approach. Thus, given that our ultimate aim is to inform decision making in practice, attention must be accorded to these practicalities.

In addition the discussion of factors which are important *for* the use of MCDA in practice, we reflect more widely than in previous chapters on the nature of insights arising *from* the use of MCDA in practice. This account is entirely anecdotal, based on our own experiences, and is intended to be illustrative rather than comprehensive.

292   *MULTIPLE CRITERIA DECISION ANALYSIS*

Finally, we returned once again to an issue which threads through this book and the subject of MCDA as a whole, that of the notion of importance and the associated concepts of criteria weights or scaling factors, focusing here on the practical perspective.

Chapter 10

MCDA IN A BROADER CONTEXT

10.1. INTRODUCTION

Our discussions so far have focused on methods for multicriteria analysis and their use in practice, at times looking at how we can enhance the practice of MCDA by drawing on methods and practice from other fields, in particular in relation to problem structuring and implementation issues. These two aspects emphasize the importance of viewing MCDA in a broader context – of seeking to be informed and supported in what we do by theory, method and experience drawn from the wider fields of Management Science and other related disciplines. At the same time, we should take the initiative to explore how MCDA can inform and support other areas.

There are many ways in which the use of MCDA in conjunction with other methods of OR/MS can lead to mutual enhancement. Exploring these possibilities, with the aim of expanding the influence of MCDA and its acceptance as a valuable tool for OR/MS practitioners, has been a particular interest of one of the authors. This work has identified a number of ways in which MCDA can work synergistically with other approaches: Belton, Ackermann and Shepherd (1997) discuss the use of the SODA methodology together with MCDA to give an *integrated* approach from problem structuring through to evaluation; Belton and Elder (1996) describe a visual interactive DSS for production scheduling in which multicriteria analysis is *embedded* in the scheduling algorithm; Belton and Vickers (1993) and Belton and Stewart (1999) discuss the *parallels* between MCDA and DEA and put forward the suggestion that an MCDA interpretation of DEA can facilitate understanding.

294   *MULTIPLE CRITERIA DECISION ANALYSIS*

In this chapter we discuss a number of actual and potential synergistic combinations, as well as opportunities to learn from developments outwith the field that have a strong multicriteria focus. We group these according to the following four broad categorizations:

a. Analytic methods of OR/MS and Statistics that have strong parallels with MCDA, allowing for cross fertilization of ideas and potential for mutual learning and benefit. In this category we consider in detail Game Theory and associated methods, Data Envelopment Analysis and methods of multivariate statistical analysis.

b. Management Science approaches which can be used to good effect in conjunction with MCDA. We have already discussed, in Chapter 3, the potential benefits of using problem structuring methods such as cognitive / cause mapping and aspects of Soft Systems Methodology (SSM) and Strategic Choice. Other links that will be discussed are with discrete event simulation, system dynamics, scenario analysis and conflict analysis.

c. Management Science and other managerial methods which have a strong multicriteria component but which have been developed without any formal links with or reference to MCDA. Examples are: the Kepner Tregoe method, Value Engineering, Kaplan and Norton's Balanced Scorecard, Quality Function Deployment and the EFQM Excellence model.

d. Application areas with a naturally strong multicriteria element, some of which have given rise to specialised analytic approaches, such as Environmental Impact Analysis, portfolio analysis, risk analysis, and quality assurance.

In the following sections we consider these in greater detail.

10.2. LINKS TO METHODS WITH ANALYTICAL PARALLELS

Although the development of MCDA as an identifiable field of OR/MS placed formal emphasis on representation of decision problems in terms of conflicting criteria, the concept is implicit in a number of other branches of OR/MS, such as:

- Game theory (e.g., Owen, 1982), and extensions thereof such as hypergame theory (e.g., Bennett et al., 1989a,b), which deal with conflicts between different "players" rather than "criteria", although the distinctions are not always clear-cut;

- Data Envelopment Analysis (DEA) (e.g., Charnes et al., 1994), which seeks to compare the efficiency of organizational units in terms of numbers of different "input" and "output" measures – The decision context is not always clear in DEA, but the presumption is that the organizational units are being compared with some ultimate action in mind.

MCDA can both learn from and contribute new insights to these areas of OR/MS, and we shall shortly deal with each in turn.

The field of multivariate statistical analysis developed to assist interpretation of multidimensional data. In many instances, these techniques are purely descriptive (for example in psychometrics or market research), and are not designed to aid decision making *per se*. Nevertheless, MCDM problems are by definition "multivariate" in nature, and the descriptive tools of multivariate statistical analysis can be exploited to give decision makers considerable insight into the available tradeoffs and the points at which hard decisions have to be made. This, too, we will deal with later in this section.

10.2.1. Game and Hypergame Theory

The earliest forms of game theory dealt with "two-person, zero-sum" games, which considered optimal strategies for two players in direct conflict, in the sense that the players act independently of each other, and that gains to one player can only be achieved by corresponding losses to the other. MCDM problems are seldom "zero-sum", however, in that ideally the search is for "win-win" situations, or at least for solutions in which no one criterion is seriously disadvantaged.

The branch of game theory which is closest in spirit to MCDA is that of cooperative non-zero-sum games. By "cooperative" is meant that the players need to act together to select a joint strategy, presumably to their mutual benefit. The emphasis is thus on group decision making, and if the interests of each player are viewed as distinct "criteria", then the correspondence between MCDA and cooperative non-zero-sum games is evident. There are, nevertheless, some interesting distinctions between the approaches adopted by MCDA and game theorists.

Much as in the usual MCDA formulation, a non-zero-sum game between m players can be described in terms the "utility" $u_i(a)$ accruing to each player i if joint strategy a is implemented. In the analysis of cooperative games introduced initially by Nash (1953), the assumption is that each player has a non-cooperative action which they can take independently of the other players if no agreement is reached, yielding a bottom-line utility irrespective of what other players do. This corre-

296 MULTIPLE CRITERIA DECISION ANALYSIS

sponds to the BATNA ("best alternative to a negotiated agreement") concept in negotiation theory. It is assumed that no player will agree to a joint strategy which offers them a lower utility than that guaranteed by their non-cooperative action. This provides a natural zero on the utility scale for each player, in that a worst case giving a zero utility (or at least a zero gain in utility from BATNA) is clearly defined and understood.

In many, but not necessarily all cases, the BATNA may be the "do-nothing" or status quo situation. In implementing MCDA, the status quo alternative is often included, but it may be useful to spend time during the problem structuring process to consider specifically whether the impacts on each criterion if no decision can be reached may be different to the current status quo. If so, the inclusion of such "BATNA"-type worst case scenarios may facilitate the process of scoring alternatives in terms of individual criteria in the light of a well-understood zero-point (allowing questions to be asked in ratio terms which seem often to be intuitively appealing).

With the above background, Nash proposed a set of axioms characterizing a desirable cooperative solution. These were initially described for the two-person game (see, for example, Owen, 1982, pp. 141-142) as follows:

Individual rationality: No player will accept a strategy which gives them a lower utility than given by their BATNA;

Pareto optimality: If for two strategies a and b, $u_i(a) \geq u_i(b)$ for both players, then a is preferred to b;

Independence of irrelevant alternatives: If a is the most preferred of a set A of strategies, then it will remain the most preferred if one or more of the remaining strategies are deleted from A;

Independence of linear transformations: The most preferred solution should not be dependent upon the scale used for defining the utilities $u_i(a)$ for each player;

Symmetry: Suppose that for any feasible strategy a yielding $u_1(a) = v$ and $u_2(a) = w$, there exists another strategy, say b, for which $u_1(b) = w$ and $u_2(b) = v$, so that the problem is symmetrical. Then (i.e. for such symmetrical problems) the most preferred solution a^* will satisfy $u_1(a^*) = u_2(a^*)$.

Nash proved that if the above axioms hold, then the most preferred solution will be that which maximizes the product $u_1(a) \times u_2(a)$ over all feasible joint strategies. Kalai (1977) showed subsequently that if the symmetry axiom is dropped, as would seem to be more realistic for most MCDM problems, then the most preferred solution would be that which maximizes:

$$[u_1(a)]^\alpha \cdot [u_2(a)]^{1-\alpha}$$

for some $0 < \alpha < 1$. In this case, the parameter α indicates the bargaining power of player 1 (which we might interpret as the "importance" of the first criterion).

As indicated by Weerahandi and Zidek (1981), the above principles can be extended to the m-player game for which the solution would be that which maximizes:

$$\prod_{i=1}^{m} [u_i(a)]^{\alpha_i}$$

where the α_i (typically normalized so that $\sum_{i=1}^{m} \alpha_i = 1$) represent the relative bargaining strengths of each player. In MCDA terms, this is equivalent to the multiplicative form of aggregation of partial values which was mentioned in Section 4.2. The difference lies in the assumption (in the axiom of independence of linear transformations) that "utilities" are expressed on a ratio scale with a natural zero defined by the BATNA option. Interpreted as partial values, these utilities would not necessarily satisfy the interval preference scale property as discussed in Section 4.2, and thus equal increments would not imply the same trade-offs with other criteria. As we have previously emphasized, it would be important to ensure during assessment of utilities for use in the Nash-Kalai solution, that these utilities are perceived by decision makers in the appropriate ratio sense. Assessment of the power indices α_i are to a large extent analogous to the assessment of weights in value function models, although problems of such assessment seem not to be widely discussed in the game theory literature.

It is interesting to note that the somewhat different axiomatizations of value measurement theory and cooperative game theory do in effect generate the same preference models, and there is perhaps scope for further research in comparing the two sets of axioms at a more fundamental level. Although value measurement (and MCDA in general) is expressed in terms of conflicting criteria for a single decision making entity, while cooperative game theory emphasizes conflicts between different groups in joint decision making, the distinctions are very blurred, and both fields can learn something from the other.

Bennett and co-workers (e.g. Bennett et al., 1989a) have raised an important point which arises when the problem is genuinely that of seeking compromise between conflicting interests of different groups. In such cases, it is probable that the different groups may have widely disparate

298   *MULTIPLE CRITERIA DECISION ANALYSIS*

perceptions and beliefs concerning strategies available to themselves and to other groups, the effects of alternative courses of action, and the values and goals of the other players. They may even have different perceptions as to who the relevant parties or players are! As a result of these differing perceptions and beliefs, perfectly "rational" game theoretical analysis by each party might well generate outcomes which are sub-optimal for all players (in the sense that "win-win" alternatives may well exist), or may be highly unbalanced in being highly advantageous to one party relative to the others.

Hypergame theory deals with the impacts of the different beliefs and perceptions as described above, but is essentially descriptive rather than prescriptive. As a means of decision support, hypergame models encourage parties to think more broadly and creatively not only about their own options and values, but also of those of the other players or interests. The link to MCDM problems arises when the upper-level criteria in the value tree relate to identifiably different interest groups (e.g. social, environmental, economic), and/or when the decision analyst is assisting one party or interest group in establishing their preferences and strategies in negotiating with other groups. The lesson from hypergame theory is the need for the decision analysis to take pains in seeking to identify potential values, preferences and strategies of other groups. For example, even when the primary focus of the MCDA is on the interests of one party, it is valuable to extend the value tree upwards to include consideration of the other interests, and in particular to conduct extensive sensitivity analysis on the effects of assumed values (scores and weights) for the other interests.

On the other hand, MCDA concepts in general, and perhaps multi-attribute value theory in particular, can go a long way towards communicating preferences, beliefs and value judgements between different parties and groups. We shall continue this theme in Section 10.3.4, where we explore the potential for integration of MCDA and Conflict Analysis, an overarching term which encompasses Hypergame Theory.

10.2.2.   Data Envelopment Analysis (DEA)

DEA is a field of management science which developed at much the same time as, but largely independently of, MCDM/MCDA. (See Charnes et al., 1994, and Lewin and Seiford, 1997, for some reviews.) The present authors (Belton and Vickers, 1993; Stewart, 1996a; Belton and Stewart, 1999), as well as a number of others (e.g. Joro, Korhonen and Wallenius, 1998, and Doyle and Green, 1993) have pointed to the clear associations between DEA and MCDA, although there are also features which are distinctive to DEA.

Data Envelopment Analysis seeks to compare related organizational units, (generally termed "decision making units", or DMUs) in terms of the "inputs" or resources provided or consumed, and the "outputs" produced. For example, academic departments in a university might be compared in terms of staff numbers, operating budgets and research funding provided (the "inputs"), and numbers of graduates, numbers of doctorates and numbers of research publications (the "outputs"). Other examples have included branches of a bank and hospitals. The two key concerns of DEA are in identifying best achievable practice (i.e. the boundary of the "production possibility set", defined by attainable combinations of inputs and outputs), and in rating the efficiency of each DMU in terms of how closely it matches the best achievable practice. Even with just one input and one output, we have a bicriterion decision problem, but in most cases there will be multiple inputs and outputs.

Formally, suppose that we have n DMUs, characterized by m different inputs and s different outputs. Let x_{ij} be the amount of input i provided to or used by DMU j, and y_{rj} the amount of output r produced by DMU j. In the most basic form of DEA, a measure of the efficiency of DMU k (say), is provided by the solution to the fractional programming problem maximizing (through choice of the u_r and v_i):

$$\frac{\sum_{r=1}^{s} u_r y_{rk}}{\sum_{i=1}^{m} v_i x_{ik}}$$

subject to:

$$\frac{\sum_{r=1}^{s} u_r y_{rj}}{\sum_{i=1}^{m} v_i x_{ij}} \leq 1 \quad (j = 1, \ldots, n)$$

$$u_r, \ v_i \geq 0$$

The idea is thus to ascertain whether there exist any sets of weights u_r and v_i such that the ratio of aggregate output $(\sum_{r=1}^{s} u_r y_{rk})$ to aggregate input $(\sum_{i=1}^{m} v_i x_{ik})$ for DMU k is at least as large as that for any other DMU. In view of the normalization constraint, the maximum value for this ratio is 1, in which case the DMU is said to be "efficient" and the inputs and outputs characterizing DMU k are on the production frontier (the boundary of the production possibility set). If the solution is strictly less than 1, then DMU k is classed as inefficient, and the solution provides a measure of the extent to which DMU k falls short of the best practice frontier defined by the other DMUs.

Some further insight into the solution is obtained by recognizing that the weights (i.e. the input and output weights) can be arbitrarily re-scaled by multiplying all through by a positive constant. A useful choice for the scaling of the weights when evaluating the efficiency of DMU k,

300   *MULTIPLE CRITERIA DECISION ANALYSIS*

is to select the multiplicative constant such that $\sum_{i=1}^{m} v_i x_{ik} = 1$. The fractional programming problem thus reduces to the linear programming problem:

$$\text{maximize} \quad \sum_{r=1}^{s} u_r y_{rk}$$

subject to:

$$\sum_{i=1}^{m} v_i x_{ik} = 1$$

$$\sum_{r=1}^{s} u_r y_{rj} - \sum_{i=1}^{m} v_i x_{ij} \leq 0 \quad (j = 1, \ldots, n)$$

$$u_r, \; v_i \geq 0$$

This is often termed the "input-oriented" version of the DEA formulation, as it can be shown that the solution gives a measure of the extent to which the inputs (x_{ik}) deviate from those defining the efficient production frontier at the given output levels (y_{rj}). An "output-oriented" version is generated by scaling the weights such that $\sum_{r=1}^{s} u_r y_{rk} = 1$, and then minimizing the denominator in the fractional programming formulation, namely $\sum_{i=1}^{m} v_i x_{ik}$. For simplicity of discussion here, we shall focus primarily on the input-oriented version.

An interesting feature of the LP formulation is the demonstration that DMU k is efficient if and only if there exists an additive function of the inputs and outputs, given by $\sum_{r=1}^{s} u_r y_{rj} - \sum_{i=1}^{m} v_i x_{ij}$ (i.e. having negative weights on inputs and positive weights on the outputs), such that DMU k has a value greater than or equal to that of any other DMU. (Note that in the DEA formulation, the weights are so standardized that the maximum value of this additive function is 0.) This creates a direct link between the DEA formulation and the theory of additive value functions in MCDA. This link is explored in detail in Stewart (1996a).

A further useful insight into the DEA model is obtained by examining the dual of the above LP which is:

$$\text{minimize} \quad E$$

subject to:

$$E x_{ik} - \sum_{j=1}^{n} \lambda_j x_{ij} \geq 0 \quad (i = 1, \ldots, m)$$

$$\sum_{j=1}^{n} \lambda_j y_{rj} \geq y_{rk} \quad (r = 1, \ldots, s)$$

$$\lambda_j \geq 0 \quad (j = 1, \ldots, n)$$

This gives rise to the concept of "data envelopment", as DMU k is now in effect being compared with a (generally) hypothetical DMU constructed as a weighted linear sum of other DMUs. The weights (λ_j) used in constructing the hypothetical DMU are chosen such that the aggregate outputs $(\sum_{j=1}^{n} \lambda_j y_{rj})$ are at least as much as that produced by DMU k. The efficiency of DMU k, now denoted by E, is shown to be the maximum ratio (taken across all inputs $i = 1, \ldots, m$) of the aggregate input for the hypothetical DMU $(\sum_{j=1}^{n} \lambda_j x_{ij})$ to that of DMU k (x_{ik}). The weighted sum of the other DMUs thus provides a benchmark against which the performance of DMU k can be assessed. Joro et al. (1998) demonstrate a close link between this data envelopment formulation and the solution of a multiple objective linear programming problem using the reference point approach (cf. Section 7.3). They work in terms of the output-oriented formulation, in which the data envelopment LP is shown to be essentially equivalent to a multiobjective maximization of outputs from linear combinations of DMUs, subject to the inputs being constrained to those of DMU k, and with the outputs of k as a "reference point".

As an illustration of the basic DEA model, consider the example described in Table 10.1, taken from Belton and Vickers (1992). The example related to a hypothetical academic department in which the three inputs related to numbers of lecturing, research and support staff respectively, the first three outputs to numbers of undergraduate, postgraduate and doctoral students, and the final output to number of publications per annum. All inputs and outputs have been re-scaled so that the maximum is 100 in each case. Although such re-scaling is not conventional in DEA, it facilitates the interpretation of weights, especially when subjective bounds on the weights need to be specified (see later).

Solutions to the above LPs for each of the 9 DMUs yield full efficiency $(E = 1)$ for all except DMUs 5, 8 and 9. For future reference, we record the results for two of the inefficient DMUs:

DMU 5: The efficiency is 0.724, and the non-zero λ_j's are $\lambda_1 = 0.0489$, $\lambda_2 = 0.0950$, $\lambda_3 = 0.8469$ and $\lambda_4 = 0.0646$. It is easily confirmed that $0.0489 y_{r1} + 0.0950 y_{r2} + 0.8469 y_{r3} + 0.0646 y_{r4}$ is respectively 30, 30, 50 and 60.56 for the four output measures (i.e. equal to DMU 5 for the first three outputs and greater for the fourth). Similarly, it can be confirmed that $0.0489 x_{i1} + 0.0950 x_{i2} + 0.8469 x_{i3} + 0.0646 x_{i4}$ is respectively 28.94, 21.71 and 30.98 for the three inputs. The ratios of these to the inputs for DMU 5 are 0.724 for the first two and 0.516 for the third, confirming the efficiency measure.

302 MULTIPLE CRITERIA DECISION ANALYSIS

Table 10.1. Performance of academic departments taken from Belton and Vickers, 1992

DMU	x_{1j}	x_{2j}	x_{3j}	y_{1j}	y_{2j}	y_{3j}	y_{4j}
1	20	50	40	20	40	75	75
2	60	100	100	100	60	25	30
3	24	10	20	20	50	50	60
4	30	20	40	40	0	25	50
5	40	30	60	30	50	50	40
6	80	25	90	55	85	10	20
7	70	75	10	30	25	0	50
8	50	55	55	0	25	10	40
9	100	80	100	90	100	0	25

DMU 9: The efficiency is 0.864, and the non-zero λ_j's are $\lambda_2 = 0.2659$, $\lambda_3 = 1.5907$, $\lambda_4 = 0.6546$ and $\lambda_7 = 0.1803$. The aggregate inputs and outputs can again be computed, to confirm that these dominate the corresponding values for DMU 9 and that the efficiency is as stated.

A limitation imposed by the simple definition of efficiency according to the basic DEA model may be seen from the above dual formulation. The problem is that the benchmark allows for arbitrary scaling up or down of inputs and outputs, which may not always be achievable in practice. For example, the benchmark for DMU 9 includes an upwards rescaling of the inputs and outputs of DMU 3 by a factor of nearly 60%, at which level the ratios of outputs to inputs might not be sustainable in practice. The permitting of such re-scaling confounds two sources of inefficiency, namely scale (or design) inefficiency, and technical (or operational) inefficiency. Scale inefficiency results from operating at a sub-optimal scale, i.e. too large or too small. In comparing existing DMUs, however, the scale may well be imposed by external or historical forces and be out of the control of management, in the short term at least. The important question relates to how well the DMU is performing at the its current scale of operation, namely its technical efficiency. The technical efficiency of DMU k can be assessed by solving the dual LP formulated above with the inclusion of the further constraint:

$$\sum_{j=1}^{m} \lambda_j \leq 1$$

which prevents arbitrary re-scaling of the benchmark DMUs. In the case of DMU 9, the technical efficiency turns out to be 1. For DMU 5, the technical efficiency is 0.875, with the benchmark defined by $\lambda_1 = \lambda_2 = 0.125$ and $\lambda_3 = 0.75$.

Up to this point, the DEA model as we have described it is essentially objective in the sense that no value judgements or preference information is included. In MCDA terms, the process does no more than to identify the non-dominated solutions which is often not very helpful as many DMUs may be categorized as efficient on the basis of very extreme sets of weights. Standard DEA implementations do generally include weight restrictions of the form $u_r \geq \epsilon$, $v_i \geq \epsilon$ for some $\epsilon > 0$. This is, however, a somewhat ad hoc response, only eliminating the most extreme cases, and it is not clear how the magnitude of ϵ is to be specified. More comprehensive weight restrictions are equivalent to the assessment of weights for additive value function models; thus the care needed for such assessments as discussed in Chapter 5 apply just as well to DEA (although this is not always well recognized in the DEA literature). In particular, it is important to ensure that inputs and outputs are measured on interval preference scales, that a consistent scaling is used, and that weights have the appropriate "swing weight" interpretation. This has been discussed in detail in Belton and Vickers (1993), Stewart (1996) and Belton and Stewart (1999). Halme et al. (1999) point to the known difficulties in MCDA of assessing importance weights in a meaningful manner, and suggest alternative means of incorporating judgemental data in the form of "most-preferred" combinations of inputs and outputs which restrict the magnitudes of the DEA efficiencies (which they then term value-efficiency in contrast to the scale and technical efficiencies referred to above).

Stewart (1996) also describes a variation to the DEA model in which weights are randomly generated within a defined set (rather than optimized for each DMU), and efficiencies calculated for each DMU for each set of weights generated. This yields a distribution of possible efficiency vales for each DMU. If the distribution includes an efficiency of 1, then the DMU is "efficient" by the usual definition, but the robustness of this assessment is demonstrated by the spread of possible efficiencies which can be displayed graphically for example by "box-and-whisker" plots. For the example of the academic departments described above, the box-and-whisker plots for DMUs 4, 5 and 6 are displayed in Figure 10.1. As noted above, DMUs 4 and 6 are efficient, but not DMU 5. The range of efficiencies for DMUs 4 and 6 do stretch to 100%, but are much more widely spread than those of DMU 5, indicating that although DMU 5 does not achieve full efficiency for any set of weights, neither does it register very low efficiency scores, in contrast to DMUs 4 and 6.

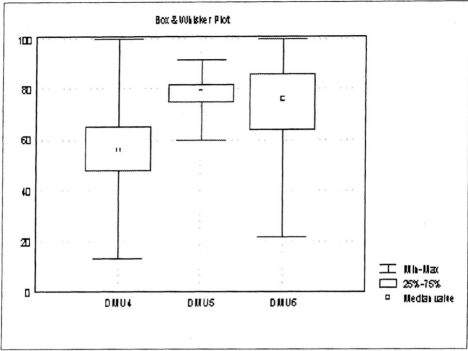

Figure 10.1: Ranges of efficiencies for three DMUs

Although DEA involves comparisons of objects (DMUs) on the basis of multiple criteria (inputs and outputs), it differs from conventional MCDA in the sense that the emphasis is not on selecting or ranking alternatives. The aim is to evaluate DMUs against benchmarks in order to suggest areas in which improvements in efficiency should be possible. Nevertheless, there appear to be instances in which some form of rank ordering may be sought, while some writers (e.g. Doyle and Green, 1993) have suggested that the DEA concept may be useful as a more general MCDA tool. The more objective DEA formulation may in particular be useful for preliminary screening of alternatives in order to generate a shortlist along the lines discussed for the land use planning case study in Section 3.5.3. We have elaborated on this concept in Belton and Stewart (1999), where the advantages of using the monte carlo approach are emphasized. A further useful property of the monte carlo implementation is that the results (viewing the DMUs as variables, and each monte carlo repetition as a case) can be subjected to factor analysis to reveal which DMUs (i.e. alternatives) tend to have high efficiencies for similar sets of weights. It is then possible to extract representatives with high median efficiency from each "factor" (similar sets of DMUs), to generate a shortlist of substantially different but potentially good alternatives.

10.2.3. Multivariate Statistical Analysis

Multivariate statistical analysis developed as a means of identifying patterns and structures in multivariate data. Typically the data are described in terms of "cases" or "subjects", for each of which a number of different pieces of information (the "variables") are recorded. For example, the cases may represent students assessed according to a number of different tests or courses; or companies assessed according to various performance indices. Some of the goals of multivariate analysis are to identify underlying "factors" which differentiate cases most clearly, or according to which cases are most similar, leading to some form of clustering or ordering of cases. Many statistical packages exist on which these analyses are easily carried out, and we refer the reader to such packages and associated documentation for further details.

The basic data describing an MCDM problem can also be viewed in multivariate terms, with the alternatives as cases and the criteria as variables. The value of the descriptive tools of multivariate analysis may be appreciated by reflecting on a simple MCDM problem with only two criteria. It is then simple to plot the alternatives on a pair of axes representing performance on the two criteria as illustrated in Figure 10.2 (where each alternative is represented by a ■). There is hardly any need for much formal MCDA as described elsewhere in this book! The dominated alternatives are obvious. The efficient frontier is represented by the alternatives joined by the dotted lines, and it is relatively simple for the decision maker to evaluate whether the tradeoffs represented by each segment of the dotted lines are acceptable or not. The alternatives can be compared to the generally unattainable "ideal" outcome indicated on the figure, so that the proximity of the non-dominated alternatives to the ideal can be evaluated directly.

The problem is that graphical displays such as that of Figure 10.2 are considerably less easy to interpret for three criteria, and cannot adequately be displayed for more than three criteria. By use of multivariate statistical analysis techniques, we can however generate approximate two-dimensional representations of the problem, allowing us to obtain similar insights for higher dimensional problems as for the case of two criteria.

The use of two multivariate statistical techniques for this purpose, namely correspondence analysis and factor analysis, has been described in Stewart (1981). Since software support for factor analysis is generally available in all statistical packages, while the use of correspondence

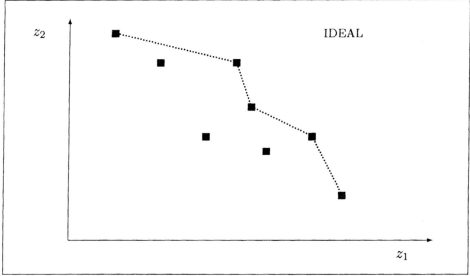

Figure 10.2: Graphical display of a simple two-criterion problem

analysis as described in Stewart (1981) involves some non-standard manipulations, we shall restrict our discussion here to the use of factor analysis. This approach is best explained by working through a specific example, rather than by a description in general theoretical terms. For this purpose, we return to the business location case study which was extensively analysed in Chapter 5. The basic input data describing this decision problem, prior to any substantial preference modelling, is given in Table 5.4. For some of the criteria, the evaluations in Table 5.4 are given on constructed or qualitative scales, while for others the performance levels are represented by values of objectively determined attributes. No attempt has yet been made at this stage to convert these evaluations into partial values, and certainly no aggregation of values across criteria has been carried out. In this sense we have essentially "raw" data, simply describing the problem as it presents itself to the decision makers.

In multivariate statistical terms, as previously indicated, the criteria (or the attributes used as surrogates for these criteria) can be viewed as the "variables", while each alternative represents a "case". Note that in this example, the criterion measurements are all defined in an increasing sense (i.e. larger values are preferred). This is convenient for interpretation of the multivariate statistical output, and can always be ensured without loss of generality by appropriate definition of the criteria.

The data in Table 5.4 can be subjected to a standard factor analysis using any appropriate statistical package. It is useful to apply "varimax" rotation to the principal components in order to obtain maximum separation of criteria into associated sets. In the case of the business location case study data, the first two factors account for approximately 75% of the statistical variation between the alternatives. In other words, the six attribute values can to a high level of precision be approximated by expressions of the form:

$$z_i(a) = \lambda_{1i} y_1(a) + \lambda_{2i} y_2(a)$$

where $y_1(a)$ and $y_2(a)$ are performance measures for alternative a in terms of the two "factors". The coefficients λ_{1i} and λ_{2i} are termed the "factor loadings" associated with attribute i, i.e. measures of the extent to which the associated criterion is associated with each factor.

Although the performance measures $y_1(a)$ and $y_2(a)$ are not directly observed, these "scores" are estimated in the factor analysis from expressions of the form $\sum_{i=1}^{m} \alpha_{ij} z_i(a)$, where the α_{ij}, termed the "factor score coefficients" for factor j, are estimated directly from the data. By plotting the estimated scores for each alternative on the factor axes, as illustrated in Figure 10.3 (where the position of each alternative is indicated by the first three letters of its name, e.g. Par, Bru, Ams, etc.), a two dimensional representation of the relationships between the alternatives is obtained. It is useful to include on the plot the position corresponding to the "ideal", namely the hypothetical alternative which achieves the best outcome on each criterion. The corresponding factor scores, generated using the same factor score coefficients, are represented by the position labelled IDEAL in Figure 10.3. Strictly speaking, the IDEAL point should be viewed as a projection on to the plane of an unattainable point lying well outside of the plane. Nevertheless, the relationship between this point and those corresponding to the real alternatives is an indication of their relative closeness to the ideal.

Interpretation of the relationships between the alternatives as displayed in Figure 10.3 is facilitated by also plotting the factor loadings $(\lambda_{1i}, \lambda_{2i})$ for each criterion on to the same set of axes. These points are connected to the origin by lines which represent in effect the projection of the 6-dimensional coordinate system in criterion space on to the 2-dimensional factor plot. The origin in Figure 10.3 corresponds to an alternative which is average on all criteria, so that the directions to the plotting points of each criterion indicate improving performance in terms of that criterion. In fact, the cosine of the angle between the directions to a particular alternative and to a particular criterion respectively indicates the extent to which this alternative is associated with above

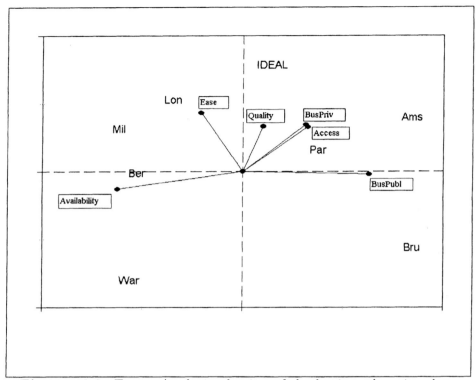

Figure 10.3: Factor Analysis plotting of the business location data

average performance on the given criterion. For example, we observe that London and Milan are associated with good performance on the *ease of set up and operations* criterion, as all of these lie in a direction to the "north-west" of the origin.

The following properties of the decision problem may then be identified as follows:

- The factor corresponding to the horizontal axis in Figure 10.3 represents primarily a contrast between performance in terms of *availability of staff* on the one hand, and *business potential – public sector* on the other hand (with some small contrasts between the other criteria). Thus we see directly that good performance on *availability of staff* can only be achieved in the current set of alternatives at the expense of *business potential – public sector*, identifying an important tradeoff assessment that the decision maker must make (unless another alternative can be found which performs well on both of these criteria).

- In similar fashion, the factor on the vertical axis indicates that good performances on *ease of set up and operations, quality of life, business potential – private sector* and *accessibility from US* are only attainable in the current set of alternatives by accepting lower than average performances on the other two criteria.

- No alternative lies particularly near to the projection of the ideal, but London, Paris and Amsterdam are perhaps the closest, and are thus deserving of more detailed investigation.

The factor analysis thus provides us with visual insights into the ranges of outcomes which are achievable, and the necessary value trade-offs which have to be made (unless further alternatives can be identified). In some examples, alternatives are found naturally to form into clusters on the factor analysis plot. In such cases, it may be useful to direct decision makers first to make comparisons between clusters (by evaluating a representative alternative from each cluster), as this focuses attention on to the most critical tradeoff decisions. Particularly if the number of alternatives is relatively large, this provides a shortlist of alternatives to which other MCDA methods, such as value measurement or outranking, may more easily be applied. Once the preferred cluster is identified, alternatives within the cluster may be compared at a finer level of detail, in order to reach the final decision.

As discussed in Section 8.6, the idea of using principal components analysis (unrotated factor analysis) has also been suggested as an extension to the PROMETHEE approach, termed GAIA (Brans and Mareschal, 1990). Recall that in GAIA, the principal components analysis is applied to the "normed flows", defined for each alternative a and criterion i by:

$$\phi_i(a) = \frac{1}{n-1} \sum_{b \neq a} \{P_i(a, b) - P_i(b, a)\}$$

where $P_i(a, b)$ is the "intensity of preference" for a over b in terms of criterion i, defined in terms of one of the functional shapes illustrated in Figure 8.3, and n is the number of alternatives. In comparing the plot displayed in Figure 10.3 (derived from factor analysis applied to the raw data describing the alternatives) with the GAIA plot shown in Figure 8.4, we note that the revealed patterns are essentially the same. Except for a minor shift of *business potential – private sector* and *accessibility from US* from *quality of life*, Figure 10.3 can be seen as a mirror image of Figure 8.4 rotated through approximately 90 degrees. Interpretations of the plots are, however, invariant

310   *MULTIPLE CRITERIA DECISION ANALYSIS*

to rotations and mirror imaging, so that the plots are approximately equivalent. Stewart (1992) observes that the two plots become in fact virtually indistinguishable when applied to the example (relating to siting of a power plant) used by Brans and Mareschal to illustrate the GAIA approach. The introduction of the PROMETHEE preference functions into the analysis does not thus appear to add much to the geometrical representation of the alternatives and criteria and their interrelationships. The real value of GAIA lies in the idea of projecting a weighted preference direction on to the plane, as illustrated in Figure 8.4. With equal weights, this preference direction points more-or-less towards the "ideal" displayed in Figure 10.3. Within GAIA, the user can interactively adjust the weights on the criteria, which rotates this vector around the origin , so that the user can obtain a direct sense of what relative weights would "point" in the direction of specific alternatives such as Paris or London.

10.3.   LINKS TO OTHER OR/MS APPROACHES (INTEGRATING AND EMBEDDING)

10.3.1.   Problem Structuring Methods

In Chapter 3 we discussed in some detail the use of problem structuring methods such as cognitive / cause mapping and soft systems methodology (SSM) in the process of eliciting and capturing information in the initial stages of an investigation, preceding the building of a multicriteria model. The output of such a process is a rich and detailed qualitative representation of the issue, possibly in the form of a cause map or a rich picture, from which the required form of multicriteria model may be extracted. However, the qualitative model does not cease to be useful at this point. In comparison to the detail captured in this type of model most multicriteria models are very sparse, capturing the essence of a problem, but not representing the detail. The continued use of the detailed qualitative model throughout the multicriteria analysis can serve as a reminder of the detailed context and explanations of structure of the derived multicriteria model. This is particularly useful when the investigation of an issue continues over a significant length of time and thus does not command the continual attention of the problem owners, or if new people become involved and need to understand the model as the investigation progresses. It can also be useful, as suggested by Belton et al. (1997), as a means of capturing the discussion leading to and justification for quantitative elements of a model, such as relative weights of criteria.

10.3.2. Discrete Event Simulation and System Dynamics

Discrete event and continuous simulation (system dynamics) are tools to support the exploration of the impact of alternative system configurations. A simulation is comprised of a logical model of the system being investigated, incorporating as many rules governing the behaviour of the system as are necessary to give an acceptable representation.

Discrete event simulation focuses on modeling the progress of individual entities in a system of queues, capturing the stochastic nature of the system through the use of theoretical or empirical statistical distributions to represent variability. Users tend to be concerned with relatively short-term performance measures such as queue lengths and resource utilization. System dynamics is concerned with aggregate flows rather than the movement of individual entities, and, as the name suggests, focuses more on the dynamics of a system, in particular the impact of feedback loops on the system's evolution over time.

Simulation models more often than not provide the user with multiple performance measures that are typically handled intuitively by the decision makers. MCDA is a natural extension of the analysis that can draw together the output of a simulation with other relevant factors and enable the decision makers to fully appreciate the choices open to them. The linking of simulation and multicriteria analysis provides a decision support tool linking decision space, solution space and value space as described by Belton and Elder (1994). For example, an organization looking to set up a telephone call center to deal with customer queries has to make decisions about the number of lines to operate, how long to spend with each customer, and so on (decision space). A simulation can be developed to explore the implications of these decisions on output measures such as, how long callers are kept waiting, how many give up, how many calls are handled, etc. (solution space). This information has then to be considered, together with other factors, in the light of the organizations objectives, such increasing volume of activity and increasing quality of service (value space) in arriving at a decision – a process that can be supported by MCDA.

The overall analysis promotes understanding and creativity by allowing the user rapidly to try out new ways of doing things, through the simulation to be immediately able to see the consequences and to explore these through the multicriteria analysis.

Examples of the combination of discrete event simulation and multicriteria analysis are: Gravel et al. (1991) which uses multicriteria analysis to evaluate production plans developed using simulation; Belton and El-

312   *MULTIPLE CRITERIA DECISION ANALYSIS*

der (1998) which describes the simulation and evaluation of an airport check in system; and Spengler and Penkuhn (1996) describing a DSS which combines a flowsheet-based simulation with multicriteria analysis. Early work combining system dynamics and MCDA was carried out by Gardiner and Ford (1980) using the SMART approach, and by researchers in Social Judgement Theory (Hammond et al., 1977 and Mumpower et al., 1979). More recently, the use of PROMETHEE has been embedded as a control mechanism in systems dynamic models of socio-economic systems, as described by Brans et al. (1998), while Santos and Belton (2000) explore the combined use of systems dynamics and multicriteria analysis for performance measurement and management.

10.3.3. Scenario Planning

We referred to scenario planning approaches in Chapter 3, in the context of exploring uncertainty during problem structuring for MCDA. In this section we explore in a little more detail the potential for synergy between the two approaches. Van der Heijden (1996) describes scenario planning as a process of organizational learning – an approach to strategic planning than is distinguished by its emphasis on the explicit and ongoing consideration of multiple futures. A key component of that process is clearly the construction of the scenarios, stories which describe the current and plausible, but challenging, future states of the business environment. Van der Heijden (1996, p5–7) stresses that the scenarios constructed are *external* to the organization, they will arise from events which are out of its control, and they are value-free in the sense that they are not construed as good or bad. In the broader context of strategic planning the scenarios provide alternative perspectives that will challenge an organization in viewing the future and in evaluating its strategies and action plans.

The process of scenario construction has developed very much as a craft, which has evolved from and is informed by practice, rather than theoretical underpinnings. As a consequence there are many individual methods that differ, for example, in the number of scenarios constructed and whether or not extreme scenarios are utilized. In essence, all of these seek first to understand the past and the present, to identify constant factors, key trends, driving forces, possible one-off events, interrelationships between elements, the uncertainties inherent in each of these and their potential impact on the organization's strategic agenda. This information is then utilized to create a set of scenarios, each of which is written up as a story, that encapsulate the future. Van der Heijden (p187) lists five principles which should guide scenario construction:

MCDA in a Broader Context   313

- At least two scenarios are required to reflect uncertainty, but more than four has proved to be impractical

- Each scenario must be plausible, meaning that it can be seen to evolve in a logical manner from the past and present.

- Scenarios must be internally consistent.

- Scenarios must be relevant to the client's concerns and they must provide a useful, comprehensive and challenging framework against which the client can develop and test strategies and action plans.

- The scenarios must produce a novel perspective on the issues of concern to the client.

For detailed expositions on the process of scenario development, the reader is referred to van der Heijden (1996) or to Eden and Ackermann (1998).

Once constructed, the scenarios can be used to inform strategy design, to explore the strength of an organization, or to evaluate alternative strategies – a process sometimes referred to as wind-tunnelling. This process ensures that all strategies are "tested" against the full range of plausible conditions. The potential for synergy with MCDA is particularly strong at this stage. Van der Heijden (p232–235) rejects "traditional rationalistic decision analysis" as an approach which seeks to find a "right answer"; however, as argued elsewhere in this book, our view is that MCDA has much more to offer than this, and is entirely consistent with the philosophy of organizational learning embraced by scenario planning. The main tool for presenting the result of the wind-tunneling process is the scenario-option matrix depicting a holistic evaluation of each option in the light of each scenario against a 7 point scale ($-\ -\ -$ through 0 to $+\ +\ +$), as illustrated in Figure 10.4. The aim of MCDA in this context would be to enrich this evaluation, leading to a better understanding and more explicit statement of the strengths and weaknesses of the options and further informing the creation of options which are more robust. This view is also expressed by Goodwin and Wright (1997), who illustrate the use of SMART in conjunction with scenario planning.

It is not possible to describe the process of scenario planning in full detail, or to do justice to the richness of the approach. Nevertheless, we hope that the following simple hypothetical example, based on the Land Use Planning case study introduced in Section 2.2.3, will give a sense of how it could be used in conjunction with MCDA. The problem context, that of determining policies for land use in a developing country, is one

314   *MULTIPLE CRITERIA DECISION ANALYSIS*

which is rife with uncertainty and lends itself well to the use of scenario planning. In addition to the construction of the internal *policy scenarios* described in Section 3.5.3, it would be necessary also to devote time to the creation of *external scenarios*. It is essential that the group tasked with scenario development is able to " ... suspend disbelief, think the unthinkable, and let intuition and premonitions flow freely" (van der Heijden, p183). Processes for problem structuring described in Chapter 3, such as post-it exercises and cognitive mapping are equally useful in the scenario construction process. In creating scenarios for the Land Use Planning problem issues such as climatic change, globalisation, economic development, political stability, demographic issues, technological developments, etc. may be taken into account. (Within the context of this case study, it is interesting and relevant to make reference to the development of scenarios for South Africa, which is described in some detail by van der Heijden, pages 199–207.) Let us imagine that two scenarios are elaborated; these will be referred to as "Widespread Westernization" and "The Expanding Desert", the labels being intentionally evocative of the detail. Suppose further that initially three policy scenarios (options) have been identified for evaluation; these are labeled "conserve at all costs", "commercialise" and the "status quo". An illustrative scenario-option matrix is shown in Figure 10.4; this indicates whether an option performs well or badly under each scenario, using the 7-point scale previously mentioned.

	A Conserve at all costs	B Commercialise	C Status Quo
Widespread Westernisation	0	++	−−
The expanding desert	+	−−	−

Figure 10.4: Scenario-option matrix for illustrative example

MCDA can further inform this analysis by making explicit the basis for evaluation and further enhancing debate. The holistic evaluations shown in Figure 10.4 could have been derived from the simple multi-criteria evaluation, illustrated here using a multi-attribute value model, depicted in Figure 10.5. In this figure, three key factors are shown to contribute to the overall evaluation: Economic Return; Conservation of Ecosystems; and Water Supply and Quality. The performance of the options against these three criteria, in each of each of the scenarios is illustrated by the profile graphs (using a local scale where best and worst performances against each criterion across all scenarios define the upper and lower limits of the scale). These values are aggregated to give an overall evaluation for the options in each scenario, giving greater weight to Economic Return in the Widespread Westernisation scenario, and greater weight to Water Supply and Quality in the Expanding Desert scenario.

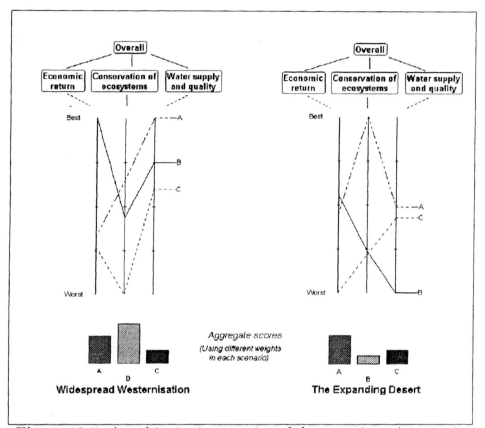

Figure 10.5: A multi-criteria extension of the scenario-option matrix

The more detailed presentation of the evaluation of options highlights the strengths and weaknesses of those currently being considered and should inform debate about the creation of more robust options. In the example here, attention is focused on option A; could it be possible to find ways of improving economic performance within this policy scenario – perhaps by developing tourism rather than more traditional economic activities such as forestry?

10.3.4. Conflict Analysis

The analysis of interdependent individual decisions, and the potential conflicts which ensue, is an evolving methodological approach which

316   *MULTIPLE CRITERIA DECISION ANALYSIS*

stems from classical game theory as described in Section 10.2. The broad term, conflict analysis, embraces Metagame Analysis (Howard, 1971,1989a,1989b), Hypergame Analysis (Bennett et al., 1989) which was also mentioned in Section 10.2, and the more recent extension of these methods known as drama theory or soft game theory (Bennett and Howard, 1996; Bryant, 1997; Bennett, 1998; Howard 1999). As discussed earlier, there are many parallels between game theory and MCDA; it is also possible to consider their use in an integrated way.

An initial problem structuring identifies the parties to the conflict, the actions open to them, and the scenarios resulting from their interactions. In this context, a scenario represents a possible pathway into the future, i.e. an anticipation thereof; it is the result of a specific combination of actors' actions (or, in our previous terminology, policy scenarios for each actor). Usually not all combinations are considered feasible. Among the feasible scenarios some have particular relevance, for example representing a player's "position" – the future they are trying to persuade other players to accept – and are the subject of more detailed interpretation and analysis. On the basis of a holistic determination of parties' (ordinal) preferences over the relevant scenarios, their stability is investigated. A scenario is said to be *stable* if there are no incentives for any party to depart from it (there are no *unilateral improvements*), or if each incentive is deterred by a credible *sanction* by the other party, i.e. a change in his/her action(s) making the first party worse off. A stable scenario represents a possible resolution of the game.

The table below shows a very simple situation involving two business partners; one of the partners, Jo, has recently not been contributing her full share of work to the business activity, having taken up a demanding hobby and the second partner, Chris, is threatening to leave unless Jo gives up her hobby. To try to resolve the situation the two partners have agreed to meet next week. Chris wants to do the most possible to save the business and has decided to carry out some formal analysis of the situation, thus the analysis which follows pertains to Chris's perspective.

The parties to the conflict (players in the game) and the actions open to them are shown down the left hand side of table. In the body of the table a 1 or 0 indicates if a player takes a particular action (1) or not (0). Each column indicates a possible scenario.

MCDA in a Broader Context 317

	Scenario A (Status Quo)	Scenario B (Compromise)	Scenario C (Collapse)	Scenario D (Disaster)
Jo:				
Give up hobby	0	1	0	1
Chris:				
Leave	0	0	1	1

The possible future scenarios are:

A: Status Quo – Chris carries the business while Jo continues with her hobby

B: Compromise – This represents Chris's position; Chris does not leave but Jo has to give up her hobby

C: Collapse – Jo does not give up her hobby, so Chris leaves and the business collapses

D: Disaster – Jo gives up her hobby but Chris leaves anyway – this scenario is considered infeasible

Scenario C is the "threatened" future, which will result if both players implement their "fallback" position (i.e. the action, or set of actions, they threaten to implement if their position is not accepted).

Conflict analysis then seeks to determine the two parties' preferences, at a holistic level, for the three feasible scenarios, to identify any sanctions or threats available to each of the parties, and from this to determine whether or not an equilibrium (stable scenario) exists.

Chris's preferred outcome is scenario B, followed by C and then the Status Quo (A). Chris hypothesizes that Jo's preference is for the Status Quo with B second and C third. This gives rise to the stability analysis shown in the following table, where "I" indicates that a player has an improvement from that particular scenario ("I^*" indicating that it can be achieved unilaterally) and "S" indicates a related sanction.

	Scenario A Status Quo	Scenario B Compromise	Scenario C Collapse
Jo	–	I	I
Chris	I^*	S	I
Outcome	unstable	stable	unstable

The analysis reveals that the status quo does not represent a stable scenario, because Chris will be better off by leaving (as he prefers scenario C to A) and this change could not be hampered by any sanction

318   *MULTIPLE CRITERIA DECISION ANALYSIS*

coming from Jo. Scenario B is a stable scenario because Jo's incentive to move to a better position (the improvement by choosing not give up her hobby) is effectively sanctioned by Chris threatening to quit (resulting in scenario C, which is the worst for Jo). The collapse scenario is also unstable; although neither actor can unilaterally force a change to another feasible scenario, both would prefer scenario B to C. Thus, scenario B appears to represent the equilibrium solution of the confrontation.

The analysis of stability introduces the issue of the credibility of promises and threats, and emotion, which is the new key element of drama theory. Promises and threats are used by actors to convey the intention to make an improvement or to carry out a sanction (Howard, 1989a, 1989b). Sometimes a threat or promise is inherently credible, namely when it leads to a clearly preferred outcome for the actor who proposes it; in another situation it may not be credible because the actor is thought not to be willing to carry it out. Drama theory argues that when individuals try to make unwilling threats and promises credible, for example in order to force the other party to accept their position, they are acting irrationally (i.e. against their preferences). This gives rise to emotional pressure, which can lead to a significant change in the situation. The situation could be resolved through a redefinition of some of the elements (e.g. the development of new actions which lead to more severe threats or more rewarding promises) or through a change in individual preferences that leads to the unwilling threat or promise becoming willing. According to Drama Theory, emotions play a central role in triggering (in a predictable way) changes in individual preference structures. Six basic situations causing strong emotional tensions, referred to as *dilemmas* or *paradoxes of rational choice*, are identified (see Howard, 1999, or Bennett, 1997, for details).

One such dilemma illustrated by the example is the *trust dilemma*. Chris faces a *trust dilemma* in relation to Jo's promise to give up her hobby (scenario B). This promise is not credible, because Jo actually prefers the Status Quo to scenario B. Thus, if the players were to agree to compromise on Scenario B, how can Chris trust Jo to give up? Chris could try to think of a more severe sanction to deter Jo from defecting, or an interesting reward if she does not defect. Alternatively, driven by his emotions, he could gradually begin to think differently, eventually changing his preferences and coming to accept the new situation (i.e. the current status quo). Another possible dilemma, known as the *inducement dilemma*, arises if Chris's preference for Scenario C over the Status Quo is actually very weak, meaning that he would be reluctant to carry through his threat to leave.

The analysis is primarily descriptive in that it does not purport to resolve the conflict; however, as the case studies described by Bennett et al. (1989) and Bennett (1997) reveal, the insights that derive from the process can assist in resolving some of the issues and discovering a way to move forward. Such an analysis gives an indication of what might be expected to happen in reality, and insight into the possible ways of defining actions (threats and promises) and strategies that enable an actor to improve their position.

Clearly a simple example such as this can only give a taste of the detailed methods of conflict analysis; the reader is referred to the four chapters in Rosenhead (1989) authored by Howard, Bennett, Cropper and Huxham, for a fuller account of metagames, hypergames and their integration with other soft OR methods and to the more recent papers cited earlier for a discussion of drama theory.

But what might be the links to MCDA? We feel that there is potential, in certain circumstances, for each approach to inform the other, as discussed in greater depth by Losa and Belton (2001). Within a conflict analysis a more detailed multicriteria evaluation of outcomes could lead to a better understanding of the situation, for example, shedding more light on each party's preferences, on their stability and thereby the credibility of threats and the resolvability of dilemmas. In our simple example Chris's preferences may be influenced both by financial prosperity and the value placed on time with family. Family time is threatened by Jo's current actions (i.e. the Status Quo), but financial prosperity would seem to be threatened by resignation and collapse of the business (i.e. Scenario C). These two outcomes are depicted in the multicriteria preference profile shown in Figure 10.6(a). In most multicriteria analyses the preference for one over the other would depend on the relative importance of the two criteria, possibly shedding some doubt on the credibility of the threat to leave (i.e. of the strength of preference for scenario C over the Status Quo) and giving rise to the inducement dilemma described above. Realization of this may lead Chris to further explore alternative actions and to discover a more credible threat, such as a scenario X indicated in Figure 10.6(b), which would reinforce his position and may simultaneously resolve the trust dilemma, by making the non-cooperation of Jo even more "expensive" for her (depending on her preference for X). Another way of changing the current situation might arise if Chris delves more deeply into his own preferences and realises the importance of a third factor which makes the Status Quo look more attractive than Scenario C, as illustrated in Figure 10.6(c), and reverses the order or preference for these two outcomes. Equally possible is the realization of a third criterion which strengthens Chris's preference for

Scenario C over the Status Quo, thereby increasing the credibility of his threat to leave.

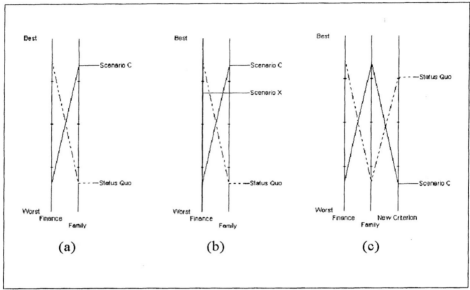

Figure 10.6: Alternative evaluations of Chris's preferences

Looking to a link in the other direction – how conflict analysis might inform MCDA – we feel that the evaluation of a situation involving multiple parties, such as that described in the Land Use Planning case study, from the perspective of conflict analysis may lead to a better understanding of the different parties' preferences and the likelihood of a compromise solution being accepted and implemented.

10.4. OTHER METHODOLOGIES WITH A MULTICRITERIA ELEMENT

There are many managerial tools that incorporate a strong multicriteria element. In the main these appear to have been developed without the involvement of analysts or researchers from the MCDA field, but recently links have been established as people active in MCDA have started to become involved in these tools and their application, as evidenced for example by conference papers presented in 2000 (e.g., Scheubrein and Bossert; Kaylan et al.; Santos and Belton; Bana e Costa and Correa). Two well established general purpose problem solving methods are: the Value Engineering approach (also known as Value Management or Value Analysis) and the Kepner-Tregoe method. More recent developments are: the Balanced Scorecard of Kaplan and Norton (1996); aspects of

Quality Function Deployment which originated in Japan in the 1960's with the ideas of Yoji Akao (Akao, 1998); the EFQM (European Federation for Quality Management) Excellence Model and framework for benchmarking. These are just some of the more widely known models; we are also aware of many in-company procedures that adopt a multi-criteria framework. Once again, it is not possible in the space we have available to give a comprehensive description of all of these approaches. Our aim is to give a brief overview of each in order to raise awareness of MCDA related work in the broader managerial context.

10.4.1. Value Engineering

Value Engineering had its roots in the General Electric Company in the 1940's, developed by L.D. Miles, whose book "Techniques of Value Engineering and Analysis" (2nd edition, 1971) gives a full description of the approach. An overview is given in Van Gundy's (1988) book on "Techniques of Structured Problem Solving". Also referred to as Value Management, Value Analysis or by other similar labels, the approach has been developed over the past 50+ years and continues to be widely used today. The initial use of the method was in improving product design with a particular focus on reducing cost while achieving the same level of quality and functionality; however, applications became more widespread. Cases recently reported in "Project" (March 1998), the newsletter of the Association for Project Management, included the prioritization of funds for the restoration of a medieval palace at Eltham in the south of England, and addressing how to maintain the functionality of planned redevelopment of the historic Sheffield City Town Hall, despite a shortfall in funding arising as a consequence of escalating costs.

Value Management is a comprehensive, facilitated approach that incorporates recommendations on all aspects of the process including team composition, workshop design and conduct. At the heart of Value Management is the Value Study, an approach to problem analysis which incorporates: specification of objectives through a process called Function Analysis; creative thinking focussing on key aspects identified by the function modeling; evaluation using a simple weighted scoring model; and the development of an action plan. Function analysis, or modeling, is a well elaborated process; functions are categorized as basic – relating to primary purpose, secondary – being essential to a design/solution but not directly related to the client's goals, and supporting – meeting the client's needs, but not essential for the basic function. Many ways of mapping the relationships between functions have been developed. Function analysis parallels the model building phase of a multicriteria

322   *MULTIPLE CRITERIA DECISION ANALYSIS*

analysis and provides an interesting alternative perspective which may be very useful in certain types of problem.

The scoring process involves prioritization of value criteria/objectives and the use of 1-10 scales for the evaluation of options against those criteria. This seems to be rather ad hoc, with little attention paid to the meaning or validity of the numbers elicited.

10.4.2.   Kepner-Tregoe Approach

The approach is described in full in the two books by Kepner and Tregoe, "The Rational Manager" (1976) and "The New Rational Manager" (1981). Also, it is outlined in Van Gundy's book on "Techniques of Structured Problem Solving" (1988). The approach comprises two phases, problem analysis and decision making, which correspond to the process of problem structuring and model building and analysis as described in the earlier chapters of this book. Each phase is broken down into more specific elements. Decision making includes the definition of objectives, the specification of alternatives and their evaluation against objectives using a simple weighted sum model. Objectives are classified as "musts" and "wants" and the process allows for weighting of the "wants" to reflect the relative importance of these objectives. Alternatives that do not satisfy the "musts" are eliminated from further consideration; alternatives which pass this screening are ranked with respect to performance against each of the "wants" or evaluated on a 10 point scale. The alternative with the highest weighted score is tentatively selected. The process then explicitly includes a review of the possible adverse consequences of a selected alternative and the careful development of an action plan to implement the chosen alternative, minimizing possible adverse consequences.

As with Value Studies, the process of weighting and scoring is not well defined.

10.4.3.   The Balanced Scorecard

The Balanced Scorecard is a relatively recent approach that has created substantial interest in many organizations. It has its origins in a study conducted in 1990 for KPMG by Kaplan and Norton (1996) on "Measuring Performance in the Organization of the Future". The study recommended that an organization's performance should be assessed on multiple dimensions, focussing on the four key areas of finance, customer, internal, and innovation and learning. With further experience the scorecard developed from a measurement tool to a strategic management system.

MCDA in a Broader Context 323

Financial
> Return on investment
> Economic value added
> (achieved through: increasing revenue, improving cost and productivity, enhancing utilization of assets and reducing risk)

Customer
> Customer satisfaction
> Customer retention
> Market share
> Customer acquisition
> Customer profitability

Internal
> Innovation
> Time
> Quality
> Cost

Learning and Growth
> Employee capabilities
> Motivation, empowerment and alignment
> Information systems capabilities

Figure 10.7: Generic Scorecard Measures

Kaplan and Norton have identified a number of generic measures in each of the four key areas, which tend to feature in most organization's scorecards. These are indicated in Figure 10.7. They stress, however, the importance of relating these to an organization's own strategy and not accepting them as a recipe for success.

Kaplan and Norton also stress two other key points. Firstly, the list included in Figure 10.7 should not be seen as an ad hoc collection of measures which must be traded-off against each other, but as an integrated and interlinked set of objectives and measures as illustrated in Figure 10.8. Secondly, the Balanced Scorecard must be developed with an explicit purpose in mind; not only is this essential to guarantee its use, it will also influence the detail of the scorecard.

Why is the Balanced Scorecard of relevance to MCDA? Firstly, its development and growing acceptance indicates an awareness of the need to explicitly consider and measure multiple factors. It is essentially a generic multicriteria model at a strategic level, built to support ongoing measurement and management. The generic set of objectives/measures

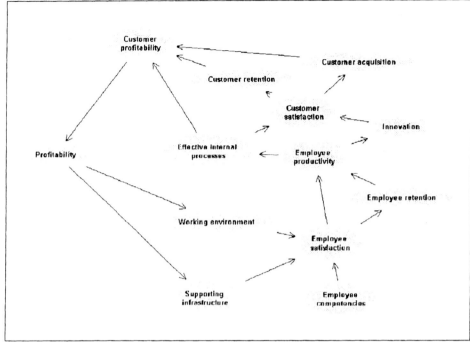

Figure 10.8: An illustrative interlinked scorecard

illustrates the power of a template in engaging organizations in such a process. Also, the emphasis on inter-relationship between objectives should perhaps sound a cautionary note for MCDA. We see the same emphasis in the work of Senge (see the "Fifth Discipline", 1992) another popular contemporary management thinker and practitioner. The issue was also highlighted in the work by Belton et al. (1997) who comment on the problem of moving from an initial problem representation in the form of a cause map, which allows complex inter-relationships including feedback loops, to a value tree. To date there has been little work in this area within the field of MCDA, exceptions being the integration of MCDA and System Dynamics described earlier and Saaty's (1996) Analytic Network Process. It is an area that merits further research.

10.4.4. Quality Function Deployment

The Quality Function Deployment (QFD) methodology has its origins in Japan in the late 1960's with the work of Akao (Akao, 1998), but is now in use worldwide. It is a team process which seeks to incorporate customer opinion in the design of products and services through a four stage process of product planning, product design, process planning and process control. The first phase, product planning, utilizes the House of Quality. This incorporates two multicriteria models: one which captures customer requirements and their perceived importance (on a scale of 1 to 5) and a second which captures a technical specification of the product. A relationship matrix captures the strength of the relationship (on a 9 point scale) between the customers' needs and the technical descriptors. The relationship matrix is multiplied by the vector of importances assigned to requirements to give a vector of importances for the technical descriptors – to give an indication of which technical aspects of a product contribute most to the customers' perceived value. Competitors' products may be evaluated according to both customer requirements and technical descriptors, leading to better understanding of the competition.

In common with the Balanced Scorecard, QFD has incorporated multicriteria modeling in a broader framework focussing on a particular managerial issue. In our view it is innovative in a multicriteria sense, and therefore of particular interest, because of the way in which it seeks to merge two different perspectives on a problem. In common with others of the tools described it makes use of importance ratings in a way that is not clearly defined.

10.4.5. The EFQM Excellence Model

The European Federation for Quality Management (EFQM) Excellence model is another performance measurement and management tool that is internationally recognized and is coming to be widely accepted in Europe. It provides an overarching framework for implementing Total Quality Management within an organization, integrating the benefits of Investors in People (IiP), which focuses on employee development and ISO9000 which focuses on process control. The EFQM model specifies a generic framework as illustrated in Figure 10.9; in common with the Balanced Scorecard it highlights the dependence of the different elements.

Each of the nine components in the model is subdivided into more detailed parts (ranging from 2 to 6, giving 32 in total) and then within each of these sub-criteria a framework referred to as RADAR (Results, Approach, Deployment, Assessment and Review) is used to as the basis for rating an organisation's performance. Simple word models (e.g. no evidence or anecdotal, some evidence, evidence, clear evidence, comprehensive evidence) define 0 to 100 scales. These scores are aggregated using pre-specified or organization-specific weights to give an overall score for the Enablers and Results elements of the model. A self-assessment tool (a detailed questionnaire) is available or an organization can employ a specialist advisor to assist them in their evaluation. The model

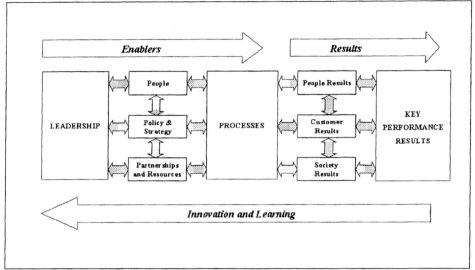

Figure 10.9: The EFQM Excellence Model (adapted from the EFQM website: www.EFQM.org)

provides a framework that organizations can use to understand the way in which their organization works, to develop strategy and a basis for continuous improvement. Should they wish, an organization can submit their self-assessment in consideration for the European Quality Award. The EFQM also provides a Benchmarking service, closely linked to the Excellence model, whereby organizations can compare themselves with role models.

The reasons for including a description of the EFQM Excellence model are similar to those for the Balanced Scorecard. It is another generic multicriteria model to support performance measurement and management that is widely used in organizations and indeed there are strong similarities between the two frameworks. The EFQM model is more tightly prescribed and has at its heart a model based on scoring and weighting that is akin to commonly used multicriteria models. However, in common with other approaches reviewed the basis for these values is not clearly expounded.

10.4.6. General comments

It would be easy to criticize the approaches discussed above from the perspective of multicriteria analysis, to dismiss them on the grounds of being simplistic, theoretically unfounded, ignorant of the principles of measurement, and so on. However, the approaches have gained widespread acceptance – they are used in many organizations – and if that is a primary goal of MCDA (which we believe it ought to be) then there must be valuable lessons to be learned. In our view, a key attraction of a number of these approaches is that they provide an "off-the-shelf" approach, a framework that is appealing and whose generic nature means that it can be easily applied by any organization. The approaches are simple and easy to use. They are also vigorously promoted by management consultants.

10.5. OR/MS APPLICATION AREAS WITH A MULTICRITERIA ELEMENT

In many OR/MS studies, there may not be explicit reference to MCDM theories or models, but there is often nonetheless strong implicit recognition of the multicriteria nature. In this section, we shall briefly identify some of these models, with the comment that an explicit recognition of the multiple criteria may help to enhance their value, and to give guidance regarding the manner in which relevant preference information (trade-offs etc.) may be elicited.

Quality management is one of the earliest applications of quantitative methods in industrial management. Traditional methods identified clearly the conflict between "consumer risk" and "producer risk", to be balanced against the cost of sampling. For a fixed sample size, the consumer and producer risks are in direct conflict for efficient sampling plans, but both can be reduced with larger sample sizes. This identifies three criteria, but the trade-offs between cost and the two forms of risk are generally left to unaided judgement or negotiation between the parties. In more recent quality assurance work, Taguchi methods for process improvement have become popular. The concern here is to design process experiments in order to identify operating conditions which yield both a good average quality and robustness (i.e. low variability in quality). The primary aim is efficiency in MCDM terms (e.g. the best robustness for a given average quality, or *vice versa*), but tradeoff along the efficient frontier is largely left to judgement, which becomes analogous to portfolio optimization discussed in the next paragraph.

Financial risk analysis and portfolio theory place considerable emphasis on the consideration of both (expected) return and risk in making investment decisions. In Markowitz theory, an important visual tool is the graphical plot of risk, generally measured by standard deviation of returns, against expectation of return. Considerable statistical and computational effort goes into characterizing the efficient frontier, but choice of desired position along the frontier may be left largely to direct judgement. As we have noted while discussing multivariate statistics,

328   MULTIPLE CRITERIA DECISION ANALYSIS

such judgement is easily applied when there are just two criteria (risk and return in this case), but rapidly becomes much more difficult for larger numbers of criteria. Even in the financial world of profits, it is not at all clear that "risk" is a single dimensional concept, while the existence of many specialized funds (limiting investments to companies with acceptable social or environmental records, for example) implies that investors do in fact have many other criteria, which tend to be taken into consideration in more *ad hoc* ways, often by application of specified constraints.

An area of public sector decision making which involves multiple conflicting criteria is that of environmental impact assessment (EIA). Many countries legislate for some form of EIA before approval is granted for any major development projects, and these may well specify a whole host of criteria relating to soil, water and atmospheric contaminants. This defines a clear multicriteria decision making problem. The approaches used often violate important principles of MCDA, however, for example:

- Attention is often focused on two alternatives only, the proposed plan and (*de facto*) the *status quo*, whereas a wider range of alternatives can lead both to more meaningful expression of values (as opposed to polarization into two camps), and to greater opportunity for discovering consensus.

- Fixed required levels of achievement may be specified at the outset, often without clear understanding of what is practically achievable, making any consideration of desirable tradeoffs almost impossible. The required performance levels may in some cases be so undemanding that the EIA does little to prevent socially undesirable developments, and in other instances may be so unrealistically demanding that all development gets bogged down in lengthy litigation and quite *ad hoc* compromises.

The above examples have attempted to illustrate the manner in which multicriteria decision making problems pervade many areas of management science. In many cases, it is probable that the intuitive or *ad hoc* treatment of the multiple criteria still generate perfectly acceptable answers. The approaches used, however, tend to emphasize two primary criteria which can easily be visualized, with remaining criteria managed by means of setting required levels of performance. Some concerns are that setting of fixed performance levels (for example) may be susceptible to manipulation, and that many good opportunities may be missed. The challenge for research is to extend formal MCDA into these other areas, in order to facilitate more systematic, fair and justifiable consideration of all the relevant criteria.

10.6. CONCLUSIONS

In this chapter we have examined the emergence of multicriteria concepts in a wide range of management and management science theories. An important point which we have emphasized is that the existence of multiple criteria in managerial decision making is very widely recognized even if not always spelt out explicitly in these terms. Many different approaches have been proposed and adopted to cope with these multiple criteria, and in our discussion of these approaches we have stressed two important and contrasting themes:

- Although the existence of multiple criteria is acknowledged, their treatment, and that of the underlying preferences, value judgements and trade-offs of the decision maker, is often left to rather *ad hoc* processes, relying on simple qualitative rules and/or on scoring systems that do not recognize many of the principles of MCDA discussed earlier in this book.

- In spite of their simplicity, however, many of these approaches do contain insights which may not immediately emerge from standard MCDA practice, so that there is much for MCDA practitioners to learn from these other schools of thought.

The challenge to research and practice in MCDA is therefore to integrate its rich theory and extensive experience into wider management practice as represented by the various methodologies discussed in this chapter. The success of many of these approaches reveals the need of managers to find adequate means of coping with conflicting criteria in decision making. The power of MCDA can address that need only to the extent that MCDA becomes an integral part of management practice, rather than to remain a specialized sub-discipline of OR/MS. It is this recognition of the need for an "integrated approach" to MCDA which motivated this book, and forms the theme of our concluding chapter.

Chapter 11

AN INTEGRATED APPROACH TO MCDA

11.1. INTRODUCTION

In this final chapter we return to focus on and explicitly to address the title and overarching theme of the book – *An Integrated Approach to Multicriteria Decision Analysis*. This discussion is directed at both theoreticians and practitioners in the MCDA community. We believe that integration, in many different forms, is essential to the growth and success of MCDA. Hence, let us begin the discussion by expanding on what we envisage by integration.

It is our view (also expressed in Pictet and Belton, 1997) that MCDA is a practical subject, one which has no value unless it is applied, and thus that theoretical developments must be grounded in practice. At the same time we believe that practice must be informed by sound theory, in particular the theories of value measurement and preference modeling to which we devote Chapter 4. However, we should also be cognizant of those theories of behavioural decision making and organizational intervention which have a bearing on the practice of MCDA. Thus the first thread of our theme focuses on the need to integrate theory and practice.

One area in which we feel there is both the scope and the need for theoretical development is in the provision of an integrating framework for the different MCDA methods. Throughout the book we have addressed different stages in, and aspects of the process of MCDA, which in itself provides the basis for an integrating framework. Within that we have considered three key schools of MCDA thinking and some of the specific approaches associated with each: namely, value function based methods, outranking methods and aspiration-level methods. In these discussions we have sought to highlight both commonalities and

332   *MULTIPLE CRITERIA DECISION ANALYSIS*

differences. We aspire to a framework, or a meta-methodology, which integrates the methods through their commonalities, but within that, acknowledges and celebrates the diversities, recognizing the strengths and weaknesses of each, their appropriateness in different contexts and when they can usefully complement each other. We explore these issues in more detail in Section 11.2.

In keeping with the view expressed above that MCDA is a practical subject, throughout the book we have sought to emphasise the use and usefulness in practice of the methods we describe and to suggest ways in which that practical value can be enhanced by drawing on and integrating experience and expertise from outside of MCDA.

We would, however, go further than simply suggesting that the practice of MCDA can be enhanced in this way. We are of the opinion that the integration of MCDA methods within a broader framework of problem structuring and organizational intervention is essential for its widespread acceptance as a practically useful methodology. The significant attention paid to these aspects of the process of MCDA is reflected in Chapters 3 and 9.

Less critical to the implementation of MCDA, but still offering substantial potential for its enhancement as a means of informing planning and decision making, is the integration with OR/MS methodologies which have a complementary focus, for example, system dynamics which emphasises the dynamic and interconnected nature of situations, scenario planning which focuses on the need to construct and to consider alternative futures, or conflict analysis which highlights the importance of viewing a situation from the perspectives of the different parties involved. The opportunities for enhancement here are mutual and MCDA analysts and practitioners should seek both to draw on these other methodologies and to demonstrate to their proponents the value of the MCDA perspective.

To summarise, we would highlight the following threads of integration:

- The integration of theory and practice

- An integrating framework for the use of MCDA methods

- The integration of MCDA methods within a broader framework of problem structuring and organizational intervention

- The integration of MCDA with other OR/MS and managerial methodologies

As indicated above, we have already devoted substantial attention to the third of these threads in Chapters 3 and 9, while the fourth has been addressed in Chapter 10. In the next section we explore further the second thread of integration, synthesizing comments made throughout the book, and in the final section, where we look to "The way forward" for MCDA, we return to the need to integrate theory and practice.

11.2.   AN INTEGRATING FRAMEWORK

In previous chapters we have discussed the different schools of MCDA which have developed, the motivation for their development and their characteristic features and assumptions. In this section we put forward our arguments for an integrating framework for MCDA, initially establishing why we feel such a framework is necessary and would be of benefit to the field, going on to discuss what such a framework might look like and concluding with our thoughts on issues which would need to be addressed.

11.2.1.   Why is there a need for an integrating framework?

To the casual observer of the field, MCDA may give the impression of a "house divided against itself", of a field fragmented with no firm foundation or philosophy. It is certainly true that from time to time adherents of different schools do enter into heated debates and exchanges which are not always productive. There is, however, a fundamental set of underlying concepts shared by all schools, although these may be differently expressed. It is this unity within the diversity that we feel must be acknowledged and promoted if MCDA is to make meaningful contributions to managerial decision making, particularly at strategic levels. A unified presentation is important both for the promotion of and for the acceptance of MCDA as a coherent field. It is easily perceived as a collection of ad hoc methodologies which give different "answers" to the same "problem", rather than as a set of approaches which can collectively generate different insights into a complex issue. Referring back to our introduction to MCDA in Chapter 1, it is a myth that MCDA can give the "right" answer because there is no such thing; rather, the aim of good MCDA is to facilitate decision makers' learning about the many facets of an issue in order to assist them in identifying a preferred way forward. To use a medical analogy, perhaps the use of different MCDA methodologies should be perceived in the same way as the use of different diagnostic tests, each yielding particular insights into the condition of a patient, rather than as alternative remedies for the illness.

We are at pains to emphasise that we are advocating *unity* not *uniformity*. The acknowledgement of unity does not mean that the different

334 MULTIPLE CRITERIA DECISION ANALYSIS

schools of MCDA should be forced artificially into a dull uniform mould. To the contrary, we feel strongly that such uniformity would be counter-productive. As indicated above it is the synergistic accumulation of insights from the different perspectives of each school that may yield the greatest added value.

11.2.2. What would an integrating framework look like?

An integrating framework must allow for a common exposition of the multi-criteria approach without attempting to impose a methodological straightjacket which suggests, for example, "you must use this method, in this way, for this type of problem ... ". It must recognize those elements which are truly common to all approaches; it must highlight the distinctive strengths of each approach and seek to encourage analysts from other schools to embrace these in their own practice and research; furthermore it must alert us to the weaknesses of the overarching framework and of individual approaches in order to encourage research which seeks to address these.

Firstly, what are the truly common elements? The fundamental concept shared by all schools of MCDA is, by definition, that of structuring complex decision problems in terms of alternatives and criteria. Exponents of different schools might approach this structuring in different ways. Some, for example, may have stressed the *hierarchical* nature of criteria more explicitly than others as part of their methodology, but the value of hierarchical representation for purposes of problem structuring and building up of understanding of the problems can be recognized by all schools.

A further common theme across schools is that in non-trivial problems the "ideal" is generally not attainable. The gap between the "ideal" and what is achievable can and should provide the spur to think creatively about developing new alternatives, but at the end of the day compromise between the conflicting criteria is usually inevitable. Schools may differ as to how compromises are generated, but all tend to concentrate attention on alternatives which are either efficient, or at least sufficiently close to the efficient frontier so that (in the terminology of the outranking school) no efficient alternative is "strictly preferred" to it. Acceptance of a compromise is in effect the acceptance of tradeoffs between criteria. Value function methodologies focus directly on these tradeoffs, requiring decision makers to confront the sometimes difficult problem of expressing acceptable tradeoffs explicitly. In contrast, goal programming requires decision makers to speculate about the consequences of tradeoffs in defining plausible aspiration levels, while outranking meth-

An Integrated Approach to MCDA   335

ods allow incomparability and avoid the issue of assessing tradeoffs entirely by comparing pairs of alternatives in terms of evidence favouring each alternative over the other. All three of these insights contribute to understanding the multicriteria decision problem.

Within the broad commonality of approach, schools differ as to the *a priori* assumptions made and as to the information presented to the decision makers. It may be useful at this point to review briefly some of the particular contributions and insights arising from different schools, as follows.

- The standard *value function* approach stresses the value of imposing some form of discipline in the building up of preference models, by ensuring that these are consistent with simple operationally meaningful "rationality" axioms. As we have emphasized earlier, it is accepted that unaided human decision making may often be inconsistent with these axioms, but attempting to frame preferences within the context of such axioms helps decision makers to obtain greater understanding of their own values, and to justify their final decisions where required. As noted above, the value function approaches also encourage explicit statements of acceptable tradeoffs between criteria. These may be difficult to specify, but even rough indications are useful in understanding and conveying preferences in an unambiguous manner. Although these approaches derive from decision theory which adopts a normative perspective, many practitioners utilise them in a constructive spirit.

- The *outranking* school focuses on pairwise evaluation of alternatives, identifying vetoes and incomparabilities as well as assessing preferences and indifferences. It has drawn attention to the fact that preferences and values are not pre-existing, waiting to be measured, but evolve as part of the MCDA process within the context of the choices to be made. The aim should thus be neither to describe what decision makers do, nor to prescribe what they should do; the emphasis should be on giving support to decision makers in *constructing* their preferences between the alternatives under consideration. As indicated earlier, a constructivist philosophy is easily and usefully incorporated into other methodologies.

- The *goal programming* and *reference point* approaches emphasize the existence of situations in which decision makers may find it very difficult to express tradeoffs or importance weights, but may nevertheless be able to describe outcome scenarios which they would find satisfying. These would be expressed in terms of aspirations or goals for each criterion.

336   *MULTIPLE CRITERIA DECISION ANALYSIS*

- The *AHP* mode of developing preference functions has demonstrated that in many situations decision makers may be much more comfortable in expressing values on semantic scales (e.g. moderately important, highly important) rather than in terms of numerical scores. The use of pairwise comparisons and overspecification of values, giving rise to a measure of consistency, are also distinctive features of the AHP approach.

As we have indicated above, principles identified by one school can and should inform the implementation of methodologies employed by other schools, without these schools having to abandon their own principles and techniques. In this way we can move towards a meta-MCDA framework. The present authors have tended to make more use of value function methods than those of other schools. But in our application of these methods, we have adopted an explicitly "constructivist" approach in accordance with the arguments of the outranking school. In building up partial and aggregated value functions, we guide clients to think in terms of what preferences are revealed by comparison of real or hypothetical alternatives, thus *constructing* preferences consistent with the axioms of value measurement. Critics of this approach have suggested that we are evading the issue of what the decision makers' "true" preferences are and/or that we compromise the legitimacy of the value function approach. Our response is that "true" preferences do not exist "in the decision makers' head", waiting to be elicited, while the axioms are certainly not descriptively valid, and are at best conditionally prescriptively justified. The most "legitimate" use of the axioms (in our view) is thus as a set of simple guidelines which can be explained to decision makers at the outset, and which will be used to place a structure on the preferences being constructed.

Another example of the linking of concepts from different schools is provided by Pictet and Belton (2001) who seek to alert users of a multiattribute value function analysis to pairs of options which might be judged incomparable by an outranking method.

In similar vein, we note also that the use of semantic scales, pairwise comparisons and overspecification of values (which are fundamental to the standard implementation of the AHP) have been incorporated into other value function methods by techniques such as the MACBETH and related approaches discussed in Section 6.2. The main difference between these two approaches is that whereas AHP associates fixed numerical values with each qualitative preference statement (such as "weakly preferred", "strongly preferred", etc.), MACBETH translates such statements into bounds on the underlying numerical values.

11.2.3. Attention to practice

If an integrating framework is to effectively inform the good practice of MCDA it must recognize issues outside of the MCDA methods themselves, in particular the practical issues discussed at length in Chapter 9, and take cognizance of the fact that different methods are better suited to different contexts. For example:

- *The nature of the client group:* Is the group responsible for the decisions to be taken or are they acting in an advisory capacity? This will dictate the extent to which preferences can be elicited directly as opposed to a need to work with incomplete or inferred preferences.

- *The nature of contact with the client group:* Is analysis to be carried out interactively with those directly responsible for decision making, or is the majority of analysis to be done in the backroom with only key issues to be reported? Interactive involvement calls for an approach which is transparent and easily explainable as well as for robust, easy to use and flexible supporting software.

- *The organizational context:* In an organic and entrepreneurial organisation the emphasis is likely to be on learning and finding a robust, preferred way forward without a lot of attention to recognized processes; in a larger more bureaucratic organization the emphasis may be on being seen to do things correctly and the ability to justify decisions to others through the provision of an audit trail. Although the nature of all MCDA methods is to make explicit the basis of decisions, the nature of the "audit trails" provided is different, so that different approaches may be better suited in different organizational contexts. It is also important to be aware of the political context of the decision – the openness of most MCDA methods may conflict with the need to maintain a degree of equivocality.

- *The time available for analysis:* There is no doubt that a detailed multicriteria analysis can be very demanding of the decision makers – it is important to recognise that in some circumstances a "quick and dirty" analysis, working with little, or poor data may be better than no analysis at all.

- *Skills of the analysts / facilitators:* Whilst we advocate an ideal situation wherein all analysts are familiar with the broad range of methodologies and have both analytic and process skills, it has to be accepted that this may not always be the case in practice. It is thus important to make clear what skills are needed to implement effectively a particular approach to analysis.

338   *MULTIPLE CRITERIA DECISION ANALYSIS*

- *Technical support available:* The effective use of most methodologies relies on the availability of reliable, easy to use software. Use in an interactive context calls for good display facilities within the software as well as the technology to do so. Allowing users directly to input their own judgements calls for increased user friendliness and support. The use of two or more approaches in combination would be much facilitated by a standard protocol which enabled the sharing of databases. All of these are factors which in practice may influence the decision to work with particular approaches.

Furthermore, an integrating framework should recognize the tension between these practical issues on the one hand and on the other the desire for theoretically sophisticated methodologies.

Horses for courses

Although we have argued that an integrating framework should not seek to impose which approach should be used when, it is the case that the different approaches best lend themselves to use in particular contexts and phases of decision making, and the framework should thus seek to provide a catalogue of good practice. In our experience we have found the following links between methodologies and context to be particularly apt.

- Use of goal programming or reference point methods at early stages of problem analysis, for purposes of initial screening and generation of shortlists of alternatives for more detailed evaluation by groups;

- In building portfolios of options, initial detailed evaluation of individual elements followed by use of a goal programming or reference point approach which incorporates metacriteria such as balance factors;

- Use of value function methods within workshop settings, for purposes of facilitating the construction of preferences by working groups which may only represent a subset of stakeholder interests;

- Use of outranking methods, and/or 'what-if' experiments with value function models, for purposes of informing political decision makers regarding consequences of particular courses of action.

11.3.   THE WAY FORWARD

11.3.1.   What are the benefits of an integrating framework?

The major benefit is that expounded under the heading of *Why is there a need for an integrating framework* in Section 11.2, namely the articulation of a unified front which generates interest in the potential for MCDA amongst both analysts and decision makers. A growth in applications of MCDA is an essential basis for the furthering of useful research which can itself inform and lead to better practice, thereby ensuring continuing and healthy growth of the field. Furthermore, an integrating framework will provide a basis for research which directly benefits the field as a whole, it will provide a common language to facilitate communication amongst researchers and practitioners in the field, and it will provide practitioners of the future with a stronger toolbag and set of tools.

11.3.2. What are the pre-requisites?

Clearly such a framework can only effectively be realized if the right culture exists in the MCDA community. A spirit of cooperation, a recognition of the strengths of others' approaches and of the constraints imposed by others' positions, and a culture of mutual learning is needed rather than one of competition and aggressive criticism. This must exist across schools and between theory and practice.

The nature of the MCDA field to date has been that new entrants, both to theory and practice, have tended naturally to establish their strengths within the school in which they began to develop their research or to practice their craft. Furthermore, the fact that the field began to develop rapidly only in the 1970's means that the initiators and principle exponents of each of the schools still play a leading role and command significant "loyalty". Whilst we believe that such specialism is vital to the field and to the maintenance of its diversity, we believe also that as the field matures it is now important to encourage in an active way the development of cross-school expertise. The (approximately) biennial MCDA Summer Schools, which have run since 1983, seek to give new researchers and practitioners a thorough understanding of common issues and the breadth of multicriteria methods. Participants tend, however, to return to their home ground after the schools and focus on developing a specialist interest. It may be that the growth of cross-school expertise is a natural development of the field, but we hope to see those in established positions taking a proactive role in facilitating and encouraging this, for example through collaborative degrees and research programmes.

Good supporting software is in any case essential for the effective practice of MCDA, but the smooth integration of different approaches will call for common protocols which allow the sharing and easy exchange of data.

340 *MULTIPLE CRITERIA DECISION ANALYSIS*

There are also fundamental issues on which the whole field rests that must be acknowledged by an integrating framework and could be the focus of a collaborative research programme. We refer in particular to the measurement of preferences and to the ubiquitous concept of the relative "importance" of criteria, topics to which we have returned many times throughout the book.

11.3.3. Reflections on the nature of research in MCDA

In contemplating "the way forward" we return to the view put forward at the start of this chapter, namely that MCDA in practice should be informed by sound theory, the development of which should in turn be grounded in practice, comprising a classical learning loop such as proposed by Kolb et al. (1984). Embedded within this theory-practice loop we would like to emphasise three interacting foci of research; the *development of an integrating framework* which is informed by and in turn informs research into the *implementation of MCDA* and research into *methodological development*.

Development of an integrating framework

We have dwelt at length on the need for and the nature of an integrating framework. It should be evident that we see this development as the major challenge for research in MCDA. Such research may be initiated from either a practical or a theoretical standpoint. From a practical perspective, by specifying key factors which characterize actual or potential MCDA interventions we can begin to evaluate the extent to which different approaches, individually or in combination, are useful in specific contexts defined by combinations of these key factors. Factors that we consider to be relevant in characterizing interventions have been emerging throughout the book, for example:

- Stage of process (preliminary screening; initial evaluation with small task or special interest groups; direct involvement with DMs in the final decision);

- Level of contact with decision makers (Largely backroom according to broad terms of reference only; spasmodic interaction to obtain some partial preference information; intensive contact);

- Size and bureaucratization of the organization or group, and the associated need for audit trail (small group, no such need; intermediate levels; politicized with high demand for justification/motivation);

- Familiarity of problem setting (similar to many previous decisions; somewhat unusual; totally new strategic ground);

- Time and resources available, i.e. the time available in absolute and elapsed terms (in conjunction with the level of contact with decision makers);

- Integration with other approaches to modeling (ranging from qualitative models such as cognitive/cause maps to quantitative models e.g. for simulation or scheduling).

From a theoretical standpoint, considerations of a particularly attractive feature of one approach, or a weakness of another may suggest ways to develop a more powerful or flexible hybrid methodology.

As we indicated earlier, the development of an integrating framework extends beyond a simple identification of "horses for courses"; it should not only to provide a catalogue of good practice but an understanding of why this is so and how it can be built on further in a particular context. The further development of such a framework will depend upon a coherent programme of implementation research and methodological development, to which we now turn.

Implementation research

An important component of the development of MCDA must be the active pursuit of empirical research to explore the use and usefulness of methods. Such research should encompass, for example: the ability and the willingness of decision makers to engage in the process and to provide the preference information or value judgements required of them; the extent to which they are able to provide such judgments in a consistent and coherent manner; the extent to which they understand the process and to which genuine learning occurs as a consequence; the extent to which confidence in and ownership of decisions is enhanced; the extent to which shared understanding and agreement is achieved in groups; the extent to which perceived decision quality is enhanced. Empirical research encompasses both experimental investigations and action research.

- *Experimental investigation* has been used extensively in MCDA and in decision analysis generally. Typically a large group of subjects is set simple hypothetical tasks or choices under varying experimental conditions. Outcomes may be evaluated using quantitative or qualitative research tools and would be expected to shed light on how decision makers might react / respond / behave when faced with a real issue. Examples of such research are: the work by Kahneman

342 *MULTIPLE CRITERIA DECISION ANALYSIS*

and Tversky (1979) investigating decision making under uncertainty; work by Mousseau (1992), von Nitzsch and Weber (1993) and Weber and Borcherding (1993) on the interpretation of importance weights; a comparison of interactive methods by Buchanan (1994); a comparison of AHP and value function models by Belton (1986).

- *Action research* involves the careful design and monitoring of actual implementations of MCDA, incorporating a process of data capture to record the important aspects of the intervention or participants' responses to it, and a process of active reflection on the outcomes in order to develop theory and inform future interventions (e.g., Eden and Huxham, 1995). It is only through action research that implementation issues can be genuinely investigated, although as commented above, experimental investigations can yield valuable insights in a less "risky" environment, which should inform action research.

Methodological research

As already indicated, the development of a comprehensive integrating framework for MCDA must inform and be informed by both implementation research and methodological development. There will always be considerable scope for the further development of individual methodologies. It is our own experience of using MCDA in practice that each new intervention poses new challenges, some of which require careful attention to underlying theory. In the spirit of our ambition for an integrating framework we hope that many future methodological developments will be fuelled by the feedback and by the tensions between theory and practice as well as by cross-fertilisation of ideas between the different schools of MCDA, and by developments outside the field. It is not our intention here to attempt a complete review of possible research areas, but consider it worth noting two broad categories of research:

- Analysis of *special demands for particular application areas*, where the general assumptions of any single model may not hold very well. Examples of such application areas may be those of finance and investment analysis, and of environmental impact assessment.

- Identification of *general weaknesses in MCDA models* and extension of models to address these. For example, as we have mentioned elsewhere, methodologies for the treatment of risk and uncertainty in MCDA models appear still to be inadequately developed.

11.4. CONCLUDING REMARKS

The focus throughout this book has been on the presentation of MCDA methodologies, a collection of formal approaches which seek to take explicit account of multiple criteria in helping individuals or groups explore decisions that matter, and their use in practice. In this final chapter we have presented our vision of how we would like to see the field of MCDA develop in the future, in particular of how the "collection of methodologies" might benefit from an increased unity through an integrating framework.

However, a truly integrated approach to MCDA must look to outward as well as to internal integration. It must seek to be informed by and to inform developments in other fields.

Of course, the need for integration is not the only challenge which will face the MCDA community in future years. The nature and pace of global change present opportunities and challenges for everyone, including this field. Increasingly sophisticated and more widely available technology, increasingly globalised activity, increasing awareness of the possible consequences of these and reaction against them all potentially impact on the practice of MCDA and thereby on the need for supporting methodological development. This may be through the nature of support provided (e.g. more virtual groups), the nature of problems tackled (more complex) and the nature of "client" groups (e.g. the need to involve a broader community of stakeholders). It is our view that an integrated community will be better placed to meet these challenges.

Appendix A
MCDA Software Survey

As has been emphasized on a number of occasions throughout the text, the application of MCDA methods generally needs to be supported by appropriate computer software, so as to free the facilitator / analyst and decision maker from the technical implementation details, allowing them to focus on fundamental value judgements. It is certainly possible in many cases simply to set up macros in a spreadsheet to achieve this end, especially when linked to the graphics functions of the spreadsheet. It can be more convenient to make use of specially designed software, however. In this Appendix, we provide a listing of some software products, based on information available in December 2000. We make no claims that the list is complete (and it will, of course, rapidly become dated), and apologize in advance to suppliers whose products we have not included. Apart from our own personal contacts with workers in the MCDA field, the list has been compiled primarily from responses to a request to the MCRIT-L list server, and from the decision analysis software surveys which appear regularly (every two years) in the journal *ORMS Today* published by INFORMS (The Institute for Operations Research and the Management Sciences).

The table which follows on the next pages summarizes information on the various software products and their availability. In most cases a web-site is indicated, and it is suggested that the reader consults these for updated information on availability and pricing. We note that there is little explicit reference to software for goal programming and related methods. This is because standard linear programming packages (including the optimizer routines within spreadsheets) can be easily be adapted to solve goal programming problems (as indicated by the examples in Chapter 7). In particular, the XPRESS-MP software contains an explicit goal programming option, details concerning which can be obtained via the internet from www.dash.co.uk.

Package	Categories of decision supported	Name and address of supplier	Methodology
Criterium Decision Plus	Discrete problems (up to 200)	InfoHarvest Inc., 4511 Shilshole Avenue, Suite 2000, Seattle WA 98107-4714, USA http://www.infoharvest.com	Value function model based on tradeoff analysis
DataScope	Discrete and continuous problems	Laszlo Bernatsky, Cygron R&D Ltd., P O Box 727, H-6724 Szeged, Hungary http://www.cygron.com	Multiattribute value functions
Decision Explorer	Problem structuring for general decision contexts	Banxia Software Ltd., 141 St James Road, Glasgow G4 0LT, Scotland http://www.banxia.com	Qualitative data analysis, linking concepts through cognitive or cause maps
Decision Lab 2000	Discrete choice problems (with provision for comparison under several scenarios)	Bertrand Mareschal, Free University Brussels, Dept of Statistics and Operations Research, Boulevard du Triomphe - CP210/01, B-1050 Brussels, Belgium http://www.visualdecision.com	PROMETHEE/GAIA
DEFINITE	Discrete choice problems (with emphasis on impact studies)	Institute for Environmental Studies, Free University of Amsterdam, De Boelelaan 1115, 1081 HV Amsterdam, The Netherlands fax: +31 20 4449555 email: marjan.van.herwijnen@ivm.vu.nl	Multiattribute value functions including option for imprecise preference information; cost-benefit analysis; outranking

Table A.1. Special purpose MCDA software

Package	Categories of decision supported	Name and address of supplier	Methodology
ELECTRE III, IV, IS & TRI	Discrete choice problems	LAMSADE Softwares, University of Paris-Dauphine, Place de Lattre de Tassigny, 75775 Paris Cedex 16, France http://www.lamsade.dauphine.fr	ELECTRE outranking methods
EQUITY	Resource allocation and project portfolio selection	Enterprise LSE Ltd., Houghton Street, London WC1A 2AE, England http://www.enterprise-lse.co.uk	Multiattribute value functions
High Priority	Discrete choice problems	Krysalis Ltd., 28 Derwent Drive, Maidenhead SL6 6LB, England http://www.krysalis.co.uk	Multiattribute value functions
HIVIEW	Discrete choice problems	Enterprise LSE Ltd., Houghton Street, London WC1A 2AE, England http://www.enterprise-lse.co.uk	Multiattribute value functions
Logical Decisions	Discrete choice problems	Logical Decisions, 1014 Wood Lily Dr., Golden, CO 80401, USA http://www.logicaldecisions.com	Multiattribute value functions and AHP

Table A.1. (Cont.)   Special purpose MCDA software

Package	Categories of decision supported	Name and address of supplier	Methodology
MACBETH	Discrete choice problems	Jean-Claude Vansnick, University of Mons-Hainaut, Faculte Warocque des Sciences Economiques, 20 Place du Parc, 7700 Mons, Belgium http://www.umh.ac.be/vansnick/macbeth.html	Multiattribute value functions using pairwise comparisons on semantic scales
On Balance	Resource allocation and project portfolio selection	Krysalis, Ltd, 28 Derwent Drive, Maidenhead SL6 6LB, England http://www.krysalis.co.uk	Multiattribute value functions
PRIME	Discrete choice problems (Group decisions)	System Analysis Laboratory, Helsinki University of Technology, P O Box 1100, 02015 HUT, Finland http://www.hipre.hut.fi/Downloadables/	Multiattribute value functions with imprecise preference information
Team Expert Choice	Discrete choice problems (Group decisions)	Expert Choice Inc., 5001 Baum Blvd. #430, Pittsburgh PA 15213, USA http://www.expertchoice.com	AHP
V·I·S·A	Discrete choice problems	SIMUL8 Corp., 141 St James Rd., Glasgow G4 0LT, Scotland http://www.SIMUL8.com/visa.htm	Multiattribute value functions

Table A.1. (Cont.)   Special purpose MCDA software

350   *MULTIPLE CRITERIA DECISION ANALYSIS*

Package	Categories of decision supported	Name and address of supplier	Methodology
Web-HIPRE	Discrete choice problems (Group decisions)	System Analysis Laboratory, Helsinki University of Technology, P O Box 1100, 02015 HUT, Finland fax: +358 9 4513096 http://www.hipre.hut.fi	Multiattribute value functions and AHP
WINPRE	Discrete choice problems (Group decisions)	System Analysis Laboratory, Helsinki University of Technology, P O Box 1100, 02015 HUT, Finland fax: +358 9 4513096 http://www.hipre.hut.fi/Downloadables/	Interactive multiattribute value functions with imprecise preference information
WWW-NIMBUS	Multiobjective mathematical programming problems	Kaisa Miettinen, University of Jyvaskyla, Dept of Mathematical Information Technology, P O Box 35 (Agora), FIN-40351, Jyvaskyla, Finland http://nimbus.mit.jyu.fi	Interactive and reference point methods

Table A.1. (Cont.)   Special purpose MCDA software

Appendix B
Glossary of Terms and Acronyms

GLOSSARY

Attribute   A quantitative measure of performance associated with a particular criterion according to which an alternative is to be evaluated.

Compact set   A closed and bounded set, i.e. a set of finite extent and such that the boundary is included in the set. For example the set of values x such that $0 \leq x \leq 1$ is compact, but not sets defined for example by $x \geq 0$ (since this is unbounded above), or $0 < x < 1$ (since the boundary values of 0 and 1 are excluded).

Complete order   A preference relationship ($\succ$) on a set of objects (e.g., decision alternatives) is said to be a complete order if (1) for any pair of objects a and b, either $a \succ b$ (a is "preferred to" b) or $b \succ a$ (but not both); and (2) the relationship is transitive, i.e. if $a \succ b$ and $b \succ c$ then $a \succ c$.

Concave function   Let $f(\mathbf{x})$ be a real-valued function defined on points in an n-dimensional vector space. The function is said to be concave if $f(\alpha \mathbf{x}^1 + (1 - \alpha)\mathbf{x}^2) \geq \alpha f(\mathbf{x}^1) + (1 - \alpha)f(\mathbf{x}^2)$ for any two arbitrarily chosen points $\mathbf{x}^1$ and $\mathbf{x}^2$, and for any $0 < \alpha < 1$. In other words, if the function is evaluated at three points lying on a straight line, then its value at the intermediate point is never less than the corresponding value on the straight line joining the values at the two end points.

Convex function   Let $f(\mathbf{x})$ be a real-valued function defined on points in an n-dimensional vector space. The function is said to be convex if $f(\alpha \mathbf{x}^1 + (1 - \alpha)\mathbf{x}^2) \leq \alpha f(\mathbf{x}^1) + (1 - \alpha)f(\mathbf{x}^2)$ for any two arbitrarily chosen points $\mathbf{x}^1$ and $\mathbf{x}^2$, and for any $0 < \alpha < 1$. In other words, if the

352 *MULTIPLE CRITERIA DECISION ANALYSIS*

function is evaluated at three points lying on a straight line, then its value at the intermediate point is never greater than the corresponding value on the straight line joining the values at the two end points.

Convex set A set $\mathcal{C}$ is said to be convex if for any two arbitrarily chosen elements $c^1, c^2 \in \mathcal{C}$, $\alpha c^1 + (1 - \alpha)c^2$ is also in $\mathcal{C}$. For example, the set of integers is not convex (since 1 and 2 are integers, but not 1.5), but the set of values in the $X - Y$ plane lying in the inside of a circle is convex.

Criterion A particular perspective according to which decision alternatives may be compared, usually representing a particular interest, concern or point of view.

MAUT Multiattribute utility theory. See Section 4.3

MAVF Multiattribute value function. See Chapters 5 and valindint

MAVT Multiattribute value theory, i.e. the theory of preference modelling underlying the construction of multiattribute value functions

MOLP Multiple objective linear programming

Preorder This is also termed a weak order, and represents a weakening of the concept of a complete order to allow for the possibility that both $a \succ b$ and $b \succ a$ (in which case a and b are often said to be "indifferent")

Pseudoconcave function Let $f(\mathbf{x})$ be a real-valued function defined on points in an n-dimensional vector space, and define $\nabla f(\mathbf{x}^1)$ as the gradient vector of $f(\mathbf{x})$ at the point $\mathbf{x}^1$ (i.e. the vector of partial derivatives). The function is said to be pseudoconcave if $f(\mathbf{x}^2) \leq f(\mathbf{x}^1)$ for any point $\mathbf{x}^2$ such that $\nabla f(\mathbf{x}^1) \cdot (\mathbf{x}^2 - \mathbf{x}^1) \leq 0$. In other words, any move from a point $\mathbf{x}_1$ in a direction of locally decreasing values of $f(\mathbf{x})$ will never lead to a reversal of this downward slope. Concave functions are also pseudoconcave, but pseudoconcave functions are not necessarily concave.

SSM Soft systems methodology

References

Ackoff, R. L. (1978). *The Art of Problem Solving*. John Wiley & Sons, New York.

Agrell, P. J., Lence, B. J., and Stam, A. (1998). An interactive multicriteria decision model for multipurpose reservoir management: the Shellmouth Reservoir. *Journal of Multi-Criteria Decision Analysis*, 7:61–86.

Akao, Y. (1998). *Quality Function Deployment, Integrating Customer Requirements into Product Design*. Productivity Press.

Angehrn, A. A. and Luthi, H. J. (1990). Intelligent decision support systems : A visual interactive approach. *Interfaces*, 20(6):17–28.

Bana e Costa, C. A., editor (1990). *Readings in Multiple Criteria Decision Aid*. Springer-Verlag, Berlin.

Bana e Costa, C. A. and Correa, E. C. (2000). Construction of a total quality index using a multicriteria approach. Paper presented at the EURO XVII Conference, Budapest, July 2000.

Bana e Costa, C. A., Ensslin, L., Corrêa, E. C., and Vansnick, J. (1999). Decision support systems in action: integrated application in a multicriteria decision aid process. *European Journal of Operational Research*, 113:315–335.

Bana e Costa, C. A. and Vansnick, J. C. (1994). MACBETH – an interactive path towards the construction of cardinal value functions. *International Transactions in Operational Research*, 1:489–500.

Bana e Costa, C. A. and Vansnick, J. C. (1997). Applications of the MACBETH approach in the framework of an additive aggregation model. *Journal of Multi-Criteria Decision Analysis*, 6:107–114.

Bana e Costa, C. A. and Vansnick, J. C. (1999). The MACBETH approach: basic ideas, software and an application. In Meskens and Roubens, 1999, pages 131–157.

354   *MULTIPLE CRITERIA DECISION ANALYSIS*

Barda, O. H., Dupuis, J., and Lencioni, P. (1990). Multicriteria location of thermal power plants. *European Journal of Operational Research*, 45:332–346.

Barzilai, J., Cook, W. D., and Golani, B. (1987). Consistent weights for judgements matrices of the relative importance of alternatives. *Operations Research Letters*, 6:131–134.

Bell, D. E., Raiffa, H., and Tversky, A., editors (1988). *Decision Making: Descriptive, Normative and Prescriptive Interactions.* Cambridge University Press.

Bellman, R. E. and Zadeh, L. A. (1970). Decision making in a fuzzy environment. *Management Science*, 17:B141– B164.

Belton, V. (1985). The use of a simple multiple criteria model to assist in selection from a shortlist. *Journal of the Operational Research Society*, 36:265–274.

Belton, V. (1986). A comparison of the analytic hierarchy process and a simple multi-attribute value function. *European Journal of Operational Research*, 26:7–21.

Belton, V. (1993). Project planning and prioritisation in the social services – an OR contribution. *Journal of the Operational Research Society*, 44:115–124.

Belton, V., Ackermann, F., and Shepherd, I. (1997). Integrated support from problem structuring through to alternative evaluation using COPE and V·I·S·A. *Journal of Multi-Criteria Decision Analysis*, 6:115–130.

Belton, V. and Elder, M. D. (1994). Decision support systems – learning from visual interactive modelling. *Decision Support Systems*, 12:355–364.

Belton, V. and Elder, M. D. (1996). Exploring a multicriteria approach to production scheduling. *Journal of the Operational Research Society*, 47:162–174.

Belton, V. and Elder, M. D. (1998). Integrated MCDA: a simulation case study. In Stewart and van den Honert, 1998, pages 347–359.

Belton, V. and Gear, A. E. (1982). On a shortcoming of Saaty's method of analytical hierarchies. *OMEGA: International Journal of Management Science*, 11:228–230.

Belton, V. and Gear, A. E. (1985). The legitimacy of rank reversal – a comment. *OMEGA: International Journal of Management Science*, 13:143–144.

Belton, V. and Hodgkin, J. (1999). Facilitators, decision makers, D.I.Y users: Is intelligent multicriteria decision support for all feasible or desirable? *European Journal of Operational Research*, 113:13–26.

Belton, V. and Pictet, J. (1997). A framework for group decision using a MCDA model: Sharing, aggregating, or comparing individual information? *Journal of Decision Systems*, 6:283–303.

Belton, V. and Stewart, T. J. (1999). DEA and MCDA: Competing or complementary approaches? In Meskens and Roubens, 1999, pages 87–104.

Belton, V. and Vickers, S. P. (1992). VIDEA: Integrated DEA and MCDA - a visual interactive approach. In *Proceedings of the Tenth International Conference on Multiple Criteria Decision Making*, volume II, pages 419–429.

Belton, V. and Vickers, S. P. (1993). Demystifying DEA – a visual interactive approach based on multi criteria analysis. *Journal of the Operational Research Society*, 44:883–896.

Benayoun, R., de Montgolfier, J., Tergny, J., and Larichev, O. (1971). Linear programming with multiple objective functions: step method (STEM). *Mathematical Programming*, 1:366–375.

Bennett, P. (1998). Confrontation as a diagnostic tool. *European Journal of Operational Research*, 109:465–482.

Bennett, P., Cropper, S., and Huxham, C. (1989a). Modelling interactive decisions: the hypergame focus. In Rosenhead, 1989.

Bennett, P. and Howard, N. (1996). Rationality, emotion and preference change: drama-theoretic models of choice. *European Journal of Operational Research*, 92:603–614.

Bennett, P., Huxham, C., and Cropper, S. (1989b). Using the hypergame perspective, a case study. In Rosenhead, 1989.

Benson, H. P., Lee, D., and McClure, J. P. (1997). A multiple-objective linear programming model for the citrus rootstock selection problem in Florida. *Journal of Multi-Criteria Decision Analysis*, 6:283–295.

Bouyssou, D., Perny, P., Pirlot, M., Tsoukiàs, A., and Vincke, P. (1993). A manifesto for the new MCDA era. *Journal of Multi-Criteria Decision Analysis*, 2:125–127.

Brans, J. and Mareschal, B. (1990). The PROMETHEE methods for MCDM; the PROMCALC, GAIA and BANKADVISER software. In Bana e Costa, 1990, pages 216–252.

Brans, J. P., Macharis, C., Kunsch, P. L., and Schwaninger, M. (1998). Combining multicriteria decision aid and system dynamics for the control of socio-economic processes. an iterative real-time procedure. *European Journal of Operational Research*, 109:428–441.

Brans, J. P., Mareschal, B., and Vincke, P. (1984). PROMETHEE: a new family of outranking methods in multicriteria analysis. In Brans, J. P., editor, *Operational Research '84*, pages 477–490. North Holland, Dordrecht.

356 *MULTIPLE CRITERIA DECISION ANALYSIS*

Brans, J. P., Mareschal, B., and Vincke, P. (1986). How to select and how to rank projects: The PROMETHEE method. *European Journal of Operational Research*, 24:228–238.

Brans, J. P. and Vincke, P. (1985). A preference ranking organization method: the PROMETHEE method for multiple criteria decision-making. *Management Science*, 31:647–656.

Briggs, T., Kunsch, P. L., and Mareschal, B. (1990). Nuclear waste management: An application of the multicriteria PROMETHEE methods. *European Journal of Operational Research*, 44:1–10.

Brownlow, S. A. and Watson, S. R. (1987). Structuring multi-attribute value hierarchies. *Journal of the Operational Research Society*, 38:309–318.

Bryant, J. (1997). The plot thickens: understanding interaction through the metaphor of drama. *OMEGA: International Journal of Management Science*, 25:255–266.

Buchanan, J. T. (1994). An experimental evaluation of interactive MCDM methods and the decision making process. *Journal of the Operational Research Society*, 45:1050–1059.

Buede, D. M. (1986). Structuring value attributes. *Interfaces*, 16:52–62.

Buede, D. M. and Choisser, R. W. (1992). Providing an analytical structure for key system design choices. *Journal of Multi-Criteria Decision Analysis*, 1:17–27.

Butterworth, N. J. (1989). Giving up 'The Smoke'; a major institution investigates alternatives to being sited in the city. *Journal of the Operational Research Society*, 40:711–718.

Buzan, T. (1993). *The Mind Map Book*. BBC Books, London.

Charnes, A. and Cooper, W. W. (1961). *Management Models and Industrial Applications of Linear Programming*. John Wiley & Sons, New York.

Charnes, A., Cooper, W. W., Lewin, A. Y., and Seiford, L. M., editors (1994). *Data Envelopment Analysis: Theory, Methodology and Application*. Kluwer Academic Publishers, Boston/Dordrecht/London.

Checkland, P. (1981). *Systems Thinking, Systems Practice*. John Wiley & Sons, Chichester.

Checkland, P. (1989). Soft systems methodology. In Rosenhead, 1989.

Cook, W. D. and Kress, M. A. (1991). Multiple criteria decision model with ordinal preference data. *European Journal of Operational Research*, 54:191–198.

De Bono, E. (1985). *De Bono's Thinking Course*. Ariel Books.

De Bono, E. (1990). *Six Thinking Hats*. Penguin Books, London.

de Keyer, W. and Peeters, P. (1996). A note on the use of PROMETHEE multicriteria methods. *European Journal of Operational Research*, 89:457–461.

Delbecq, A. L., van de Ven, A. H., and Gustafson, D. H. (1975). *Group Techniques for Program Planning: A Guide to the Nominal Group and Delphi Processes*. Scott, Foresman and Company, Glenview.

Doyle, J. R. and Green, R. H. (1993). Data envelopment analysis and multiple criteria decision making. *OMEGA: International Journal of Management Science*, 21:713–715.

Dyer, J. S. (1990a). A clarification of "Remarks on the analytic hierarchy process". *Management Science*, 36:274–275.

Dyer, J. S. (1990b). Remarks on the analytic hierarchy process. *Management Science*, 36:249–258.

Eden, C. (1988). Cognitive mapping: a review. *European Journal of Operational Research*, 36:1–13.

Eden, C. (1989). Using cognitive mapping for strategic options development and analysis (SODA). In Rosenhead, 1989.

Eden, C. (1990). The unfolding nature of group decision support – two dimensions of skill. In Eden and Radford, 1990, pages 48–52.

Eden, C. and Ackermann, F. (1998). *Making Strategy: The Journey of Strategic Management*. SAGE Publications, London.

Eden, C. and Huxham, C. (1995). Action research for the study of organisations. In Clegg, S., Hardy, C., and Nord, W., editors, *Handbook of Organisational Studies*. Sage, Beverly Hills.

Eden, C. and Jones, S. (1984). Using repertory grids for problem construction. *Journal of the Operational Research Society*, 35:779–790.

Eden, C. and Radford, J., editors (1990). *Tackling Strategic Problems: The Role of Group Decision Support*. Sage Publications, London.

Eden, C. and Simpson, P. (1989). SODA and cognitive mapping in practice. In Rosenhead, 1989.

Edwards, W. and Barron, F. H. (1994). SMARTS and SMARTER: Improved simple methods for multiattribute utility measurement. *Organizational Behavior and Human Decision Processes*, 60:306–325.

Farquar, P. H. and Pratkanis, A. R. (1993). Decision structuring with phantom alternatives. *Management Science*, 39:1214–1226.

French, S. (1988). *Decision Theory: An Introduction to the Mathematics of Rationality*. Ellis Horwood, Chichester.

French, S. (1989). *Readings in Decision Analysis*. Chapman and Hall, London.

French, S. (1995). Uncertainty and imprecision: modelling and analysis. *Journal of the Operational Research Society*, 46:70–79.

358   *MULTIPLE CRITERIA DECISION ANALYSIS*

Friend, J. (1989). The strategic choice approach. In Rosenhead, 1989, pages 121–157.

Friend, J. and Hickling, A. (1987). *Planning Under Pressure*. Pergamon Press.

Gal, T., Stewart, T. J., and Hanne, T., editors (1999). *Multicriteria Decision Making: Advances in MCDM Models, Algorithms, Theory, and Applications*. Kluwer Academic Publishers, Boston.

Gardiner, L. R. and Steuer, R. E. (1994). Unified interactive multiple objective programming. *European Journal of Operational Research*, 74:391–406.

Gardiner, P. C. and Ford, A. (1980). Which policy is best, and who says so? In Legasto, A. A., Forrester, J. W., and Lyneis, J. M., editors, *System Dynamics*, volume 14 of *TIMS Studies in the Management Sciences*, pages 241–257. North Holland, Amsterdam.

Geoffrion, A. M., Dyer, J. S., and Feinberg, A. (1972). An interactive approach for multicriterion optimization with an application to the operation of an academic department. *Management Science*, 19:357–368.

Gershon, M., Duckstein, L., and McAniff, R. (1982). Multiobjective river basin planning with qualitative criteria. *Water Resources Research*, 18:193–202.

Goodwin, P. and Wright, G. (1997). *Decision Analysis for Management Judgement*. John Wiley & Sons, Chichester, second edition.

Gravel, M., Martel, J. M., Nadeau, R., Price, W., and Tremblay, R. (1991). A multicriterion view of optimal resource allocation in job-shop production. Presented to IFORS SPC1 on Decision Support Systems, Bruges.

Greco, S., Matarazzo, B., and Slowinski, R. (1999). The use of rough sets and fuzzy sets in MCDM. In Gal et al., 1999, chapter 14.

Halme, M., Joro, T., Korhonen, P., Salo, S., and Wallenius, J. (1999). A value efficiency approach to incorporating preference information in data envelopment analysis. *Management Science*, 45:103–115.

Hammond, J. S., Keeney, R. L., and Raiffa, H. (1999). *Smart Choices: A Practical Guide to Making Better Decisions*. Harvard Business School Press, Boston, Mass.

Hammond, K. R., McClelland, G. H., and Mumpower, J. (1980). *Human Judgement and Decision Making: Theories, Methods and Procedures*. Praeger, New York.

Hammond, K. R., Mumpower, J. L., and Smith, T. H. (1977). Linking environmental models with models of human judgement: a symmetrical decision aid. *IEEE Transactions on Systems, Man and Cybernetics*, SMC-7:358–367.

References 359

Harker, P. T. and Vargas, L. G. (1990). Reply to "Remarks on the analytic hierarchy process" by J.S. Dyer. *Management Science*, 36:269–273.

Hodgson, A. M. (1992). Hexagons for systems thinking. *European Journal of Operational Research*, 59:220–230.

Howard, N. (1971). *Paradoxes of Rationality: Theory of Metagames and Political Behaviour*. MIT Press, Cambridge, Mass.

Howard, N. (1989a). The CONAN play: a case study illustrating the process of metagame analysis. In Rosenhead, 1989.

Howard, N. (1989b). The manager as politician and general: the metagame approach to analysing cooperation and conflict. In Rosenhead, 1989.

Howard, N. (1999). *Confrontation Analysis: How to Win Operations Other than War*. CCRP Publication Series.

Howard, R. A. (1992). In praise of the old time religion. In Edwards, W., editor, *Utility Theories: Measurements and Applications*, pages 27–55. Kluwer, Boston.

Huxham, C. (1990). On trivialities of process. In Eden and Radford, 1990.

Ignizio, J. P. (1976a). *Goal Programming and Extensions*. Lexington Books, Lexington, Mass.

Ignizio, J. P. (1976b). *Introduction to Linear Goal Programming*. Sage.

Ignizio, J. P. (1983). Generalized goal programing: an overview. *Computers and Operations Research*, 10:277–289.

Islei, G. and Lockett, A. G. (1988). Judgement modelling based on geometric least squares. *European Journal of Operational Research*, 36:27–35.

Islei, G., Lockett, G., Cox, B., Gisbourne, S., and Stratford, M. (1991). Modelling strategic decision making and performance measurements at ICI Pharmaceuticals. *Interfaces*, 21(6):4–22.

Jacquet-Lagrèze, E. (1990). Interactive assessment of preferences using holistic judgements: the PREFCALC system. In Bana e Costa, 1990, pages 335–350.

Jacquet-Lagrèze, E. and Siskos, Y. (2001). Preference disaggregation: 20 years of MCDA experience. *European Journal of Operational Research*, 130:233–245.

Jordi, K. C. and Peddie, D. (1988). A wildlife management problem: a case study in multiple-objective linear programming. *Journal of the Operational Research Society*, 39:1011–1020.

Joro, T., Korhonen, P., and Wallenius, J. (1998). Structural comparison of data envelopment analysis and multiple objective linear programming. *Management Science*, 44:962–970.

360   *MULTIPLE CRITERIA DECISION ANALYSIS*

Kahneman, D. and Tversky, A. (1979). Prospect theory: An analysis of decision under risk. *Econometrica*, 47:263–291.

Kahneman, D., Tversky, A., and Slovic, P., editors (1982). *Judgement under Uncertainty: Heuristics and Biases*. Cambridge University Press, Cambridge.

Kalai, E. (1977). Nonsymmetric Nash solutions and replications of 2-person bargaining. *Int. Journal of Game Theory*, 6:129–133.

Kaplan, R. S. and Norton, D. P. (1996). *The Balanced Scorecard*. Harvard Business School Press, Boston, Mass.

Kaylan, A. R., Bayraktar, E., and Yazgaç, T. (2000). Integrating AHP and FMEA tools with QFD for supply chain system design. Paper presented at the 15th International Conference on MCDM, Ankara, July 2000.

Keeney, R. L. (1981). Analysis of preference dependencies among objectives. *Operations Research*, 29:1105–1120.

Keeney, R. L. (1992). *Value-Focused Thinking: A Path to Creative Decision Making*. Harvard University Press, Cambridge, Massachusetts.

Keeney, R. L. and Nair, K. (1977). Selecting nuclear power plant sites in the Pacific Northwest using decision analysis. In Bell, D. E., Keeney, R. L., and Raiffa, H., editors, *Conflicting Objectives in Decisions*, pages 294–322. John Wiley & Sons, Chichester.

Keeney, R. L. and Raiffa, H. (1972). A critique of formal analysis in public sector decision making. In Drake, A. W., Keeney, R. L., and Morse, P. M., editors, *Analysis of Public Systems*, pages 64–75. MIT Press, Cambridge, Mass.

Keeney, R. L. and Raiffa, H. (1976). *Decisions with Multiple Objectives*. J. Wiley & Sons, New York.

Kelly, G. (1955). *The Psychology of Personal Constructs*. Norton, New York.

Kepner, C. H. and Tregoe, B. B. (1976). *The Rational Manager*. Princeton Research Press, Princeton, NJ.

Kepner, C. H. and Tregoe, B. B. (1981). *The New Rational Manager*. Princeton Research Press, Princeton, NJ.

Kolb, D. A., Rubin, I. M., and McIntyre, J. M. (1984). *Organisation Psychology: An Experiential Approach to Organizational Behavior*. Prentice-Hall, Englewood Cliffs.

Korhonen, P. (1988). A visual reference direction approach to solving discrete multiple criteria problems. *European Journal of Operational Research*, 34:152–159.

Korhonen, P. and Laakso, J. (1986). A visual interactive method for solving the multiple criteria problem. *European Journal of Operational Research*, 24:277–287.

Korhonen, P. and Wallenius, J. (1996). Behavioural issues in MCDM: neglected research questions. *Journal of Multi-Criteria Decision Analysis*, 5:178–182.

Korhonen, P., Wallenius, J., and Zionts, S. (1984). Solving the discrete multiple criteria problem using convex cones. *Management Science*, 30:1336–1345.

Lee, S. M. (1972). *Goal Programming for Decision Analysis*. Auerbach, Philadelphia.

Lee, S. M. and Olson, D. L. (1999). Goal programming. In Gal et al., 1999, chapter 8.

Lewin, A. Y. and Seiford, L. M. (1997). Extending the frontiers of Data Envelopment Analysis. *Annals of Operations Research*, 73:1–11.

Lootsma, F. A. (1997). *Fuzzy Logic for Planning and Decision Making*. Kluwer Academic Publishers, Dordrecht.

Losa, B. and Belton, V. (2001). Combining MCDA and conflict analysis. Working paper series, University of Strathclyde, Department of Management Science.

Lotfi, V., Stewart, T. J., and Zionts, S. (1992). An aspiration-level interactive model for multiple criteria decision making. *Computers and Operations Research*, 19:671–681.

Lotfi, V., Yoon, Y. S., and Zionts, S. (1997). Aspiration-based search algorithm (ABSALG) for multiple objective linear programming problems: theory and comparative tests. *Management Science*, 43:1047–1059.

MacCrimmon, K. R. (1973). An overview of multiple objective decision making. In Cochrane, J. L. and Zeleny, M., editors, *Multiple Criteria Decision Making*, pages 18–44. The University of South Carolina Press, Columbia, South Carolina.

Mann, L. and Janis, I. L. (1985). *Decision Making: A Psychological Analysis of Conflict Choice and Commitment*.

Martins, A. G., Coelho, D., Antunes, C. H., and Climaco, J. (1996). A multiple objective linear programming approach to power generation planning with demand-side management (DSM). *International Transactions in Operational Research*, 3:305–317.

Masud, A. S. M. and Hwang, C. L. (1981). Interactive sequential goal programming. *Journal of the Operational Research Society*, 32:391–400.

Meskens, N. and Roubens, M., editors (1999). *Advances in Decision Analysis*. Kluwer Academic Publishers, Dordrecht.

Miles, L. D. (1971). *Techniques of Value Engineering and Analysis*. McGraw-Hill, New York, second edition.

362 MULTIPLE CRITERIA DECISION ANALYSIS

Miller, G. A. (1956). The magical number seven plus or minus two: some limits on our capacity for processing information. *The Psychological Review*, 63:81–97.

Millet, I. and Harker, P. (1990). Globally effective questioning in the Analytic Hierarchy Process. *European Journal of Operational Research*, 48:88–97.

Moore, C. M. (1994). *Group Techniques for Idea Building*. Sage, London.

Morgan, T. (1993). Phased decision conferencing. *OR Insight*, 6(4):3–12.

Mousseau, V. (1992). Are judgments about relative importance of criteria dependent or independent of the set of alternatives? An experimental approach. Cahier du LAMSADE 111, Université de Paris Dauphine.

Mumpower, J. L., Veirs, V., and Hammond, K. R. (1979). Scientific information, social values and policy formulation: the application of simulation models and judgement analysis to the denver regional air pollution problem. *IEEE Transactions on Systems, Man and Cybernetics*, SMC-9:464–476.

Musselman, K. and Talavage, J. (1980). A tradeoff cut approach to multiple objective optimization. *Operations Research*, 28:1424–1435.

Nash, J. (1953). Two-person cooperative games. *Econometrica*, 21:128–140.

Nunamaker, J. F. and Dennis, A. R. (1993). Group support systems research: Experience from the lab and field. In Jessup, L. M. and Valacich, J. S., editors, *Group Support Systems*. MacMillan, New York.

Owen, G. (1982). *Game Theory*. Academic Press, New York, second edition.

Perny, P. (1992). *Modélisation, agrégation et exploitation de préférences floues dans une problématique de rangement: bases axiomatiques, procédures et logiciels*. Thèse de Doctorat, Université Paris-Dauphine.

Perny, P. (1998). Multicriteria filtering methods based on concordance and non-discordance principles. *Annals of Operations Research*, 80:137–165.

Phillips, L. D. (1984). A theory of requisite decision models. *Acta Psychologica*, 56:29–48.

Phillips, L. D. (1989). People-centred group decision support. In Doukidis, G., Land, F., and Miller, G., editors, *Knowledge Based Decision Support Systems*, pages 208–221. Ellis Horwood, Chichester.

Phillips, L. D. (1990). Decision analysis for group decision support. In Eden and Radford, 1990, pages 142–153.

Phillips, L. D. and Phillips, M. C. (1991). Facilitated work groups: theory and practice. *Journal of the Operational Research Society*, 44:533–549.

References   363

Phong, H. T. and Tabucanon, M. T. (1998). Viewing a plant layout problem as a multiobjective case to enhance manufacturing flexibility. In Stewart and van den Honert, 1998, pages 360–376.

Pictet, J. and Belton, V. (1997). MCDA: What message? Opinion Makers Section, Newsletter of the European Working Group on Multicriteria Aid for Decisions.

Pictet, J. and Belton, V. (2001). ACIDE: Analyse de la compensation et de l'incomparabilité dans la décision vers une prise en compte pratique dans MAVT. To be published in a book of refereed papers from the 50th Anniversary meeting of the EURO working group on Multicriteria Aid for Decisions.

Pruzan, P. and Bogetoft, P. (1997). *Planning with Multiple Criteria.* Copenhagen Business School Press, second edition.

Ramesh, R., Karwan, M. H., and Zionts, S. (1989). Interactive multicriteria linear programming: an extension of the method of Zionts and Wallenius. *Naval Research Logistics*, 36:321–335.

Rios Insua, D. (1990). *Sensitivity Analysis in Multi-Objective Decision Making.* Springer (Lecture Notes in Economics and Mathematical Systems, Vol. 347), Berlin.

Roberts, F. S. (1979). *Measurement Theory with Applications to Decisionmaking, Utility and the Social Sciences.* Addison-Wesley, London.

Romero, C. (1986). A survey of generalized goal programming (1970-1982). *European Journal of Operational Research*, 25:183–191.

Rosenhead, J., editor (1989). *Rational Analysis for a Problematic World.* John Wiley & Sons, Chichester.

Roy, B. (1985). *Méthodologie Multicritère d'Aide à la Décision.* Economica, Paris.

Roy, B. (1996). *Multicriteria Methodology for Decision Aiding.* Kluwer Academic Publishers, Dordrecht.

Roy, B. and Bouyssou, D. (1993). *Aide Multicritère d'Aide à la Décision: Méthodes et Cas.* Economica, Paris.

Roy, B. and Hugonnard, J. (1982). Ranking of suburban line extension projects on the Paris metro system by a multicriteria method. *Transportation Research*, 16A:301–312.

Roy, B., Présent, M., and Silhol, D. (1986). A programming method for determining which Paris metro stations should be renovated. *European Journal of Operational Research*, 24:318–334.

Saaty, T. L. (1980). *The Analytic Hierarchy Process.* McGraw-Hill, New York.

Saaty, T. L. (1982). *Decision Making for Leaders.* Lifetime Learning Publications.

364   *MULTIPLE CRITERIA DECISION ANALYSIS*

Saaty, T. L. (1990a). An exposition of the AHP in reply to the paper "Remarks on the analytic hierarchy process". *Management Science*, 36:259–268.

Saaty, T. L. (1990b). How to make a decision: the analytic hierarchy process. *European Journal of Operational Research*, 48:9–26.

Saaty, T. L. (1996). *The Analytic Network Process*. RWS Publications, Pittsburgh.

Saaty, T. L. and Vargas, L. G. (1984). The legitimacy of rank reversal. *OMEGA: International Journal of Management Science*, 12:513–516.

Saaty, T. L., Vargas, L. G., and Wendell, R. E. (1983). Assessing attribute weights by ratios. *OMEGA: International Journal of Management Science*, 11:9–12.

Salo, A. A. and Hämäläinen, R. P. (1992). Preference assessment by imprecise ratio statements. *Operations Research*, 40:1053–1061.

Santos, S. and Belton, V. (2000). Integrating multicriteria analysis and system dynamics. Paper presented at the 15th International Conference on MCDM, Ankara, July 2000.

Schein, E. H. (1987). *Process Consultation (Volume 2): Lessons for Managers and Consultants*. Addison Wesley, Mass.

Schein, E. H. (1988). *Process Consultation (Volume 1): Its Role in Organisation Development*. Addison Wesley, Mass., second edition.

Schein, E. H. (1999). *Process Consultation Revisited: Bulding the Helping Relationship*. Addison Wesley, Mass.

Scheubrein, R. and Bossert, B. (2000). Balancing the Balanced Scorecard. Paper presented at the 15th International Conference on MCDM, Ankara, July 2000.

Schniederjans, M. J. (1995). *Goal Programming: Methodology and Applications*. Kluwer, Boston.

Schoner, B. and Wedley, W. C. (1989). Ambiguous criteria weights in AHP: consequences and solutions. *Decision Sciences*, 20:462–475.

Schwarz, R. M. (1994). *The Skilled Facilitator*. Jossey-Bass, San Francisco.

Senge, P. (1992). *The Fifth Discipline*. Doubleday, New York.

Sharp, J. A. (1987). Haulier selection – an application of the Analytic Hierarchy Process. *Journal of the Operational Research Society*, 38:319–328.

Shin, W. S. and Ravindran, A. (1991). Interactive multiple objective optimization: Survey I – continuous case. *Computers and Operations Research*, 18:97–114.

Simon, H. A. (1976). *Administrative Behavior*. The Free Press, New York, third edition.

References   365

Siskos, J. (1980). Comment modéliser les préférences au moyen de fonctions d'utilité additives. *RAIRO Recherche Opérationnelle/Operations Research*, 14:53–82.

Siskos, Y., Spyridakos, A., and Yannacopoulos (1999). Using artificial intelligence and visual techniques into preference disaggregation analysis: the MIIDAS system. *European Journal of Operational Research*, 113:281–299.

Spengler, T. and Penkuhn, T. (1996). KOSIMEUS – a combination of a flowsheeting program and a multicriteria decision support system. Paper presented at the SAMBA II workshop (State of the Art Computer Programs for Material Balancing), Wuppertal Institut.

Spronk, J. (1981). *Interactive Multiple Goal Programming*. Martinus Nijhoff Publishing.

Srinivasan, V. and Shocker, A. D. (1973). Linear programming techniques for multi-dimensional analyses of preferences. *Psychometrica*, 38:473–493.

Steuer, R. E. (1986). *Multiple Criteria Optimization: Theory, Computation, and Application*. John Wiley & Sons, New York.

Steuer, R. E. and Choo, E. U. (1983). An interactive weighted Tchebycheff procedure for multiple objective programming. *Mathematical Programming*, 26:326–344.

Stewart, T. J. (1981). A descriptive approach to multiple criteria decision making. *Journal of the Operational Research Society*, 32:45–53.

Stewart, T. J. (1986). A combined logistic regression and Zionts-Wallenius methodology for multiple criteria linear programming. *European Journal of Operational Research*, 24:295–304.

Stewart, T. J. (1987). Pruning of decision alternatives in multiple criteria decision making, based on the UTA method for estimating utilities. *European Journal of Operational Research*, 28:79–88.

Stewart, T. J. (1988). Experience with prototype multicriteria decision support systems for pelagic fish quota determination. *Naval Research Logistics*, 35:719–731.

Stewart, T. J. (1991). A multicriteria decision support system for R&D project selection. *Journal of the Operational Research Society*, 42:17–26.

Stewart, T. J. (1992). A critical survey on the status of multiple criteria decision making theory and practice. *OMEGA: International Journal of Management Science*, 20:569–586.

Stewart, T. J. (1993). Use of piecewise linear value functions in interactive multicriteria decision support: A monte carlo study. *Management Science*, 39:1369–1381.

Stewart, T. J. (1995). Simplified approaches for multi-criteria decision making under uncertainty. *Journal of Multi-Criteria Decision Analysis*, 4:246–258.

Stewart, T. J. (1996a). Relationships between data envelopment analysis and multicriteria decision analysis. *Journal of the Operational Research Society*, 47:654–665.

Stewart, T. J. (1996b). Robustness of additive value function methods in MCDM. *Journal of Multi-Criteria Decision Analysis*, 5:301–309.

Stewart, T. J. (1997). Scenario analysis and multicriteria decision making. In Climaco, J., editor, *Multicriteria Analysis*, pages 519–528. Springer-Verlag, Berlin.

Stewart, T. J. (1999a). Concepts of interactive programming. In Gal et al., 1999, chapter 10.

Stewart, T. J. (1999b). Evaluation and refinement of aspiration-based methods in MCDM. *European Journal of Operational Research*, 113:643–652.

Stewart, T. J. and Joubert, A. (1998). Conflicts between conservation goals and land use for exotic forest plantations in South Africa. In Beinat, E. and Nijkamp, P., editors, *Multicriteria Analysis for Land Use Management*, pages 17–31. Kluwer Academic Publishers, Dordrecht.

Stewart, T. J. and Scott, L. (1995). A scenario-based framework for multicriteria decision analysis in water resources planning. *Water Resources Research*, 31:2835–2843.

Stewart, T. J. and van den Honert, R. C., editors (1998). *Trends in Multicriteria Decision Making*. Springer-Verlag, Berlin.

Tilanus, C. B., de Gans, O. B., and Lenstra, J. K. (1983). Postscript: a survey of reasons for failure or success of quantitative methods in management. In Tilanus, C. B., de Gans, O. B., and Lenstra, J. K., editors, *Quantititative Methods in Management: Case Studies of Failures and Successes*, pages 269–273. Wiley.

Van der Heijden, K. (1996). *Scenarios: The Art of Strategic Conversation*. John Wiley & Sons, Chichester.

Van Gundy Jr., A. B. (1988). *Techniques of Structured Problem Solving*. Van Nostrand Reinhold, New York.

Vargas, L. G. (1985). A rejoinder. *OMEGA: International Journal of Management Science*, 13:249.

Vincke, P. (1992). *Multicriteria Decision-Aid*. John Wiley & Sons, Chichester.

Vincke, P. (1999). Outranking approach. In Gal et al., 1999, chapter 11.

von Neumann, J. and Morgenstern, O. (1947). *Theory of Games and Linear Programming*. Wiley, New York, second edition.

von Nitzsch, R. and Weber, M. (1993). The effect of attribute ranges on weights in multiattribute utility measurements. *Management Science*, 39:937–943.

von Winterfeldt, D. and Edwards, W. (1986). *Decision Analysis and Behavioral Research*. Cambridge University Press, Cambridge.

Walker, W. E. (1988). Generating and screening alternatives. In Miser, H. J. and Quade, E. S., editors, *Handbook of Systems Analysis: Craft Issues and Procedural Choices*. John Wiley & Sons, New York.

Watson, S. R. and Buede, D. M. (1987). *Decision Synthesis. The Principles and Practice of Decision Analysis*. Cambridge University Press.

Watson, S. R. and Freeling, A. (1982). Assessing attribute weights. *OMEGA: International Journal of Management Science*, 10:582–583.

Watson, S. R. and Freeling, A. (1983). Comment on: assessing attribute weights by ratios. *OMEGA: International Journal of Management Science*, 11:13.

Weber, M. and Borcherding, K. (1993). Behavioral influences on weight judgments in multiattribute decision making. *European Journal of Operational Research*, 67:1–12.

Weber, M., Eisenführ, F., and von Winterfeldt, D. (1988). The effects of splitting attributes on weights in multiattribute utility measurement. *Management Science*, 34:431–445.

Weerahandi, S. and Zidek, J. V. (1981). Multi-Bayesian statistical decision theory. *Journal of the Royal Statistical Society A*, 144:85–93.

Weistroffer, H. R. (1985). Careful usage of pessimistic values is needed in multiple objectives optimization. *Operations Research Letters*, 4:23–25.

Weistroffer, H. R. and Hodgson, L. (1998). Using AHP for text book selection. In Stewart and van den Honert, 1998, pages 434–447.

Wierzbicki, A. P. (1980). The use of reference objectives in multiobjective optimization. In Fandel, G. and Gal, T., editors, *Multiple Criteria Decision Making Theory and Practice*, pages 468–486. Springer, Berlin.

Wierzbicki, A. P. (1999). Reference point approaches. In Gal et al., 1999, chapter 9.

Wooler, S. (1982). A decision aid for structuring and evaluating career choice options. *Journal of the Operational Research Society*, 33:343–352.

Yu, P. L. (1990). *Forming Winning Strategies. An Integrated Theory of Habitual Domains*. University of Kansas, School of Business.

Zadeh, L. A. (1965). Fuzzy sets. *Information and Control*, 8:338–353.

Zahedi, F. (1986). The AHP – a survey of the method and its applications. *Interfaces*, 16(4):96–108.

368 MULTIPLE CRITERIA DECISION ANALYSIS

Zeleny, M. (1982). *Multiple Criteria Decision Making*. McGraw-Hill, New York.

Zimmermann, H. J. and Zysno, P. (1979). Latent connectives in human decision making. *Fuzzy Sets and Systems*, 2:37–51.

Zionts, S. and Wallenius, J. (1976). An interactive programming method for solving the multiple criteria problem. *Management Science*, 22:652–663.

Zionts, S. and Wallenius, J. (1983). An interactive multiple objective linear programming method for a class of underlying nonlinear utility functions. *Management Science*, 29:519–529.

Index

action research, 342
aggregation, 81, 143
    additive, 86, 120, 163
    multilinear, 101
    multiplicative, 94, 101
AHP, 133, 134, 151–159, 289, 336
    absolute measurement mode, 153
    consistency index, 155
    consistency ratio, 155
    debate, 157
AIM (aspiration-level interactive model),
    230
Allais' paradox, 97
alternative rank orders, 184
alternatives
    classifying, 29–31
    direct rating of, 129–131
    generating, 277
    identifying, 52–55
analysis versus intuition, 283
analytic hierarchy process, *see* AHP
anchoring and availability biases, 212
ascending order, 242
aspiration levels, 104, 210
attribute, 351

backroom analysis, 275
balanced scorecard, 322–324
BATNA (best alternative to a negotiated
    settlement), 296
Beaufort scale, 129
breakpoints, 166

cases
    aerial policing facility, 284
    game reserve planning, 20–21, 67–
        71, 205, 217, 219, 220, 224,
        225, 228
    land use planning, 22–24, 71–77, 178

maintenance and development of
    computer systems, 28–29
office location, 17–19, 64–67, 120,
    133, 144, 153, 237, 240, 248,
    256, 285
strategic futures, 284
Universities Funding Council, 25–28
CATWOE, 45
CAUSE framework, 45, 52, 65
cognitive mapping, 48–51
compact set, 351
complete order, 351
compromise programming, 218, 222
concave function, 172, 351
concordance, 109
concordance index, 235, 243
conflict analysis, 315–320
conjoint scaling, 189
constructed measurement scales, 127
constructivist approach, 35, 80, 119, 204,
    335, 336
convex cones, 201
convex function, 172, 351
convex set, 352
corresponding tradeoffs condition, 90
credibility index, 245, 249
criteria, 352
    hierarchies, 55, 58
    identifying, 55–59
Criterium Decision Plus, 347

data envelopment analysis (DEA), 298–
    305
    weight restrictions, 303
DataScope, 347
decision analyst, 14
decision conferencing, 7, 267, 269
Decision Explorer, 39, 41, 347
Decision Lab 2000, 255, 347
decision maker, 14, 262

370 *MULTIPLE CRITERIA DECISION ANALYSIS*

decision making unit (DMU), 299
decision theory, 63
decision workshop, 271–275
DEFINITE, 193, 347
descending order, 241
deviational variables, 214
discordance, 109
discordance index, 236, 244
displaced ideal, 222
distillation
    ascending, 246
    descending, 246
DIY analysis, 279–281
domination, 83
drama theory, 316

efficiency, 83
eigenvector estimation, 154, 158
ELECTRE I, 234–239
ELECTRE II, 241–242
ELECTRE III, 242–250, 348
ELECTRE IS, 348
ELECTRE IV, 251, 348
ELECTRE TRI, 251, 348
environmental impact assessment (EIA), 328
EQUITY, 348
European Federation for Quality Management (EFQM), 325
expected utility, 96
Expert Choice, 349

facilitator, 14, 268, 270, 272
factor analysis, 306–309
final order, 242
financial risk analysis, 327
fuzzy sets, 111

GAIA procedure, 255–258, 309
game theory, 295–298
gaps, 171
Geoffrion-Dyer-Feinberg method, 195
goal programming, 105, 209–232, 289, 335, 346
    Archimedean, 216
    generalized, 227–230
    interactive, 220–227
    linear, 213–219
    preemptive, 218
    Tchebycheff, 219
goals, 210
group dynamics, 270

habitual domains, 39
High Priority, 348
HIVIEW, 348
hypergames, 298, 316

ideal values, 212
ideas
    generating, 40–48
    structuring, 48–52
implementation research, 341
importance weights, *see* weights
incomparability, 107, 236
independence
    additive, 100
    preferential, 88
    utility, 100
indifference, 107
integrating framework, 333–338, 340
interaction between individuals, 262, 269
interactive methods, 193–204
interactive multiple goal programming (IMGP), 226
interactive sequential goal programming (ISGP), 227
interactive weighted-sums/filtering approach, 200
interval assessments, 166
intuition, 283, 286
inverse preferences, 188–193

JOURNEY making, 39

Kepner-Tregoe approach, 322
kernel, 238
Kipling's six thinking men, 44

λ-preference, 245
Logical Decisions, 348

MACBETH, 129, 133, 172–174, 336, 349
MAVF, 86, 119–151, 163–206
MAVT, 7, 85–95, 119
MCDA, 1
    establishing a contract, 265
    expectations, 265
    integrating framework, 333–338
    practice, 337
    process, 5, 263
    research, 340
    software, 281–282, 345–346
MCDM, 1, 13–16
    classification, 31–33
    discrete choice, 16–19
    multiobjective design, 19–21
    project evaluation, prioritization and selection, 24–29
median order, 242
metagames, 316
methodological research, 342
middle-most solution, 177
MIIDAS, 189
MINORA, 189

modelling issues, 266
multiattribute utility theory, 7, 63, 95–103
multiattribute value function, *see* MAVF
multiattribute value theory, *see* MAVT
multiple criteria decision analysis, *see* MCDA
multiple criteria decision making, *see* MCDM
multiple objective linear programming, 69–71, 196, 213
multivariate statistics, 257, 305–310

Nash solution, 295
nature of MCDA support, 262
non-dominated, 83

On Balance, 349
outranking, 106–111, 233–259, 290, 335
outranking flows, 254
outranking graph, 239
outranking relation
    building, 236, 241
    exploiting, 238, 241
    strong, 241
    weak, 241

pairwise comparisons, 132, 152, 234
Pareto optimality, 83
payoff table, 71, 212
pessimistic value, 212
piecewise linear approximation, 166
policy scenarios, 72
portfolio theory, 327
Post-It, 41, 42
potential optimality, 151
PREFCALC, 189
preference, 107
    compensatory, 218
    complete, 85
    non-compensatory, 218
    strict, 107
    transitive, 85
    weak, 107
preference function, 252
    partial, 83
preference index, 253
preference information
    categorical, 164
    holistic, 188–193
    imprecise, 165–187
    ordinal, 164–187
    trade-offs, 194
preference intensity, 252
preference regions, 150
preorder, 242, 352
PRIME, 349

principal components analysis, 255, 309
problem structuring, 35–64, 310
problematique, 15
profile graph, 145–146
PROMCALC, 253, 255
PROMETHEE, 252–258
pseudoconcave, 352
pseudoconcave function, 196, 197

quality function deployment, 324
quality management, 325, 327
quasi-criterion, 243, 246

rank reversal, 159
reference point method, 106, 218, 227–230, 335
reference set, 188
relative importance, 134
research in MCDA, 340
robustness, 148–151, 239
rough sets, 113

satisficing, 104, 209, 218
scalarizing function, 229
scale
    categorical, 166
    constructed, 127
    global, 121
    interval, 91, 121, 169
    local, 121
    qualitative, 130, 166
    ratio, 157
    semantic, 158, 164, 166
scenario planning, 63, 312–315
scoring, 121
sensitivity analysis, 148–151, 156, 239
sigmoidal function, 172
simulation, 311
small groups, 279–281
SMART, 7, 313
SMARTER, 142
SODA, 39
soft systems methodology, 39
software support, 281–282, 345–346
spray diagram, 65
spreadsheets, 345
stakeholders, 59–60
STEM (Step Method), 223
Strategic Choice, 39
swing weights, 135, 138
system dynamics, 311

Tchebycheff norm, 200, 218
Team Expert Choice, 349
Thomsen condition, 90
tradeoff cuts, 196

uncertainty, 60–64

UTA (utilité additive), 189
utility, *see* multiattribute utility theory

value difference, 169
value engineering, 321
value focused thinking, 39, 46
value function, 85, 289, 335
    additive, 86, 120, 163
    direct assessment, 123
    functional shape, 171
    indirect assessment, 124, 181–187
      bisection method, 124
      difference methods, 125
    interval assessment, 174
    marginal, 85
    multiplicative, 94
    partial, 85, 123–128, 163, 171, 181
    piecewise linear, 166
    upper and lower bounds, 175, 183
value measurement, 85–95

value tree, 55, 58, 66, 80, 139
veto threshold, 236, 244
VIG, 229
VIMDA, 229
V·I·S·A, 349

weak order, 242
Web-HIPRE, 350
weights, 92, 114–117, 157, 166, 175, 234
    cumulative, 139
    elicitation, 134–143, 288–291
    interpretation, 288–291
    normalization, 137
    relative, 139
    swing, 135, 138
WINPRE, 350
WWW-NIMBUS, 350

Zionts-Wallenius method, 197

Lightning Source UK Ltd.
Milton Keynes UK
UKOW042215250712

196562UK00007B/15/A